The Cambridge Double Star Atlas

The Cambridge Double Star Atlas is back! It is the first and only atlas of physical double stars that can be viewed with amateur astronomical instruments. Completely rewritten, this new edition explains the latest research into double stars and the equipment, techniques, and opportunities for you to discover, observe, and measure them. The target list has been completely revised and extended to 2,500 binary or multiple systems. Each system is described with the most recent and accurate data from the authoritative Washington Double Star Catalog, including the Henry Draper and Smithsonian numbers that are most useful in our digital age. Hundreds of remarks explain the attributes of local, rapidly changing, often-measured or known orbital systems. The color atlas charts by Wil Tirion have been updated, to help you easily find and identify the target systems, as well as other deep-sky objects. This is an essential reference for double star observers.

BRUCE MACEVOY is a Fellow of the Royal Astronomical Society and member of the Astronomical Society of the Pacific. He built his first telescope at age 11 and has studied double stars intensively since 2006. He has lectured on double star astronomy in California and Hawaii, developed database software to edit the Washington Double Star Catalog, and has observed over 5,300 double star systems from his observatory near Sebastopol, California. He was formerly a senior research psychologist with SRI International and director of research for Yahoo! He also curates one of the largest websites devoted to astronomical topics, www.handprint.com/ASTRO/.

WIL TIRION is a full-time uranographer. He is famous among the amateur astronomy community for the numerous atlases and star charts he has created. Among his other successful books for Cambridge University Press are *Sky Atlas 2000.0* (co-published with Sky Publishing), *The Cambridge Star Atlas*, *The Monthly Sky Guide*, and *A Walk Through the Heavens*. A full list of Wil Tirion's publications is available on his website: www.wil-tirion.com.

THE CAMBRIDGE
DOUBLE
STAR ATLAS

BRUCE MACEVOY

WIL TIRION
Based on the original concept by
James Mullaney

Shaftesbury Road, Cambridge CB2 8EA, United Kingdom

One Liberty Plaza, 20th Floor, New York, NY 10006, USA

477 Williamstown Road, Port Melbourne, VIC 3207, Australia

314–321, 3rd Floor, Plot 3, Splendor Forum, Jasola District Centre,
New Delhi – 110025, India

103 Penang Road, #05–06/07, Visioncrest Commercial, Singapore 238467

Cambridge University Press is part of Cambridge University Press & Assessment,
a department of the University of Cambridge.

We share the University's mission to contribute to society through the pursuit of
education, learning and research at the highest international levels of excellence.

www.cambridge.org
Information on this title: www.cambridge.org/9781107534209

First edition © J. Mullaney and W. Tirion 2009
Second edition © B. MacEvoy and W. Tirion 2015

First published 2009
Second edition 2015 (version 3, September 2022)

Printed in Great Britain by Ashford Colour Press Ltd., September 2022

A catalogue record for this publication is available from the British Library

Library of Congress Cataloging-in-Publication data
Names: MacEvoy, Bruce, author. | Tirion, Wil, author. | Mullaney,
James, 1940–
Title: The Cambridge double star atlas / Bruce MacEvoy, Wil Tirion ;
based on the original concept by James Mullaney.
Description: Second edition. | Cambridge : Cambridge University Press,
2015. | First edition by James Mullaney. | Includes bibliographical
references and index.
Identifiers: LCCN 2015037678 | ISBN 9781107534209 (Paperback :
alk. paper)
Subjects: LCSH: Double stars–Observers' manuals. | Double
stars–Charts, diagrams, etc.
Classification: LCC QB821 .M829 2015 | DDC 523.8/41–dc23 LC
record available at http://lccn.loc.gov/2015037678

ISBN 978-1-107-53420-9 Paperback

CONTENTS

INTRODUCTION

This new edition of the *Cambridge Double Star Atlas* is designed to improve its utility for amateur astronomers of all skill levels.

For the first time in a publication of this type, the focus is squarely on double stars as *physical systems*, so far as these can be identified with existing data. Using the procedures described in Appendix A, the target list of double stars has been increased to 2,500 systems by adding 1,100 "high probability" physical double and multiple stars and deleting more than 850 systems beyond the reach of amateur telescopes or lacking any evidence of a physical connection. Wil Tirion has completely relabeled the *Atlas* charts to reflect these changes, and left in place the previous edition's double star icons as a basis for comparison. This new edition provides a selection based on evidence rather than traditional opinion, so that the twenty-first century astronomer can explore with fresh eyes the astonishing actual variety in double stars.

Continuing the emphasis on physical systems, this *Atlas* explains the origin and dynamic properties of double stars and the role they have played in our understanding of star formation and stellar evolution. The elements of binary orbits, stellar spectral types, and methods of detecting and cataloging double stars, are explained to enrich the observer's understanding of double star astronomy. There is also practical guidance for the visual astronomer – information on optics, equipment preparation, useful accessories, viewing techniques and opportunities for amateur research. The references suggest both print and online double star resources. Finally, over 330 systems in the target list are marked with a star (★) in the left margin. These indicate "showpiece" systems of intrinsic beauty or charm, "challenge" pairs of close separation or large brightness contrast, and several systems that have been important in the history of astronomy. From most observing locations, at least three dozen of these targets will be in view at any time of night on any evening of the year.

Jim Mullaney's choice of nineteenth century double star catalog labels has been retained as a tribute both to his original *Atlas* concept and to the bygone astral explorers who discovered over 90% of the systems in the target list (see Appendix D). These labels are also a convenient link to the legacy double star literature and a compact labeling style for the *Atlas* charts. However, as a convenience to the digital astronomer, the target list provides both the Henry Draper (HD) and Smithsonian (SAO) catalog numbers for each system. The first will identify each system in the research literature and online astronomical databases, the second is a compact targeting command or search keyword recognized by most GoTo telescope mounts and planetarium software.

What are double stars?

Let's start by adapting the definition from double star astronomer Wulff Heintz:

> A double star is two or more stars that are bound by mutual gravitational attraction into an enduring (usually lifelong) dynamic system.

The fundamental unit is two stars – usually termed a *binary star* – orbiting their mutual center of gravity. But a double star may also be triple, quadruple, quintuple and so on, under the umbrella category of *multiple star*. Using the singular *star* indicates that a binary or multiple star is a single physical system, an astrophysical fact. By contrast, an *optical double star* is (as the name implies) only an optical illusion, two stars far apart in space that appear side by side on the illusory celestial sphere. Although in most cases evidence for physicality is inconclusive or entirely lacking, especially in distant pairs, advances in astronomy in recent decades have given us a new capability to distinguish fact from illusion in double star astronomy.

Double stars display an enormous range of orbital dynamics and stellar types, and produce characteristic visual patterns that the astronomer will encounter often at the eyepiece. The most common of these patterns are illustrated on the back cover. The challenge for double star astronomers is to understand how these systems were formed and how they will change over time, then to apply this knowledge to answer basic questions about our Galaxy.

Remarkably, all evidence suggests that most if not all stars in the Galaxy were formed as members of double star systems. This means that nearly all the double stars we observe have been united from birth. And most double stars will eventually die together, one after the other, like Romeo and Juliet. We know this from the many binary systems that contain a dead or dying star, and the large number of binaries that are orbiting so closely they can never be torn apart.

The old view was that double stars formed by randomly falling into mutual orbits as they circled the Galaxy, or appeared when a single massive star rotated so rapidly that it split in two. The current view is that double stars are literally born together from a single *cloud core* of gas and dust collapsing into its own gravity. The collapsing core, stressed by external shock waves and internal turbulence, divides into two or more protostars (*prompt fragmentation*). Matter that continues to fall toward a protostar swirls into an enormous accretion disk that often develops spiral arms or irregular clumps (*disk fragmentation*). These also gather mass to become low mass companion stars or planetary systems.

These collapsing cloud cores rarely form in isolation: most are found inside a much larger concentration of gas and dust known as a *star forming region* (SFR). The number of stars that form within a single SFR depends on the mass, density and turbulence of the gas and dust it contains, but a typical SFR can span tens of parsecs and produce hundreds or thousands of new stars. Inside these murky clouds, usually found churning along the arc of a galaxy spiral arm, protostars attract matter and grow hotter and more compact with the increasing pressure of gravitational contraction. Within a few million years at most, the most massive of these young stars fire up their thermonuclear cores, push back the clouds with the force of their radiation, light up the dispersing gas as an emission nebula, and unveil a young star cluster to our view.

The masses, rotational speeds and orbits of double protostars depend on the turbulence of the cloud core, the density of gas and dust in the SFR, the rate of their mass accretion and the angular momentum inherited from their accretion disks. They also depend on interactions with other stars in their cloud core and natal star cluster. Binary protostars are slowed into smaller orbits by friction with their enveloping clouds; as they grow in mass, near encounters with other stars in the natal cluster can shear apart widely separated or "soft" companions, perturb stable orbits, and bind tighter already close orbiting or "hard" binaries. The often elliptical shape of double star orbits and the extreme variation in orbital periods are the result of these cumulative influences. Even stars too far apart to have formed in the same cloud core can display *common proper motion* – parallel motion across the sky – because they were born in the same SFR and escaped in the same direction as the natal cluster dissolved. For all these reasons, double stars have been called the "fossils of star formation."

How stars form is only one of the many mysteries that double stars have illuminated in the history of astronomy. William Herschel discovered in 1802 that they moved in orbits, demonstrating that Isaac Newton's gravitational attraction governed not just our solar system but the visible universe. In the nineteenth century, systematic discovery and observation of nearby binary systems led to refined methods of measurement and orbital calculation, which allowed astronomers to "weigh" double stars and discover the enormous range in stellar masses. With accurate estimates of stellar distance from the parallax surveys of the twentieth century, mass could be compared to intrinsic brightness (*absolute magnitude*). This confirmed that the brightest stars are also the most massive and pointed to nuclear fusion as the only possible source of starlight; theoretical physics could then deduce the paths of stellar evolution. Double stars have also been essential to our understanding of star clusters, many types of variable stars, supernovae, black holes and exotic high energy sources in deep space. They are the keystone species of the Galaxy.

What are double stars? The astronomer Simon Portegies Zwart answered the question this way:

Binaries are the basic building blocks of the Milky Way as galaxies are the building blocks of the universe. In the absence of binaries many astrophysical phenomena would not exist and the Galaxy would look completely different over the entire spectral range.

The binary orbit

The essence of double stars is found in the binary orbit, which is a stable dynamic balance between mutual gravitational attraction and centrifugal orbital energy.

Let's start with the simplest example of two identical stars in a circular orbit. (Circular orbits are often found in close binaries that orbit in 10 days or less.) Each star attracts the other, so the total gravitational attraction between the two stars is proportional to their combined *system mass* ($M_1 + M_2$). But the strength of their mutual attraction varies as the *inverse square* of the distance between the stars. If the distance is multiplied by a number, the gravitational attraction is reduced by the reciprocal of the number squared. Increasing the distance by three times reduces the gravitational attraction to 1/9; reducing the distance by half increases the gravitational attraction four times.

In a circular binary the nominal *orbital radius* (r) is the distance from one star to the other, but each star actually orbits their common center of mass or *barycenter*, at the center of their shared circular orbit. The two stars are always connected by a line through this fulcrum point, which means they have the same *orbital period* (P). As the stars revolve around the barycenter, their constant gravitational attraction is offset by a constant orbital velocity. A greater system mass or smaller orbital distance would require a greater orbital velocity to offset the greater gravitational attraction.

This simplest of all possible binary orbits can be imbalanced in two ways. First, the two stars are usually of unequal mass. In that case, balance is restored by making the distance of each star from the barycenter proportional to the *mass ratio* (q), the mass of the smaller star divided by the mass of the larger (M_2/M_1). Like unequal weights on a balance beam, balance requires the larger star to be closer to the center of gravity. As a result, the heavier star

moves in a separate circular orbit inside the orbit of the less massive star, and because its orbit is smaller, its orbital velocity is proportionally less.

Second, the distance between the stars oscillates in synchrony with the orbital period, from a point of closest approach (*periastron*) to a point of farthest separation (*apastron*). This sends the stars into opposing elliptical orbits around the barycenter, now located at one focus of each ellipse (see Figure 1, top). The two stars are still connected by a line through the barycenter, the orbits have the same elliptical elongation or *eccentricity* (*e*) (see Appendix B) and the larger star still moves at a lower average velocity in a proportionately smaller orbit. Because the mutual gravitational attraction increases as the mutual distance from the barycenter decreases, the changing distance between the stars must be balanced by a changing orbital velocity, reaching peak velocity at periastron, lowest velocity at apastron. This elegant combination of system mass, mass ratio, orbital radius, eccentricity and orbital velocity around the barycenter is the *absolute orbit*, the actual physical motions of a binary star (see Figure 1, top).

Unfortunately, the barycenter of a binary system is invisible to an observer, so it cannot be used as a reference point to measure the orbital motion. Instead, the brighter and usually more massive *primary star* is made the anchor point, and the motion of the fainter, less massive *secondary star* is measured in relation to it (see Figure 4). This is the *relative orbit*. It has the same period and eccentricity as the absolute orbit, but now the average orbital radius (r) is equal to the *semi-major axis* (*a*), half the longest dimension of the orbital ellipse. (Half the shortest dimension is the *semi-minor axis, b*.)

As a final wrinkle, the relative binary orbit is almost always tilted to our direction of view. This will make a circular orbit appear elliptical, like the rim of a cup viewed from one side, and the point of periastron in an elliptical orbit may not be the point of smallest visual separation between the two stars (see alpha Centauri in Figure 1). This *apparent orbit* is what we actually observe and measure on the celestial sphere. Complex mathematics, applied to painstaking observations of position and orbital velocity over decades or even centuries, are required to derive the

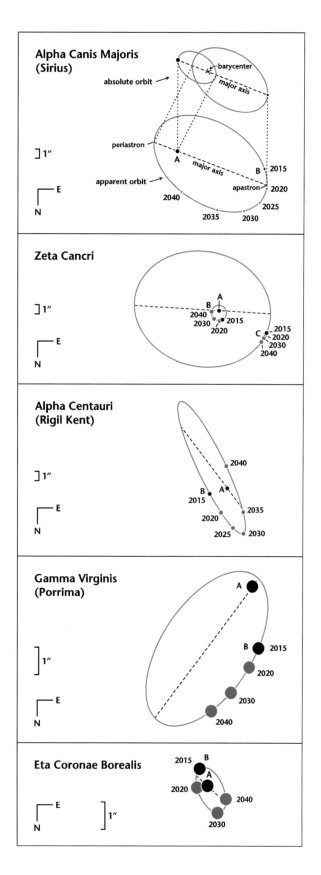

dimensions of the relative orbit and the dynamics of the absolute orbit from the distorted path of the apparent orbit.

The balance in a binary orbit between gravitational attraction and orbital energy is summarized in a proportion known as Kepler's third law. This is easiest to calculate if we measure system mass in units of *solar mass* (1 $M_{\odot}$ is the mass of the Sun), orbital radius in *astronomical units* (1 AU is the distance from the Earth to the Sun) and period in *Earth years*. Then Kepler's third law is simply:

$$(M_1 + M_2) = r^3/P^2$$

The target list indicates both the period (P) and average orbital radius (r) for all systems with an orbital solution. For these, you can use Kepler's third law to calculate the system mass.

If we don't know the orbital radius, but know the distance (d) to the double star in parsecs (denoted pc; one parsec is equal to 206,265 AU or 3.26 light years) and have measured the angular separation (ρ or *rho*) between the stars in arcseconds, then the *projected separation* (ps) between the stars, again in astronomical units, is:

$$\mathrm{ps} = \rho \cdot d.$$

This is always a minimum separation, because a tilted (foreshortened) orbit will make the distance between stars appear smaller than it actually is.

Multiple star orbits

What happens in a system of three or more stars? Here a binary orbit still prevails, but in a remarkable way – it becomes enormously larger. This creates the defining feature of a stable multiple star: a *hierarchical orbit structure* (Figure 2).

A binary pair – the building block of every double star – is bound by mutual attraction to a common barycenter. If a third star approaches too close to this

Figure 1 – Five binary orbits
The absolute orbit of Sirius (top) shows the elliptical orbits of the two components around their mutual center of gravity. The five apparent orbits shown are the orbits we actually measure. The dotted line indicates the major axis of the relative orbit with apastron and periastron at opposite ends. The current (2015) position of the components is shown with their predicted future positions out to 2040. The diagrammed star disks match the Airy disk diameter produced by a 250 mm aperture.

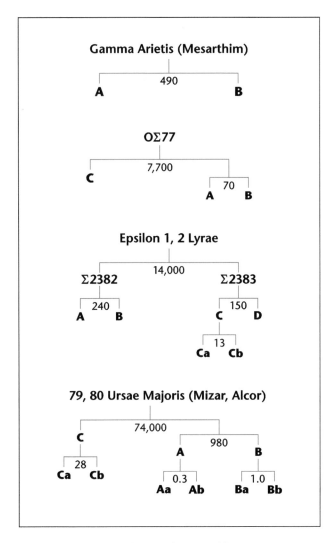

Gamma Arietis (Mesarthim)

490

A   B

OΣ77

7,700

C

70

A   B

Epsilon 1, 2 Lyrae

14,000

Σ2382   Σ2383

240   150

A   B   C   D

13

Ca   Cb

79, 80 Ursae Majoris (Mizar, Alcor)

74,000

C

980

A   B

28

Ca   Cb   0.3   1.0

Aa   Ab   Ba   Bb

Figure 2 – Hierarchical multiple star orbits
Multiple stars are composed of binaries and single stars
arranged in a hierarchy of orbits. The horizontal bars indicate
the orbits, the number below each bar the orbital radius in
astronomical units. Double stars of 2, 3, 5 and 6 components
are drawn from the target list. The examples also show how
multiple star components are labeled.

couple, the barycenter formed by all three stars
becomes unstable and their orbits unpredictable.
But a binary can join with another binary or single
star if they partner at a much greater orbital distance,
often 100 to 1,000 times the orbital radius of the
binary. At this remove a binary influences the distant
barycenter as if it were a single large star, and can
form one half of a stable "binary" unit. From the
inverse square principle, we see this can reduce to
1/1,000,000th the gravitational disruption that the
third component might exert on the binary orbit.
Yet this bond can still be strong enough to resist the

attraction from other stars in the Galaxy, making the
triple star an enduring dynamic system.

This hierarchical segregation of orbits distinguishes
a multiple star from its natal star cluster. In the
cluster, all the star systems orbit the single barycenter
formed by the entire cluster, stars are deflected into
new orbits each time they pass through the cluster,
and "evaporation" (as dispersing gas and dust reduces
the mass of the cluster, weakening the gravity that
holds the cluster together) dissolves nearly all natal
star clusters within a few 10 million years.

Multiple protostars seem to form within a single
cloud core, so at birth they will have similar,
dynamically unstable orbits. So how does a
hierarchical structure develop? Through competition.
As the protostars orbit their common center of
gravity, by chance all three can approach periastron
near the same time. When this happens the two most
massive or closest stars can join forces with their
greater mutual attraction and hurl the less massive
or more distant third star into a larger, higher
energy and higher velocity orbit; this transfer of
orbital energy allows the dominant pair to settle
into a tighter, lower energy orbit. This process can
be repeated many times within a few million years,
"hardening" the inner binary and eventually imparting
an escape velocity to the third star, ejecting it from
the system.

Although ejection is very likely, it isn't inevitable.
There are over 270 examples of "2+1" systems in the
target list – many of them an unresolved spectroscopic
binary with a wide third component – that have
found a stable dynamic configuration. The less
common "1+2" systems (a binary orbiting a more
massive primary star) and the even rarer "double
double" (2+2) systems can evolve in the same way,
binding the binary units more tightly while increasing
the orbital radius to other components. A nearby
example is the wide naked eye pair Mizar and Alcor
in Ursa Major. Mizar is also a close telescopic double,
forming a 2+1 triple system. In fact all three
components are very close binaries, forming a sextuple
system of hierarchically segregated orbits (Figure 2).

Similar quintuple and sextuple systems are rare
(only 40 are found in the target list), and hierarchical
structure seems to reach its limit in systems of about
seven stars. Beyond that, the competition among

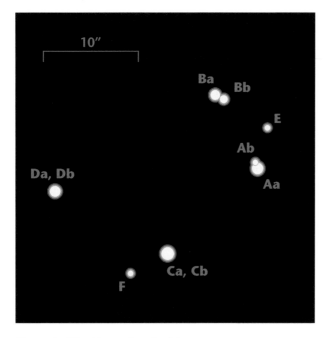

Figure 3 – The Trapezium in Orion
An unstable group of at least a dozen stars emerging from its natal cloud of gas and dust. Because the four brightest stars are competing to dominate their shared barycenter, the group is unstable and will eventually break apart.

components dissolves large stellar groups before they can develop a stable hierarchical arrangement. The 430 parsec far Trapezium, emerging from the Great Orion Nebula, is a case in point (Figure 3). A visual quartet where each of the four massive stars is already a binary or multiple system, there is no clear hierarchical ordering in their orbits or separations. Just a few million years old, this minicluster of a dozen or more stars already displays divergent motions, a sign that it is on the way to breaking apart.

Stellar mass and the binary life cycle

In stars, *mass is destiny*. The mass of a star determines how long it will live and how it will die. Because most binaries remain bound for the life of the component stars, this means mass determines the life cycle of the binary system as well.

A star is an enormous sphere of plasma, heated by the thermonuclear fusion of hydrogen or helium at its core. The fusion is ignited and contained by the enormous pressure of the star's mass, as gravity strives to collapse the star to a single point. The energy released by this fusion pushes the collapsing mass

outward in all directions, stabilizing the contraction into a spherical body with an incandescent skin or *photosphere*. This shines with a peak energy or *effective temperature* and a characteristic brightness or *absolute magnitude* (M). As stars increase in mass they become much hotter and brighter, and the color of their light, the clue to their temperature, shifts from red to blue wavelengths.

This relationship between mass, color and brightness is indicated by a star's *spectral type*. For normal or *main sequence* stars, it is simplest to think in terms of four contrasting categories of stellar mass. (1) The hottest, brightest (and rarest) *high mass* O and B type stars, such as Mintaka, Rigel or Achernar, have a mass from 120 $M_\odot$ down to about 4 $M_\odot$, and shine with a brilliant, bluish light that includes vast amounts of invisible high energy ultraviolet and X-ray radiation. (2) Less massive *A type* stars, such as Sirius, Fomalhaut and Vega, are around 3 to 1.5 $M_\odot$ and the benchmark for a bright, "pure white" star color. (3) The *solar type* stars – F, G and larger K types, like the Sun, Procyon or Rigil Kent – are about 1.5 $M_\odot$ to 0.5 $M_\odot$ and peak in the visible spectrum with a pale to distinctly yellow light. (4) The coolest, faintest (and most numerous) *low mass* stars – smaller K and M types, such as 61 Cygni or Kruger 60 – are 0.5 $M_\odot$ or less and glow with a pronounced orange light that peaks in the invisible infrared (heat) and microwave wavelengths. Arranging these spectral types in order of decreasing mass and temperature yields the sequence O B A F G K M, traditionally memorized as *Oh Be A Fine Girl, Kiss Me*. Gradations within a type are indicated with a number from 0 to 9: for example, an A0 star has double the mass and triple the brightness of an A9.

All stars eventually consume the hydrogen available to their core, and once they do they leave the main sequence of normal stars. Solar and higher mass stars switch to fusing into carbon the core of helium "ash" that results from hydrogen fusion, and the resulting surge of new energy forces the surface of the star outward to 500 or more times its normal radius. This enormously expanded *giant* or *supergiant* surface area allows the photosphere to radiate vastly more light, making it perhaps 10,000 times brighter. The rarefied surface becomes cooler, shifting the peak wavelength into the infrared and giving old solar and high mass stars a similar bloated, bright and ruddy appearance. These developments are captured in the *luminosity type* of a star. A young or midlife, main sequence star is denoted with the Roman numeral V or IV; an expanding giant star by III; and a massive and massively expanded supergiant star – the most luminous star outside a nova or supernova – by Ia, Ib or II.

Due to the extreme heat in their massively compressed thermonuclear cores, high mass (O or B type) stars feverishly consume their reserves of fuel in a few tens of million years, while relatively cool, low mass (K or M type) stars can shine frugally for tens of billion years. In binaries of unequal mass ($q < 1.0$), the more massive component will enter the giant phase first, and this can create some spectacular stellar fireworks. In close binaries ($r < {\sim}5$ AU), the dying star may expand so far that it forms a *semi-detached binary*, transferring its remaining hydrogen to the companion and giving it a life-shortening greater mass. The donor star then collapses into an incredibly compact and hot carbon remnant called a *white dwarf*. When the companion also begins to die, it expands and pours hydrogen back onto the white dwarf via an encircling accretion disk, resulting in a Type Ia supernova or X-ray binary and, eventually, the sepulchre of a matched white dwarf binary.

Most binary orbits are too large for mass transfer to occur and the giant or supergiant phase unfolds in isolation. These "giant type" binaries are not uncommon among visual double stars (the target list includes almost 560). This is because the giant and its companion are both intrinsically bright, their high system mass can sustain large orbits that can be resolved at great distances, and the giant phase can last for a billion years.

Examples of the next stage – a main sequence star with a white dwarf companion – are harder to find, even in the solar neighborhood, because white dwarfs are very faint. The three best known examples, Sirius B (Figure 1), Procyon B and 40 Eridani B, are all within 5 parsecs of the Sun, yet Procyon's 11th magnitude (m.11) white dwarf can only be glimpsed in large telescopes.

The double star population

We now can address a basic question: what is the *multiplicity ratio*, the proportion of double stars among all star systems (whether single or double stars) in the Galaxy? In the solar neighborhood (within 25 parsecs of the Sun) and considering only average or *solar type stars* (F, G and more massive K types), recent research suggests the multiplicity ratio follows a "60%–60%" allocation: *Roughly 60% of individual stars are actually members of double or multiple star systems, but roughly 60% of star systems – those individual points of light in the sky – are single stars.* The corollary to this 60%–60% rule is: *About 70% of double stars are binary.* Among the roughly 40% of local, solar type star systems identified as double stars, 72% have only two components, 21% have three, 5% have four and only 2% contain five or more components.

This 60%–60% rule is not universal because mass strongly affects the multiplicity ratio. High mass (O and B type) stars have a multiplicity ratio of 80% up to perhaps 100%. In low mass stars – small K type, M type and even smaller brown dwarfs – the multiplicity ratio is apparently less than 30%. Equal mass binaries ($q = 1.0$) seem more common in closely orbiting pairs, as components in multiple systems, and in low mass systems; unequal mass ratios ($q < 1.0$) are about equally common down to mass ratios of 0.2. Delving the extreme mass ratios, the vast majority of solar type stars, single or double, appear to support planetary systems.

The size of binary orbits, like the mass of stars themselves, covers an enormous range (see Appendix C). The closest orbiting binaries have been studied as *eclipsing variable stars*, apparently single stars that display a periodic and revealing variation in brightness as one star passes in front of the other. Some of these are contact binaries (W Ursae Majoris type variable stars), solar mass stars that circle each other in less than a day and are enclosed in a single photosphere, with a shape resembling a peanut. Other solar mass stars, in nearly circular orbits with periods of a few days or weeks, perpetually turn the same face to each other and form dramatic tidal streams or enormous star spots within a shared and tangled magnetic field (RS Canis Venaticorum variable stars). Among

A type and high mass stars, systems have been found where the tidal attraction between the stars has distorted them into an ellipsoidal shape (ellipsoidal variable stars), sometimes causing a transfer of mass from the larger star to its companion (β Lyrae variables). And stars of any mass may be the Algol type variables, with orbits of months or years – too far apart to interact, but close enough to eclipse each other along our line of sight – that let us measure the diameter of each spectral type of star.

At the other extreme, one of the widest confirmed double stars (the A type system of Fomalhaut and TW Piscis Austrini) is separated by more than 50,000 AU with an age of more than 400 million years. Multiple systems may have the heft to bind even wider orbits: Mizar and Alcor, separated by 74,000 AU, have recently been shown to be bound. The outer limit of orbits that can endure for the life of the component stars is still believed to be around 1,000 to 5,000 AU, but orbits 10 times larger are now confirmed that have survived more than one revolution around the Galaxy.

The typical binary orbit is between these extremes. Among the local, solar type double stars, the median orbital radius is about 50 AU with a median period of 250 years and a wide range of eccentricities distributed around an average of $e = 0.5$ (the semi-major axis (a) is about 15% longer than the semi-minor axis (b); see Appendix B). But Kepler's third law means the system mass will determine the orbital radius for the same 250 year orbital period: a high mass B5V binary must orbit at a radius of around 120 AU, while a low mass M5V type binary can orbit at only 30 AU, the distance of Neptune from the Sun.

Detecting double stars

By definition, a *visual double star* can be resolved into separate components with measurable relative positions. For more than two centuries, these measurements have been made with a filar micrometer: a device that lets the observer adjust the spacing between two parallel filaments in the eyepiece field of view to measure the separation between two stars, then rotate the filaments to align with and measure the position angle. Around 1975, the method of *speckle interferometry* used computers to transform atmospheric turbulence into greatly magnified star images. Two decades later, *long baseline interferometry* used computers to combine the images from widely separated telescopes to create a single high resolution aperture. Interferometry is considered a "visual" technique because it also provides measures of separation and position angle.

Most double stars in the target list were discovered by painstaking visual inspection of every star brighter than an arbitrary magnitude limit. But many were detected by other methods, and it is customary to categorize these systems by the technique used to discover and measure them.

Several hundred double stars have been discovered by analyzing the variable light from an apparently single star – those eclipsing variable stars, mentioned above. Since 1900, more than two thousand have been identified as *spectroscopic binaries*, because the two stars orbit at such high velocities that their mutual spectrum reveals Doppler shifts in the absorption lines of the much brighter star (a single line binary, denoted SB1) or of both similarly bright stars (a double line binary, SB2). Even when no Doppler shifts are apparent, *spectrum binaries* can be detected because the superimposed absorption lines of the two stars are recognizably different, and *photometric binaries* can be identified because the primary star is much brighter than its spectral type predicts. *X-ray binaries* – a white dwarf or neutron star receiving mass from a dying companion – have been identified with X-ray telescopes. Some binaries have even been discovered through a telltale stepwise (rather than instantaneous) extinction of the star's light during *occultation* by the Moon.

A small number are *astrometric binaries*, detected even though the companion is too faint or too close to the glare of the primary star to be imaged. Instead, the small elliptical motion of the primary star can be observed as a sideways wobble in the path and periodic change in the pace of its proper motion across the sky. The companion to Sirius (Figure 1) was identified in this way in 1844, nearly two decades before it was visually detected in 1862.

Finally, several thousand have been identified as *common proper motion (CPM) binaries* because they share the same speed and direction of motion across the sky. (Radial velocity toward or away from the Earth is more difficult to measure, but can be used to calculate the *true motion* in three dimensions.) These are identified by proper motion surveys that rapidly compare or "blink" matched sky photographs taken decades apart or by statistical analysis of the trajectories of stars measured by ground based telescopes and astrometric satellites. Research in the past few decades has found dozens of CPM binaries with an angular separation many times wider than the full Moon. In order to qualify as a double star, the separation of a CPM binary must be small enough to provide an enduring gravitational bond between the stars, but we've seen this limit is at least 50,000 AU in fact, and can be over 1 parsec in theory.

Stars beyond the largest binding distances can still show common proper motion: these *comoving groups*, gravitationally unbound stars with parallel orbits around the Galaxy, have emerged with similar trajectories from the same star forming region. These comoving groups can be huge. Most famous is the Ursa Major association: all but one of the stars in the "Big Dipper" asterism are at the head of an impressive stream of more than 50 stars scattered across 31 constellations.

Even with all the terrestrial and satellite instruments available to us today, visual double stars are local objects, astronomically speaking. Half the systems in the target list are within 120 parsecs of the Sun, and only high mass or high luminosity giant and supergiant systems are bright enough to be included beyond 350 parsecs. Slow positional change or highly inclined orbits prevent us from tracing long period orbits or detecting Doppler shifts; large distances can diminish even huge orbits; limited brightness obscures even neighboring low mass stars. Outside the solar neighborhood, we observe only an incomplete and biased sample of double stars and their components – low mass stars, in particular, are very difficult to detect without infrared telescopes.

As the astronomer Robert Grant Aitken complained over a century ago, a great number of optical pairs have made their way into double star catalogs. (These are retained, though recognized as optical, to prevent "rediscovery.") A repertory of

statistical tests has been developed to identify physical systems by appearance alone, and these converge on the visual profile *bright, tight, equal, similar* – the two stars should be little separated, equally bright, and have similar spectral types. The probability that an optical pair will match this profile is very small, but unfortunately this profile excludes the many unequal mass, "giant type" and visually wide CPM doubles we know exist. The real solution calls for data, and this means repeated measures of relative position made over decades or centuries of observation.

Double star catalogs

Double star observations have been painstakingly acquired and cataloged for more than two centuries by a brigade of double star astronomers, and their catalogs form a unique and irreplaceable historical record of celestial change. The target list is compiled from more than 80 of these double star catalogs dating from 1782 to the present (see Appendix D). All these catalogs (and 700 others) are now combined as the *Washington Double Star Catalog* (WDS), the authoritative and frequently updated database of visual double stars maintained since 1964 at the US Naval Observatory (USNO) in Washington, DC.

The attributes essential to include in any double star catalog (besides its celestial coordinates or location in the sky) are: (1) its catalog ID, (2) the component letter codes, (3) the position angle, (4) the separation, (5) the magnitudes of the components and (6) the epoch.

The WDS ID is currently a nine digit abbreviation of the target system's celestial coordinates, with plans to expand to 13 digits. The shorter and more easily recognized Catalog ID, used in the *Atlas* charts and in many references, may not indicate the astronomer who discovered the pair: over 400 double stars credited to F. Wilhelm von Struve (Σ) were actually discovered by William Herschel (H).

The apparent orbit is measured with just two parameters (Figure 4). *Position angle (θ)* is the "clock face" orientation of a line joining the primary (usually brighter) star to the secondary (fainter) star, measured in degrees from the line to celestial north (0°), and increasing counterclockwise through east (90°), south

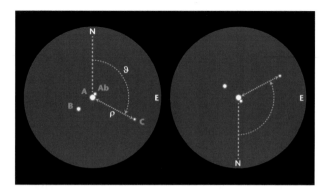

Figure 4 – Double star measurement
A binary star is described by the position angle (θ), and the separation (ρ). The orientation of the field depends on the equipment: (right) an "inverting" astronomical telescope rotates the field by 180° and position angle increases in the counterclockwise direction; (left) a mirror diagonal reverses the field left to right and position angle increases clockwise.

(180°) and west (270°). Because stars appear to drift east to west in a fixed field of view, west is traditionally referred to as *preceding* (abbreviated p.) and east as *following* (f.). These abbreviations, used in combination with *north* (n.) and *south* (s.) in the target list remarks, can point to other objects in a field of view while signaling that the direction is only approximate.

Separation (ρ) is the angular distance between the two stars, measured in arcseconds ($''$) or, in the target list, in arcminutes ($'$) if larger than 120$''$. How wide is a typical double star? The average separation of pairs in the target list is 26$''$. In comparison, the disk of Jupiter is never smaller than 29$''$.

Individual stars within multiple stars are identified by a letter *component code*, and measures of position are denoted by the component code combination. The primary star is labeled A, its companion or secondary star is labeled B, and measures of θ and ρ are listed for the pair AB; the third component is labeled C, its position relative to the primary as AC and so on. Frequently a component thought to be a single star turns out to be a close binary, so the component symbol is split by appending lowercase letters and separating the pair with a comma (C becomes the binary Ca,Cb). If one of these is also found to be a binary, the code is split by appending numbers (Ca becomes Ca1,Ca2). New components are assigned the next available letter – D, E, F and so

on. Figure 2 illustrates how this sequence of code revisions often signals the hierarchical structure of a multiple star.

In addition to positional measurements, the primary and secondary magnitudes (denoted $m1$ and $m2$) are important to calculate the system magnitude difference or *delta-m* (Δm). As the brightness contrast between the two stars, Δm can be used to estimate the mass ratio (q) of the stars when both are on the main sequence (see Appendix B).

All double stars are continually moving, in orbit around each other and in proper motion across the sky. This makes *epoch*, the year the system was measured, useful to decide if the parameters describe the current appearance of the system. In addition, the meaning of θ changes over time, because precession of the equinoxes changes the celestial direction of true north: epoch allows this to be corrected in historical measures. Although routinely omitted from most double star references, the proper motion of each component is invaluable to suggest whether two stars are moving independently or as a gravitationally bound pair. Finally, the *spectral type* and *luminosity type* are useful to understand the hierarchical structure and age of the system, estimate the system mass, and derive the absolute magnitude necessary to calculate the distance of the system from its apparent magnitude (the so-called *spectroscopic parallax*, see Appendix B).

Telescope optics

With this background understanding of physical double stars, it's time to explore the observing techniques of double star astronomy and the best use of your telescope.

The four basic optical attributes of a telescope are the *aperture* (D), the *objective focal length* (f_o), the objective focal ratio or *relative aperture* (N) of the primary mirror or objective lens and, for visual astronomy, the *eyepiece focal length* (f_e). These define the quantities of magnitude limit, resolution limit, magnification and exit pupil that determine the quality of your telescopic image. Appendix B lists the formulas used to calculate each of these quantities.

The *magnitude limit* (m_L) is an estimate of the faintest star you can detect with your telescope using averted vision. This depends on the faintest star you can detect with your naked eye, the *naked eye limit magnitude* (NELM), which is about 6.5 at a dark sky site but much less (~4.5) under light-polluted suburban skies. However, direct or foveal vision is necessary to see star colors and resolve close doubles, so these must be at least three magnitudes brighter than the magnitude limit of your telescope (wide, faint companions can still be detected with averted vision).

The λ/D or *Abbe resolution limit* (R_o) is the smallest angular width that can be imaged with your telescope aperture. It is imposed by the wave nature of light, which diffracts a perfect image into alternating ripples of light and dark that, like pixels on a computer screen, limit the smallest resolvable detail. This limit R_o is calculated (in radians) as the average light wavelength (λ) divided by the aperture (D) in millimeters. Many double star astronomers rely instead on a *resolution criterion*, the minimum separation necessary to see two distinct star images, as proposed by Lord Rayleigh using optical theory, William Dawes using telescopic observations, or C. M. Sparrow using visual experiments (see Figure 5 and Appendix B). These are found by multiplying the optical limit R_o by a perceptual adjustment factor (k). But visual acuity is an individual attribute, part of your "personal equation." Many skilled observers can detect an equal magnitude binary as an elongated, "egg shaped" or "rod shaped" star at a separation that is nearly half the Abbe limit. The optical quality and collimation of your telescope, the strength of your eyes, your viewing skill – as well as the atmospheric conditions and brightness contrast of the target system – determine your actual resolution limits.

Magnification (also denoted by M) is the visual width of the object as it appears in the image compared to its naked eye width in the sky. It is produced by the combined focal lengths of the objective (primary mirror or lens) and eyepiece. Magnification increases as the objective focal length gets longer, enlarging the image by projection distance like a film projector moved farther from a screen; and it increases as the eyepiece focal length gets shorter, enlarging the image by bringing it closer to the eye. Increased magnification has three additional effects: it reduces the angular width of your true field of view (TFOV), it makes the background sky darker, and it makes the effects of atmospheric turbulence more obvious.

Of course, to utilize fully the resolving power of your telescope, you must enlarge the resolution limit of the aperture to match the resolution limit of your eye. However, magnify as much as you like, you will never see an actual double star. Instead you see the image created as light waves from the star are disrupted by the physical limits of your telescope aperture. If the star is bright enough to be viewed directly, it will appear as a tiny drop of light called the *Airy disk*, and in brighter stars (under good seeing) this disk will be enclosed by one or more concentric ripples of light, the *diffraction rings*. Since you are not looking at an object outside the telescope but at a diffraction artifact inside the telescope, the visual size of the Airy disk must be estimated in a different way.

The *exit pupil* (d_e) of your telescope is inversely proportional to the apparent diameter of the Airy disk – as the exit pupil becomes smaller, the Airy disk will appear larger. To resolve the closest double stars or to evaluate the seeing (as explained below), you must magnify the Airy disk until it is distinctly visible. For most observers, this means an exit pupil less than 1.0 – in other words, an eyepiece focal length (f_e) that is less than the telescope's relative aperture (N) (see Appendix B). With the N=10 optics found in commercial Schmidt–Cassegrain telescopes, a 5 to 7mm eyepiece is often effective.

Preparing to observe

Astronomy requires good optics, but also careful adjustment and skillful use of your equipment. Assuming you already know how to set up and polar align your telescope, there are three preliminary steps to a night of observing: cool down, collimation and seeing evaluation.

Cool down allows the telescope to reach the same temperature as the surrounding air, which eliminates the thermal currents from the mirror and inside the telescope tube that will badly degrade the image. The routine solution is to store the telescope during the day where it can be protected from heat, set up and uncover the telescope just before sunset, and start observing when twilight ends. Cool down of at least an hour is necessary for reflecting telescopes – Newtonians and all Cassegrain formats – and large apertures may require more time. As a rule, refracting telescopes equilibrate quickly.

Collimation is the centering and parallel alignment of the optical axis of all lenses and mirrors in the telescope – including the optical axis of the eye, which must be centered over the eye lens for best views. Miscollimation can seriously degrade the quality of the telescope image, but fortunately it is easy to correct. The manuals that ship with commercial reflecting telescopes will explain how to collimate the instrument. (Refracting telescopes are already collimated and most cannot be adjusted.)

Both collimation and cool down are evaluated by looking at a bright star image defocused in the extrafocal direction (toward the observer) to show about five diffraction rings around the faint central *Poisson spot*. A warm mirror will present an undulating circumference or a wedge-shaped plume rising from the middle of the defocused star image; a miscollimated mirror will show the Poisson spot out of center.

Once the thermal wedge has disappeared and collimation is confirmed, the same bright star in crisp focus can be used to evaluate the *seeing* – the rapid distortion of the telescope image caused by thermal turbulence in the atmosphere. Many astronomers use the ten-step *Standard Scale* devised by American astronomers A. E. Douglass and E. C. Pickering to rate the seeing, but the gist of the scale can be captured in five intervals, each linked to specific features of the star image (see Figure 5). The worst seeing (1) produces an enlarged and featureless ball of rapidly seething speckles of light. As seeing improves this becomes (2) a recognizable Airy disk inside a speckle cloud, (3) a distinct Airy disk surrounded by a continuous dark ring, and (4) the Airy disk with a first ring that is unbroken at least half the time. The best seeing provides (5) a completely formed and often motionless diffraction artifact. Although turbulence originates in the atmosphere, its effect depends on the size of your telescope: on the same night at the same location, the seeing will be superior

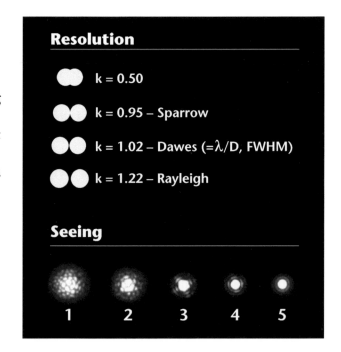

Figure 5 – Resolution and seeing
(Top) The Dawes resolution criterion is essentially the same as the Abbe or λ/D resolution limit; the Rayleigh resolution criterion is 22% larger. (Bottom): Schematic of changes in Airy disk appearance under different levels of seeing.

in apertures below 200 mm. Many double star astronomers prefer these small apertures for their pretty and vividly colored images.

The appearance of the turbulence can indicate its source and persistence. *Low frequency* seeing seems to undulate like ripples in water, causes large jumps in the star image, and often shows an upward flow in a highly defocused star image. *High frequency* seeing churns rapidly, shows little or no jumping around, and often presents a sideways flow in the defocused image. Low frequency seeing is typically in the telescope or near the ground and often abates as the telescope, ground and air cool after sunset – provided you are not observing near heated buildings or paved surfaces. High frequency turbulence often originates in changing weather or the jet stream and is likely to persist all night.

Helpful accessories

By far the most important accessory for viewing double stars – assuming you already have your thermal clothes, red light flashlight, observing chair and thermos of warming beverage – is a *standard eyepiece* that provides any combination of magnification and *apparent field of view* (AFOV) that you find routinely convenient with the telescope you are using.

Here preferences vary, depending on whether you think in terms of resolution or the field of view. I start with resolution, and prefer an eyepiece that gives an exit pupil of around 1.0: this lets me see the resolution limit of my telescope with a good sized field of view. Others prefer a magnification of 80 to 120 that presents a generous field, minimizes the effects of poor seeing in larger apertures and resolves all but the closest double stars. Try both approaches to see which one you prefer.

Eyepieces of the same focal length can have an AFOV between 40° and 100° or more, depending on optical design. I find an AFOV of 60° to 70° is optimal for the standard eyepiece: a smaller apparent field excludes too much of the surrounding skyscape, while a larger field makes it too difficult to estimate separation between wide double stars. This requires you to calculate the diameter of the *true field of view* (TFOV) of your eyepiece (see Appendix B), then use proportions of the field radius to judge angular distance. For example, if the calculated diameter of the true field is 20 arcminutes, then half the distance from the center to the edge of the field is 5 arcminutes.

Bracket this standard view with two eyepieces with half (or less) and double (or more) the standard eyepiece focal length. The lower power will deliver the most benefit with a superwide AFOV (80° or more). For the highest power eyepiece, a zoom lens across the short focal lengths is handy because you can nicely adjust the magnification to produce the best contrast and visual detail. Baader offers a very fine low power zoom; TeleVue has two excellent versions for high powers.

An *astrometric eyepiece* is available from Meade (the discontinued Celestron eyepiece can be purchased used). These show an illuminated linear scale across the field of view, used to estimate separation, and a circumference scale used to estimate position angle. The references include sources describing the use of these eyepieces. Although not as precise as a filar micrometer they are far less expensive and much easier to use.

Finally, the modern, computer-controlled GoTo mount is enormously helpful in double star astronomy. Accurate GoTo pointing makes it easy to identify an 8th magnitude target system within a field of similar stars, and it eliminates the time-consuming (although often enjoyable) challenge of star hopping with star chart and finder scope. A laptop or desktop computer provides the highest level of control, but lacking that a handset with an installed star inventory is indispensible.

Many observers prepare and print out observing checklists, finder charts for faint systems, guidebook descriptions and other aids. I print out observing lists with room for making notes or ratings of visual quality, and use the pocket-sized paper notebooks by Fabriano for making notes and drawings at the eyepiece. A compact voice-activated recorder is also handy to capture observations for later transcription.

Observing techniques

All systems in the target list can be resolved by telescopes up to 300 mm aperture under good seeing. I have not stinted on challenge doubles that will test the limits of your observing skill and observing conditions, no matter what aperture you own.

The three challenge parameters of a double star are the primary magnitude, the magnitude difference and the separation. But a $2''$ separation can be easy or difficult depending on your telescope's aperture. To eliminate this complication, just divide the separation by your telescope's resolution limit (or resolution criterion): this is the *resolution ratio* (ρ/R_o) of the binary in your telescope. A resolution ratio of 2 means the binary separation is twice as large as your resolution limit or criterion and should be easy to resolve – with sufficient magnification.

Magnitude difference is equally important, however. Their combined effect can be evaluated with a *Treanor plot* (Figure 6). The gray dots in the figure plot the Δm for doubles in the target list using the resolution ratio for a 250 mm aperture. There is clearly an area in the lower left – the combination of small resolution ratio and large Δm – where double stars are too difficult to resolve: glare or diffraction rings around the primary obscure the fainter star. The

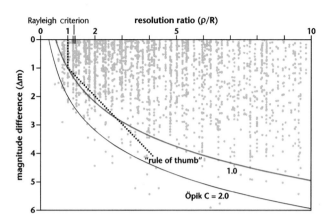

Figure 6 – A Treanor plot
A Treanor plot locates individual double stars (gray dots) by the resolution ratio (ρ/R_o) and the magnitude difference (Δm). Resolution thresholds can be drawn into the plot to indicate whether a binary will be easy or difficult to resolve: two simple examples, the Öpik limit C and the "rule of thumb," are shown.

curve ($C = 1$) in the diagram represents a resolution boundary suggested by the Estonian astronomer Ernst Öpik: double stars plotted above this curve should be easy to resolve. The shape of the curve indicates that close doubles must be similar in magnitude to be detected easily, and that glare rapidly decreases with separation. A second curve, for $C = 2$, suggests how far your observing skill and favorable seeing can increase the systems you can observe.

Unfortunately, the Öpik and similar detection formulas require a calculator to implement. A simple "rule of thumb" approximation for $C = 1$ is: if Δm is less than 1 then the pair can be resolved down to the Abbe resolution limit ($\rho/R_o = 1$); otherwise Δm must be less than the resolution ratio ($\Delta m \leq \rho/R_o$). With experience and good equipment, you will routinely detect unequal doubles beyond this simple limit.

Small separation and large brightness contrast are not the only visual difficulties you will encounter. All close doubles are difficult in bad seeing. And difficulty increases at the magnitude extremes: a bright primary star will produce more diffraction rings and cast obscuring glare at a greater distance, especially on nights when there is diffusion from atmospheric humidity or dust. A reflecting telescope with a central obstruction that is more than 30% of the diameter of the aperture pushes excess obscuring light into the diffraction rings. If the stars are too faint

for your aperture, they approach the foveal limit where your detail vision fails, and the lower resolution of your averted vision will be unable to resolve them. In all these situations, the stars must be farther apart in order to be seen. And there is occasional inaccuracy in the cataloged magnitudes or separation, making the star seem easy to resolve – on paper.

Patience is a critical ingredient in most visual astronomy. Unless the seeing is very good, you must wait for the "revelation peeps" (as astronomer Percival Lowell called them) when the air briefly steadies and you can glimpse close doubles without distortion. You must learn what a companion star looks like when it mingles with the diffraction rings of the brighter star – often appearing to be only a persistent speckle. And using sufficiently high magnification is often the key. With poor seeing, high magnification helps because the Airy disk of bright stars can be observed "through" the turbulence. (This runs counter to the practice of planetary and lunar astronomers who are taught to reduce magnification in bad seeing.) With good seeing, the right magnification can make faint components visible by suppressing the background sky brightness and make components hidden in the diffraction rings easier to detect.

How can you know if you have actually detected a close or difficult companion? The best case is that you *resolve* the binary and see two unmistakable Airy disks, separated by a sliver of darkness or barely touching at their edges. However you can *detect* a binary as a stubby rod produced by the overlap of two Airy disks, or you may see a fleeting or indistinct brightening of the diffraction rings, a fuzzy blob or persistent speckle that will not look at all like an isolated star. To confirm your observation, visually estimate the separation and position angle of the component, and compare your estimate to the target list values. This is easiest if you have a German equatorial mount (GEM). Align your focuser or star diagonal so that the eyepiece optical axis is parallel to the GEM declination axis, and the east/west (following/preceding, 90°/270°) position angles in the eyepiece field will be parallel to the optical axis of the telescope. Or just let the star drift toward the west (270°) direction. (Note that field orientation rotates 180° when you turn your telescope across the meridian, switching west and east.)

If your estimate of position angle is reasonably close to the catalog value, then detection is confirmed.

From the eighteenth century to today, star color has been one of the most carefully described and widely enjoyed aspects of double star astronomy. However, color perception varies from one person to the next and the colors you see will differ from those seen by other observers. All star colors are quite unsaturated (diluted with "white" light), which can cause star color to change the longer you look at it, and your vision will alter the apparent color of stars with complementary color contrast, for example when a bright yellow or orange star makes a white companion star appear blue or green. As stars become fainter, the same orange or red hues appear more saturated, but yellow stars will fade quickly to an apparent gray, and otherwise colorless stars can appear pale blue. The telescope aperture, the brightness of the stars, the brightness contrast and angular separation between them will all noticeably affect star color. In double stars, color really is in the eye of the beholder, with the telescope an integral part of your eye.

To examine and describe star color, think in terms of the basic incandescent or "hot metal" hue categories – *red, orange, yellow, white, blue* (abbreviated R, O, Y, W, B) and adjacent blends such as *yellow orange* (YO) or *blue white* (BW). (*Blue green, green* and *violet* are illusory.) In bright stars, color can be displayed more clearly if you slightly defocus the star in the extrafocal direction to present the color as a small disk. Notice also how much magnification affects star color: Rasalgethi, for example, will appear at low power as two yellow orange stars, but at an exit pupil of around 1.0, the beautiful yellow and blue (Y/B) contrast appears clearly. These complications limit the reliability of color as an indicator of a star's spectral type: all the more reason to just enjoy the view.

Finally, change eyepieces often. The standard eyepiece, with the target system centered in the field of view, is always your basis to compare one system to another. Compare the magnitude of the components to the magnitude of the primary, and note whether the star is in a *rich field* or a *dark field*. (Optical pairs are much more likely in a crowded field of similar

stars.) Use averted vision to search for faint components, and search the field of view for binaries, bright stars or groups of stars that may be associated with the target. Note the distance and quadrant direction of these features from the primary star.

For close pairs, use your high power eyepiece to resolve the target system. The Airy disk and the dark interval between the disk and first diffraction ring must be clearly visible. Conclude by switching to a low power eyepiece with a large TFOV. Near the galactic equator this can display splendid views in every part of the Milky Way and reveal nearby star clusters or nebulae. When you slew to the next target, this low power is ideal to locate the new system and center it in the field for viewing with the standard eyepiece.

Next steps

Double stars offer a great range of opportunities for the amateur astronomer. The best of these is simply keeping notes based on your observations. Separate from the value of a permanent record, notetaking encourages you to look longer and both study and enjoy what you see. The remarks in the target list imply a routine format for observations, but what and how you record is a highly personal aspect of double star astronomy. Whatever your style, abbreviations are especially handy to record essential or routine features efficiently.

A fun way to start is the *Double Star Observing Program* moderated by The Astronomical League. You'll be asked to observe and draw 100 famous double stars, submit your work and receive a certificate, pin and the satisfaction of having met many double stars (and optical pairs) in person.

Sampling the systems in the target list may inspire you to view all the systems in one of the important historical catalogs – an *observing campaign* similar to the Messier challenge or Herschel challenge of deep sky astronomers. The catalogs by William Herschel, James Dunlop, Charles Rumker or James South are all attractive, if liberally mixed with pair asterisms. A favorite and ambitious challenge is the list of ~ 3,000 targets compiled in the early nineteenth century by F. Wilhelm von Struve (Σ). This catalog contains a great number of interesting and spectacular objects, and for that reason has formed the backbone of nearly every double star observing list. (His supplement catalogs, denoted Σ I and Σ II, are limited to bright, wide pairs.)

Many amateurs enjoy measuring double stars, and this contributes directly to the project of confirming new physical systems and refining the orbits of those already known. This *Atlas* can be used to guide your choice of physical systems most fruitful to study, and the USNO will on request provide observers with a custom list of "neglected doubles" in dire need of attention. A micrometer eyepiece can be used to measure wide, fairly bright systems; filar and split prism micrometers are no longer manufactured commercially, and custom-made units can be very expensive. Instead, many amateurs use charge-coupled device (CCD) imaging or drift timing; articles in the references describe these techniques and the software tools necessary to produce reliable measurements. Amateurs with very large telescopes can make productive contributions to the photometry and even the spectral analysis of double stars. (Note that you will be required to demonstrate the accuracy of your equipment and methods before your measurements will be added to the WDS.) Amateurs with access to computing resources and skill in working with data have made significant contributions, aided by the high quality of astronomical resources available online. Remarkably, many double stars have been "discovered" simply by reviewing archival astrometric data or research reports.

Double stars are a marvelous esthetic inspiration. Capturing the beauty of these dynamic beacons is a worthy artistic challenge, and many observers have developed methods to draw or paint eyepiece portraits of double stars. Others meet the challenge with color video astrophotography. Photographic equipment today, especially the combination of video and lucky imaging, is capable of capturing astonishingly delicate images of the brighter double stars that were inconceivable just a few decades ago.

The *Journal of Double Star Astronomy* (JDSO) maintains an online archive of hundreds of articles that explain the amateur use of measurement methods, software and equipment. This is a great resource for browsing just to see what others have done. Although more technical in nature, the Webb Deep-Sky Society *Double Star Circulars* also describe methods of discovery and measurement and are worth exploring. Both these electronic publications (see the references) show that double stars are still being discovered by amateurs working with amateur equipment.

But don't just read about it. The surest way to expand your knowledge of double stars is to collaborate – through your local astronomy club, star parties, astronomical conferences, and through classes by professors at your high school, city college or university. Astronomy is often a dark and solitary activity, and perhaps for that reason astronomers are often eager to share their experience and encouragement with newcomers and colleagues in this fascinating and rapidly evolving field.

References

Books

R. G. Aitken, *Binary Stars* (1938). A classic, with good historical and technical chapters. Available used as a Dover reprint (1964).

R. Argyle, *Observing and Measuring Visual Double Stars*, 2nd edn., Springer (2014). The primary resource for the amateur double star astronomer, with chapters on all types of observing, measuring and imaging techniques and equipment. Absolutely essential.

R. Burnham Jr., *Burnham's Celestial Handbook, Revised and Expanded Edition*, volumes 1, 2 and 3, Dover Publications (1977). A useful basic resource on many astronomical topics and deep sky objects, but poorly indexed and badly out of date. A perennial rainy day diversion.

P. Couteau, *Observing Visual Double Stars*, MIT Press (1981). Reader-friendlier than Heintz, with many topics of interest to the double star observer. Available used.

S. Haas, *Double Stars for Small Telescopes*, Sky Publishing (2006). A field guide by an active amateur observer with an affection for star color poetry. Recently reissued.

W. Heintz, *Double Stars*, Springer (1978). A classic in the field, compact and informative, often technical or missing recent research but still irreplaceable. (You can download a free digital copy from https://archive.org/details/DoubleStars.)

R. Kent Clark, ed., *The Double Star Reader: Selected Papers From the Journal of Double Star Observations*, Collins Foundation Press (2007). A convenient selection of 45 amateur papers on all aspects of double star astronomy. (Papers after 2007 are available from the JDSO online archive.)

B. Kepple and G. Sanner, *Night Sky Observer's Guide*, volumes 1 and 2, Willmann-Bell (1998). A comprehensive guide to northern hemisphere deep sky objects, including a small selection of double stars, compiled by a cadre of dedicated amateur observers. The southern hemisphere volume by I. Cooper, J. Kay and G. Kepple was published in 2008.

I. Ridpath, ed., *Norton's Star Atlas and Reference Handbook*, 20th edn., Penguin Group Inc. (2004). In print for a century, this is perhaps the best single distillation of celestial charts, informative text, and clearly organized reference tables. The double star lists are great for casual observing.

K. Robinson, *Starlight: An Introduction to Stellar Physics for Amateurs*, Springer (2009). An overview of light, spectroscopy, stellar spectral and luminosity types and stellar evolution.

Royal Canadian Astronomical Society, *Observer's Handbook* (Annually). This invaluable publication is equal parts astronomical textbook, ephemeris and observing guide, and includes observing lists of double stars and colorful double stars.

J. B. Sidgwick, *Amateur Astronomer's Handbook*, Dover Publications (1971). A classic trove of useful and generally still accurate information on optics, visual observation and measurement (some chapters on equipment are outdated).

Magazines

Sky & Telescope is for some an essential resource because it has been in continuous publication since 1946, and all back issues up to 2009 are available as a boxed set of CD-ROM disks with a rudimentary search function. *Astronomy Magazine* began publishing in 1973 and is equally respected among amateur astronomers. In Britain, *Astronomy Now* appeared in 1987 and is of comparable style, polish and editorial focus.

All these publications are targeted to amateur astronomers, provide reports on cutting edge astronomical research, hobbyist activities, amateur equipment reviews; all are available online. They also each publish a selection of books, atlases and digital or video materials, and participate in astronomical conferences and trade shows.

Articles available online

Note: Most of these articles, and hundreds more, can be downloaded in pdf format from the SAO/NASA Astrophysics Data System website (referenced below): simply search for the author last name(s) and date of publication. JDSO articles can be downloaded in PDF format from the JDSO website at http://www.jdso.org/archives.xml. Webb Deep-Sky Society Double Star Circulars can be downloaded at http://www.webbdeepsky.com/double-stars/double-star-section-circulars.

R. G. Aitken, "The Definition of the Term Double Star" (1911). An early proposal to exclude optical pairs from double star catalogs, including an early description the "Aitken test" (as proposed by E. C. Pickering!).

R. Anton, "On the Accuracy of Double Star Measurements from 'Lucky' Images, a Case Study of Zeta Aqr and Beta Phe" (JDSO, 2009). A useful introduction to the method of "lucky imaging" to improve CCD image clarity.

R. Caballero, "Finding New Common Proper Motion Binaries by Data Mining" (JDSO, 2009). Describes how to use NOMAD data and Aladin sky survey imagery to identify CPM pairs.

R. M. Caloi, "Estimation of Double Star Parameters by Speckle Observations Using a Webcam" (JDSO, 2008). Explains the methods of speckle interferometry, which exploits atmospheric turbulence to obtain high resolution images.

J. Daley, "A Method of Measuring High Delta m Doubles" (JDSO, 2007). Describes construction and use of a stellar coronagraph in the CCD measurement of bright double stars with faint components.

S. P. Goodwin, "Binaries in Star Clusters and the Origin of the Field Stellar Population" (2009). Recent, relatively non-technical summary of the way the "birth population" of double stars is transformed by the natal cluster.

R. Harshaw, "Using VizieR to Measure Neglected Double Stars" (JDSO, 2013). Step by step explanation for using sky survey photographs to measure telescopic double stars.

C. Lada, "Star Formation in the Galaxy, An Observational Overview" (2005). An informative summary of star formation processes with emphasis on solar mass double stars.

S. Larson, "Binary Stars" (2010). A concise and lucid summary of the binary orbital dynamics. (http://sciencejedi.com/professional/classes/astrophysics/lectures/lec08_binaries.pdf)

B. Mason, "Various Orbital Solutions and Double Star Statistics" (2011). Compact introduction to speckle interferometry and issues of orbital calculation.

R. Mollise, "Photographing and Measuring Double Stars with Unconventional Imagers" (JDSO, 2005). A basic overview of cameras and software; first of three articles.

D. Raghavan, H. McAlister, T. Henry, *et al.*, "A Survey of Stellar Families: Multiplicity of Solar Type Stars" (2010). The latest and most complete analysis of multiplicity in solar type stars, a landmark study.

B. Reipurth and S. Rikkola, "Formation of the Widest Binaries from Dynamical 'Unfolding' of Triple Systems" (2013). Computer simulations show how wide binaries and "hard" binaries evolve from the dynamic decay of close triple systems.

E. Shaya and R. P. Olling, "Very Wide Binaries and Other Comoving Stellar Companions" (2011). The discovery of almost 800 wide binaries and comoving pairs through a computer analysis of the Hipparcos catalog.

D. Sinachopoulos and P. Mouzourakis, "A Statistical Approach for the Recognition of Physical Visual Double Stars" (1991). Calculates all sky 95% probability limits for the maximum angular separation by magnitude.

M. Sterzik and R. Durisen, "Are Binary Separations Related to their System Mass?" (2004). A Monte Carlo study exploring the effect of system mass on the multiplicity fraction and binary orbital elements.

P. J. Treanor, "On the Telescopic Resolution of Unequal Binaries" (1946). An early evaluation of the effect of unequal magnitude on double star detection.

R. Wasson, "Measuring Double Stars with a Dobsonian Telescope by the Video Drift Method" (JDSO, 2014). A thorough explanation of this measurement method, here using a GoTo altazimuth telescope.

B. Wilson and W. Hartkopf, "The US Naval Observatory Double Star Program: Frequently Asked Questions" (JDSO, 2011). Useful overview of the WDS database and the requirements for submitting measurements.

Websites

Astronomical Files from Black Oak Observatory: http://www. handprint.com/ASTRO/ – Extensive information on double stars and visual double star astronomy.

Astronomical League Observing Programs: https://www. astroleague.org/observing.html – Observing campaigns suited to all levels of skill and supervised by the umbrella organization for USA amateur astronomy societies. A great way to start!

Calibration Candidates: http://ad.usno.navy.mil/wds/orb6/ orb6c.html – A list of binaries that can used for calibrating astrometric instruments.

Henry Draper Catalogue at SkyMap.org: http://server6.sky-map.org/group?id=23 – Separate listings of information and related research for most HD objects. The index is a useless multipage shambles, so google the HD number to find a page directly.

Multiple Star Catalog: http://www.ctio.noao.edu/ ~atokovin/stars/index.php – Last update April 2010. A. Tokovinin's online database of multiple stars,

showing hierarchical structure, spectral types and masses of all components.

SAO/NASA Astrophysics Data System: http://adsabs. harvard.edu/ – Portal for a digital library of research papers in astronomy and physics, operated by the Smithsonian Astrophysical Observatory (SAO). A national treasure.

SIMBAD Astronomical Database: http://simbad.u-strasbg. fr/simbad/ – The comprehensive repository for all published astronomical data and survey photography maintained at the University of Strasbourg (FR).

Sixth Catalog of Orbits of Visual Binary Stars: http://www. usno.navy.mil/USNO/astrometry/optical-IR-prod/wds/ orb6 – A compilation of visual (including interferometric) binary star orbits, updated frequently.

Stelle Doppie: http://stelledoppie.goaction.it/ – Gianluca Sordiglioni's comprehensive and easy to use database of double star data, compiled from many sources.

Washington Double Star Catalog (WDS): http://www.usno. navy.mil/USNO/astrometry/optical-IR-prod/wds/ WDS – The primary database of double and multiple star information, maintained at the US Naval Observatory; the "summary" version, separated into four 4Mb text files, can be downloaded for free.

Acknowledgements

Although it is true that "this research has made use of the *Washington Double Star Catalog* maintained at the US Naval Observatory," that is meager recognition for the generous support and guidance provided to double star astronomers by William Hartkopf and Brian Mason of USNO. I also have relied on the research services of SIMBAD and VizieR with assistance from Gilles Landais and Cécile Loup (University of Strasbourg), and the online archives of physical science papers at SAO/ NASA (Harvard University) and arXiv.org (Cornell University). Special tribute is due to Gianluca Sordiglioni, curator of the *Stelle Doppie* website, for preparing the target list data, vetting the astronomical data sources and correcting many errors in the process. Dr. Hartkopf, James Barnett, Phil Sullivan, Ross Gould, Jan Randall and our copyeditor Zoë Lewin provided many useful comments on the text, and Chloé Harries skillfully guided the manuscript to press. Finally, a heartfelt thanks to Wil Tirion, my tireless and talented coauthor, who made extensive changes to the charts and created the illustrations for the text, and to Vince Higgs, our patient and encouraging editor at Cambridge University Press, for agreeing to embark on a new edition. My gratitude for their collaboration must speak my regret for any unforced errors.

I dedicate this edition to my father John MacEvoy who helped me build my first telescope and taught me to appreciate the beauty of our world.

Bruce MacEvoy, FRAS
Sebastopol, California, USA

Wil Tirion wants to thank James (Jim) Mullaney for his original idea to create a special atlas devoted to double stars. For this Second Edition, a lot of time went into updating the star maps and I want to express my thanks to Bruce MacEvoy for creating special lists to make this editing process as easy as possible. And, last but not least, my (our) thanks goes to our editior Vince Higgs, for his valuable guidance of the project.

Wil Tirion – Uranographer
Capelle aan den IJssel, The Netherlands

STAR CHARTS

CHART INDEX
Northern hemisphere

CHART INDEX
Southern hemisphere

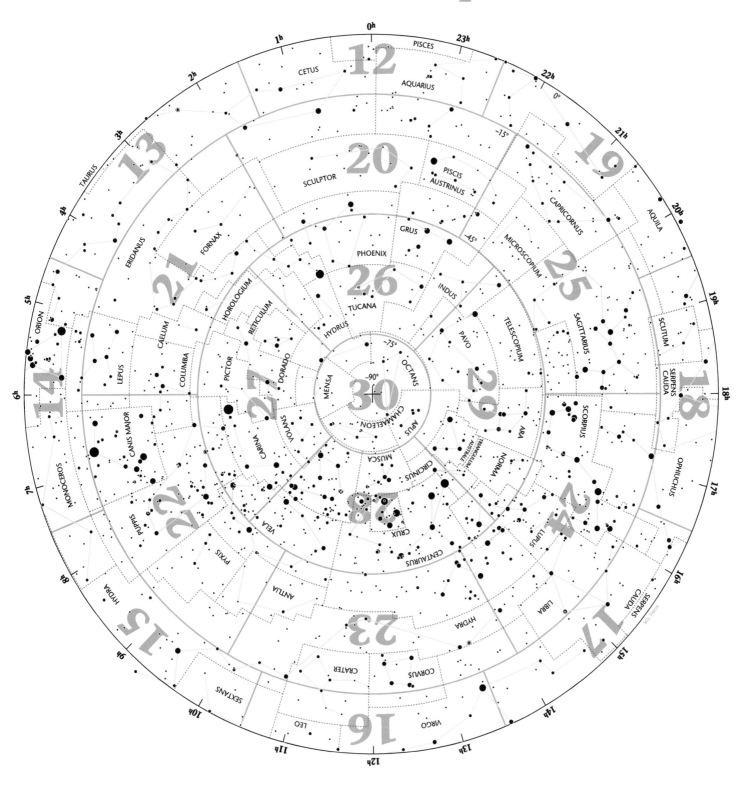

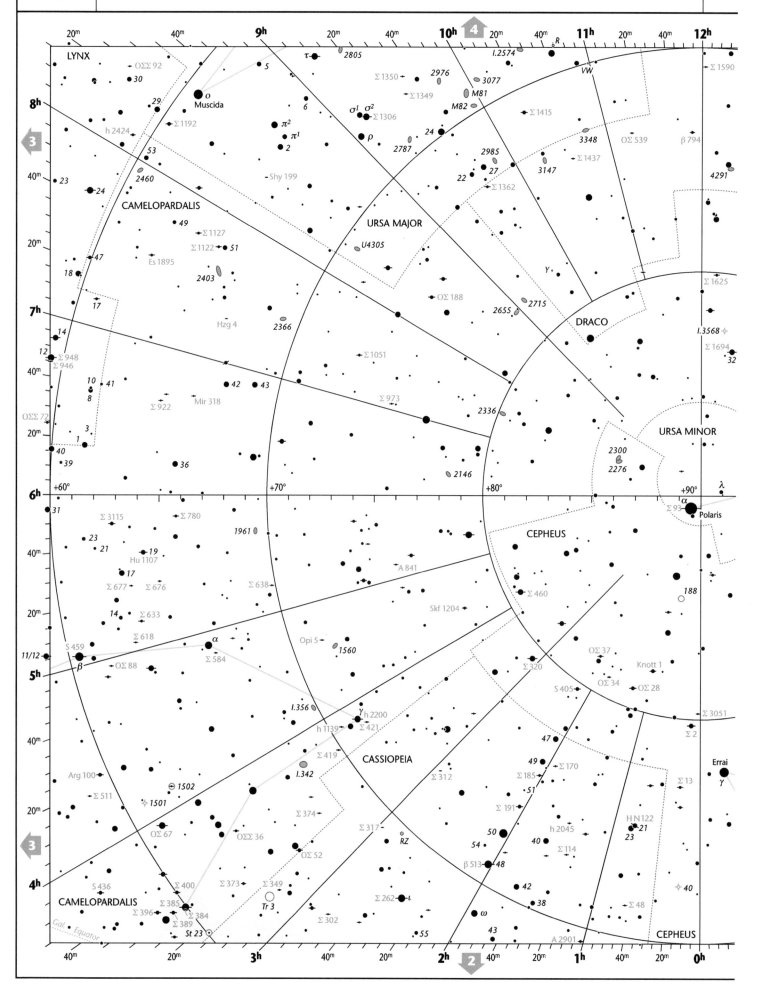

1

Magnitudes
● ● ● ● ● ● ● ● ● ● ● ● ● · · ·
brighter than 0.0 0.0–0.5 0.5–1.0 1.0–1.5 1.5–2.0 2.0–2.5 2.5–3.0 3.0–3.5 3.5–4.0 4.0–4.5 4.5–5.0 5.0–5.5 5.5–6.0 6.0–6.5 6.5–7.0 7.0–7.5 fainter

LYNX

20ᵐ   40ᵐ   **9ʰ**   20ᵐ   40ᵐ   **10ʰ**   20ᵐ   40ᵐ   **11ʰ**   20ᵐ   40ᵐ   **12ʰ**

OΣΣ 92
30
29
8ʰ
h 2424   Σ 1192
53
2460
23   24
CAMELOPARDALIS
49   Σ 1127
Σ 1122   51
18   47
17
2403
Hzg 4   2366
7ʰ
14
12   Σ 948
Σ 946
10   41
8
OΣΣ 72   Σ 922   Mir 318
40ᵐ
3
1
40   40ᵐ
39   36
6ʰ **+60°**   **+70°**   **+80°**   **+90°**   λ

τ   2805
5
6
π²
π¹
2
σ¹ σ²
ρ   Σ 1306
Shy 199   2787
24
22   U4305
OΣ 188
2655   2715
Σ 1051
Σ 973
2336
2146

URSA MAJOR

R
I.2574   Σ 1590
Σ 1350   2976
3077
Σ 1349   M81
M82
Σ 1415
3348   OΣ 539   β 794
2985   Σ 1437   4291
27   3147
Σ 1362
γ   Σ 1625
VW
DRACO   Σ 1694
I.3568   32
URSA MINOR
2300
2276
α   Σ 93
Polaris

31
Σ 3115   Σ 780
23
21   19
Hu 1107
17
Σ 677   Σ 676
14   Σ 633
Σ 618
S 459
11/12   β   OΣ 88
5ʰ
I.356
Arg 100   h 1139   Σ 419
1502
Σ 511   1501
OΣ 67   OΣΣ 36
CAMELOPARDALIS
S 436   Σ 400
Σ 385
Σ 396   Σ 384
Σ 389
Gal. Equator   St 23

40ᵐ   20ᵐ   **3ʰ**   40ᵐ   20ᵐ   **2ʰ**   40ᵐ   20ᵐ   **1ʰ**   40ᵐ   20ᵐ   **0ʰ**

1961
A 841
Σ 638
Skf 1204
Σ 460
Opi 5   1560
α   Σ 584
γ h 2200
Σ 421
CASSIOPEIA
Σ 312
I.342   Σ 191
Σ 374
Σ 317   RZ
OΣ 52
Σ 373   Σ 349
Tr 3
Σ 262   ι
Σ 302
55
ω
43
A 2901

CEPHEUS
OΣ 37
Knott 1
OΣ 34   OΣ 28
S 405
Σ 320   Σ 3051
47   Σ 2
49   Σ 170
51
50   h 2045
40
54   Σ 114
β 513   48
Errai
γ   Σ 13
H N 122
21
23
42
38   Σ 48
40
CEPHEUS

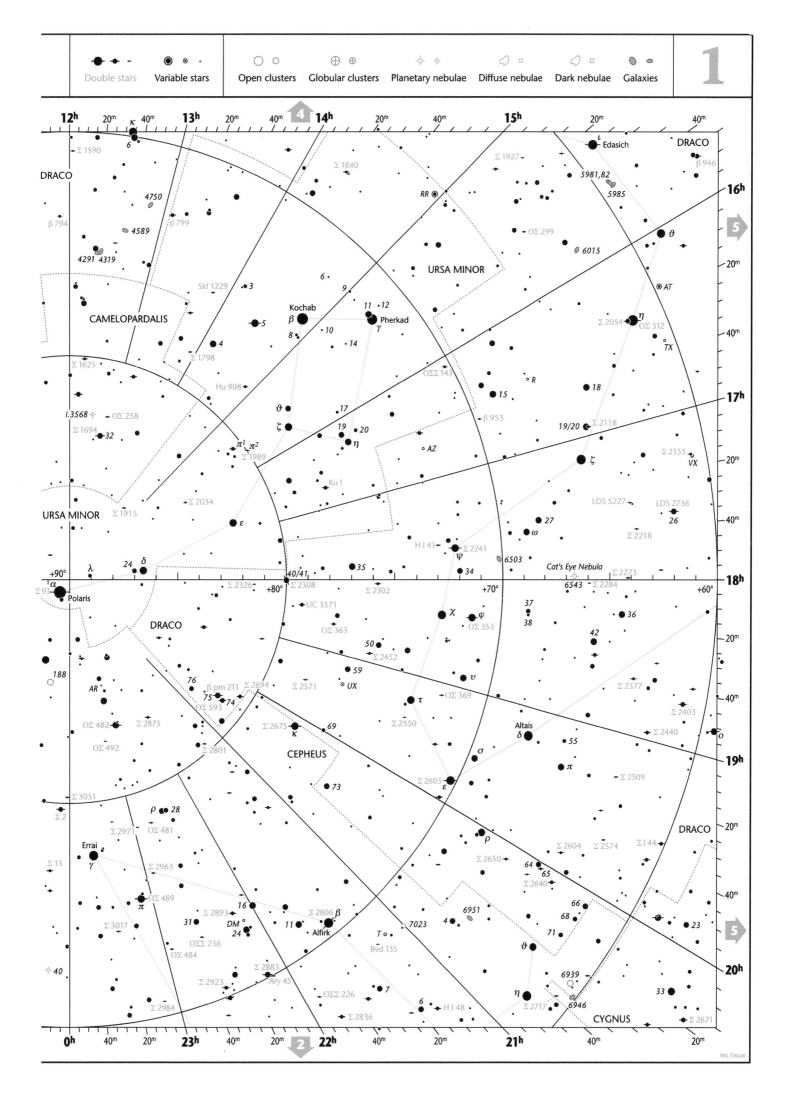

Double stars   Variable stars   Open clusters   Globular clusters   Planetary nebulae   Diffuse nebulae   Dark nebulae   Galaxies

WIL TIRION

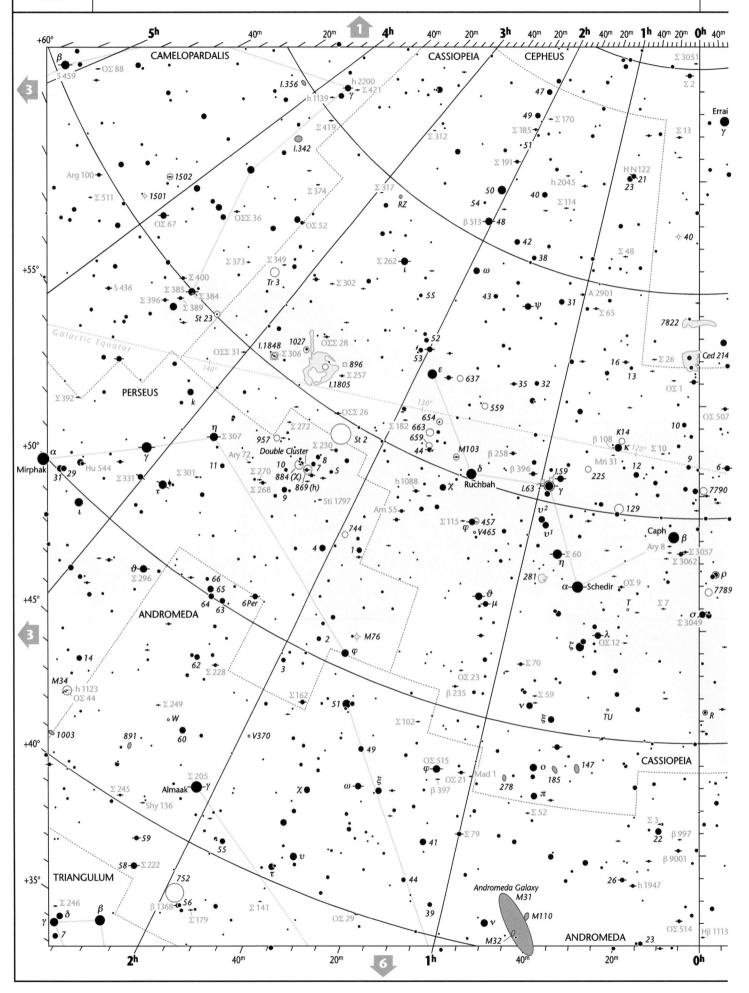

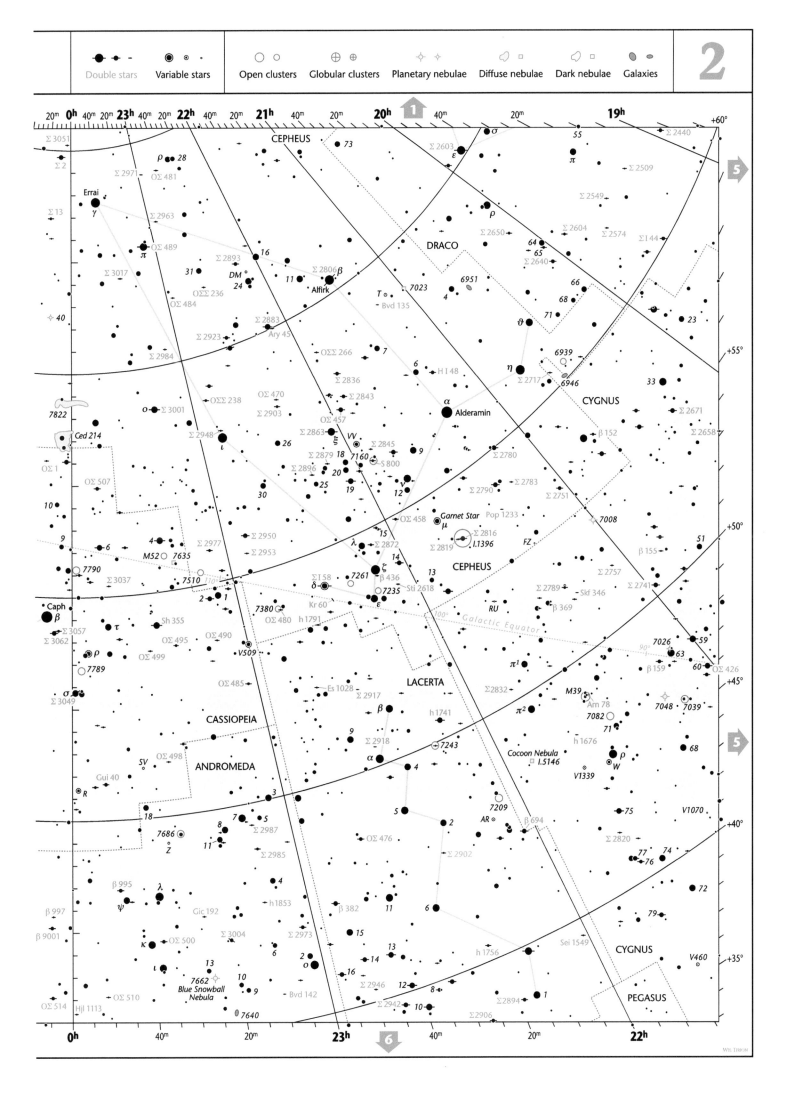

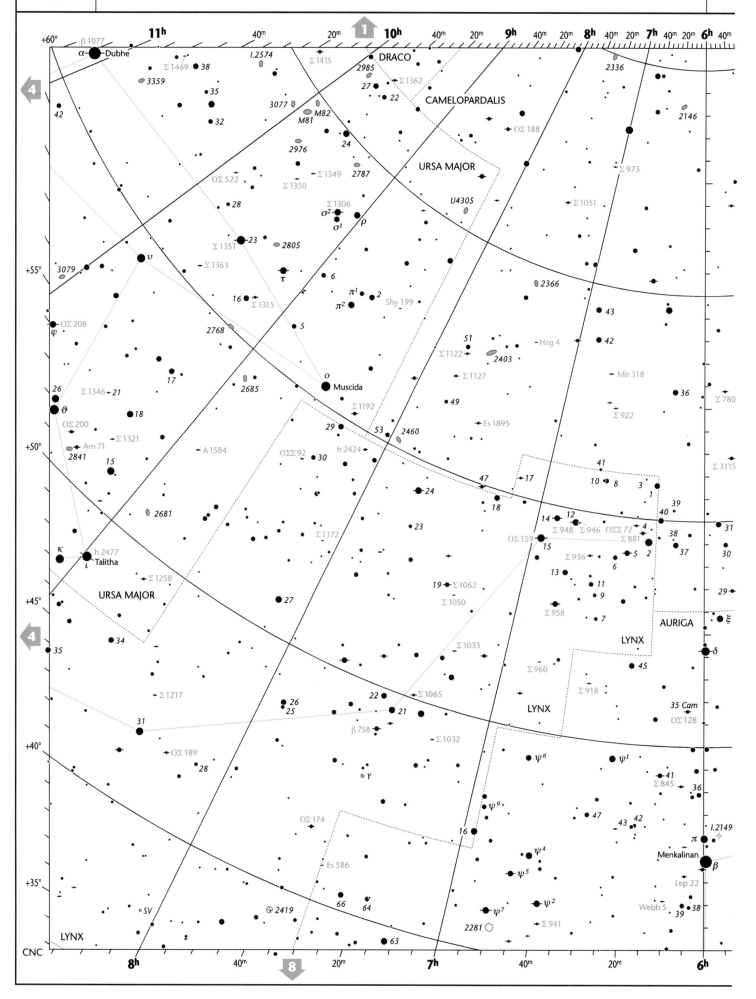

3

Magnitudes

brighter than 0.0   0.0–0.5   0.5–1.0   1.0–1.5   1.5–2.0   2.0–2.5   2.5–3.0   3.0–3.5   3.5–4.0   4.0–4.5   4.5–5.0   5.0–5.5   5.5–6.0   6.0–6.5   6.5–7.0   7.0–7.5   fainter

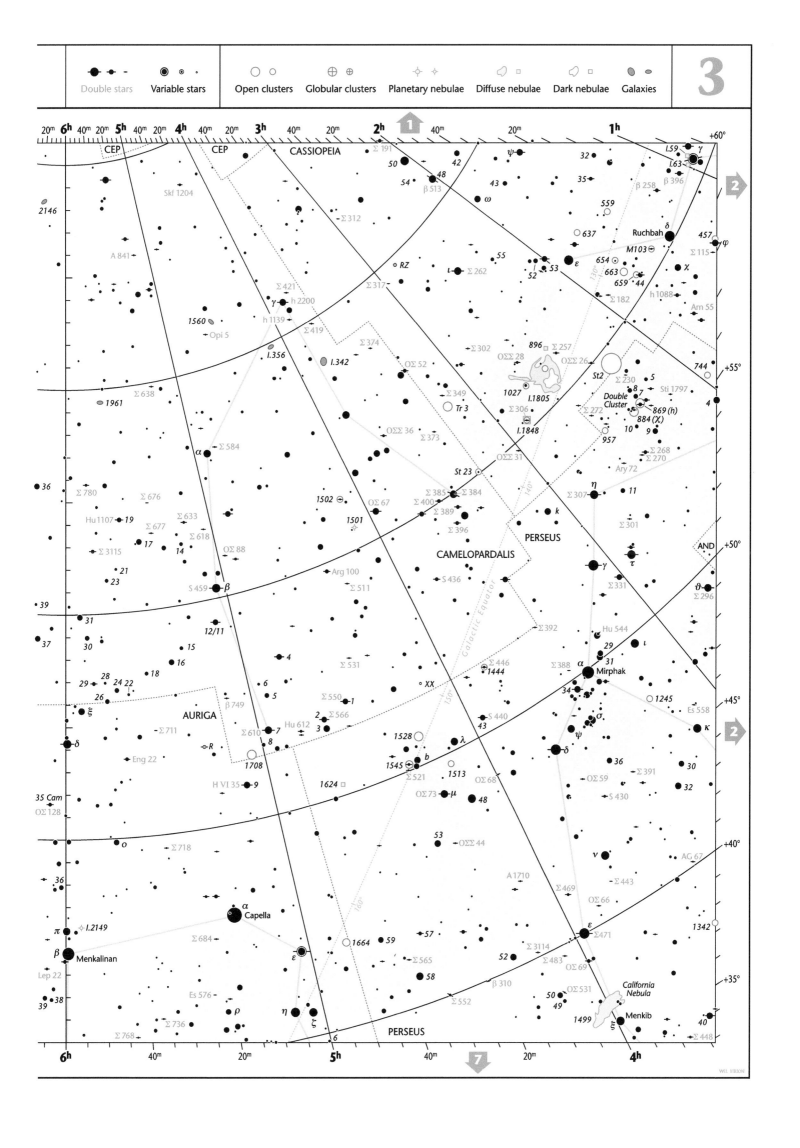

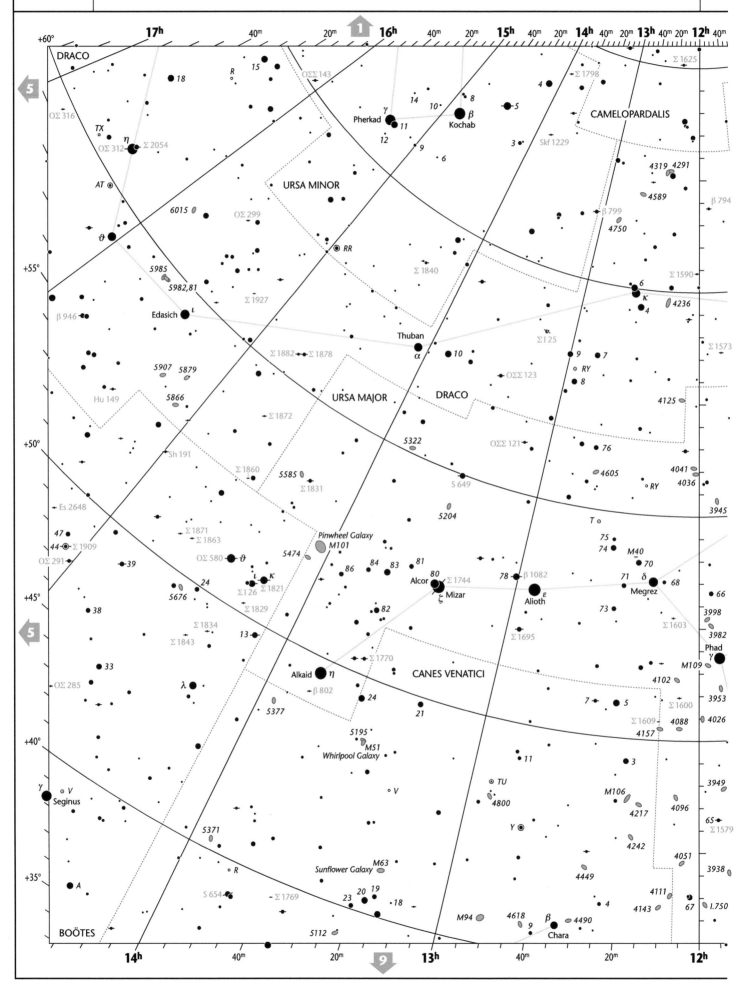

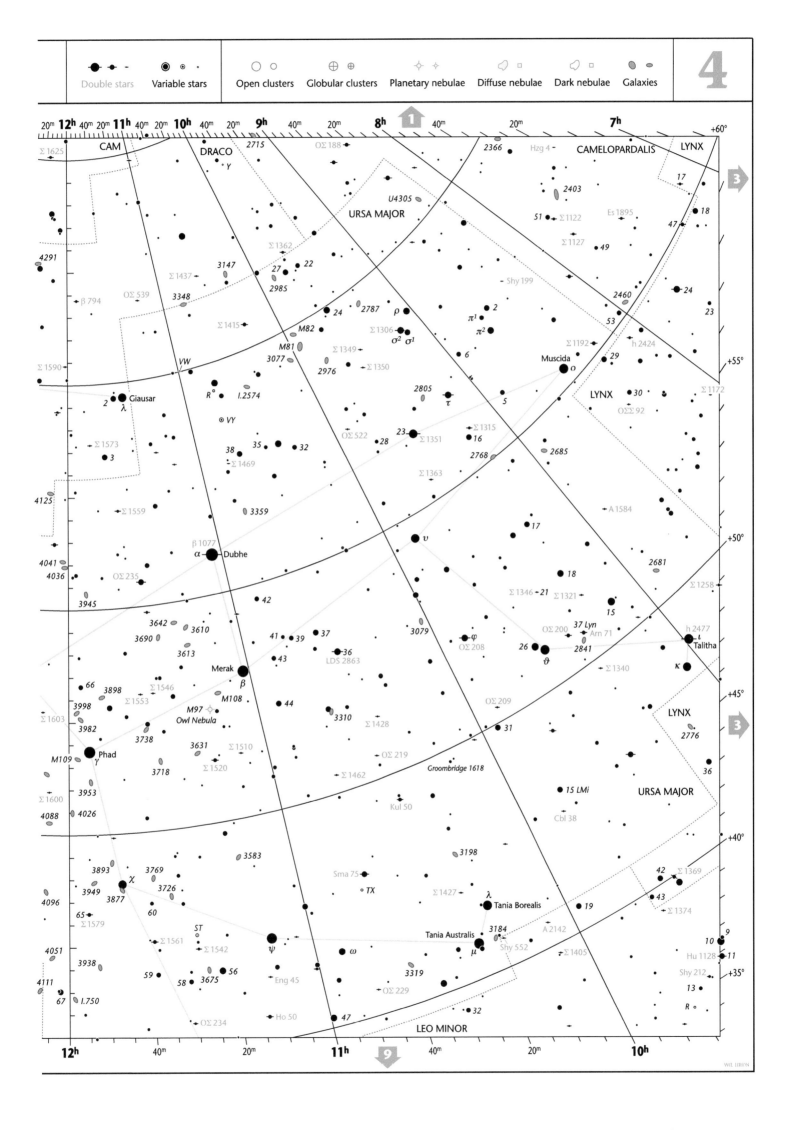

Double stars   Variable stars   Open clusters   Globular clusters   Planetary nebulae   Diffuse nebulae   Dark nebulae   Galaxies

20ᵐ 12ʰ 40ᵐ 20ᵐ 11ʰ 40ᵐ 20ᵐ 10ʰ 40ᵐ 20ᵐ 9ʰ 40ᵐ 20ᵐ 8ʰ 40ᵐ 20ᵐ 7ʰ +60°

Σ 1625
CAM
DRACO
γ
2715
ΟΣ 188
2366   Hzg 4   CAMELOPARDALIS   LYNX

17
2403
Es 1895
18
51 Σ 1122
47
U4305
49
24

4291
URSA MAJOR
2460
53
23

Σ 1362
Shy 199
3147
27 22
ΟΣ 539
Σ 1437
2985
3348

β 794
3348
24   2787   ρ
π¹   2
M82   Σ 1306   π²
σ² σ¹
Muscida
Σ 1192
h 2424
29

Σ 1415
M81
3077   Σ 1349
2976   Σ 1350
6
5
ο
LYNX
30
ΟΣ 92
Σ 1172

Σ 1590
VW
2805
τ
μ

2 Giausar
λ
R°   I.2574
VY
ΟΣ 522
28
23   Σ 1351
16
Σ 1315
Σ 1363
2768
2685

Σ 1573
3
38   35   32
Σ 1469

4125
Σ 1559
3359
A 1584

4041
β 1077
α   Dubhe
ΟΣ 235
4036
17
18
2681
Σ 1258

3945
42
υ
3079
Σ 1346 21 Σ 1321
15

3642   3610
3690
3613
41 39 37
43
36
LDS 2863
φ
ΟΣ 208
26   ϑ
37 Lyn Arn 71
2841
ΟΣ 200
h 2477
ι
Talitha
κ

66
3898
Σ 1546
Σ 1553
Merak
β
M108
44
3310
Σ 1428
ΟΣ 209
31
Σ 1340
LYNX
2776

3998
3982
M97
Owl Nebula
3738
3631   Σ 1510
Σ 1520
M109
γ Phad
3718

Σ 1603
Groombridge 1618
36
Σ 1600
3953
4088   4026
Kul 50
15 LMi
Cbl 38
URSA MAJOR

4111
3893
χ
3769
3726
3583
3198
42 Σ 1369
43
Σ 1374

3949
3877
60
Sma 75
TX
Σ 1427
λ Tania Borealis
19
9
10
Hu 1128 11

4096
65
Σ 1579
ST
Σ 1561
Σ 1542
ψ
ω
3184
Shy 552
A 2142
Shy 212

4051
3938
59
58 3675 56
Eng 45
3319
Tania Australis
μ
Σ 1405
13
R

67 I.750
ΟΣ 234
Ho 50
47
32
ΟΣ 229
LEO MINOR

20ᵐ 12ʰ 40ᵐ 20ᵐ 11ʰ 40ᵐ 20ᵐ 10ʰ

WIL TIRION

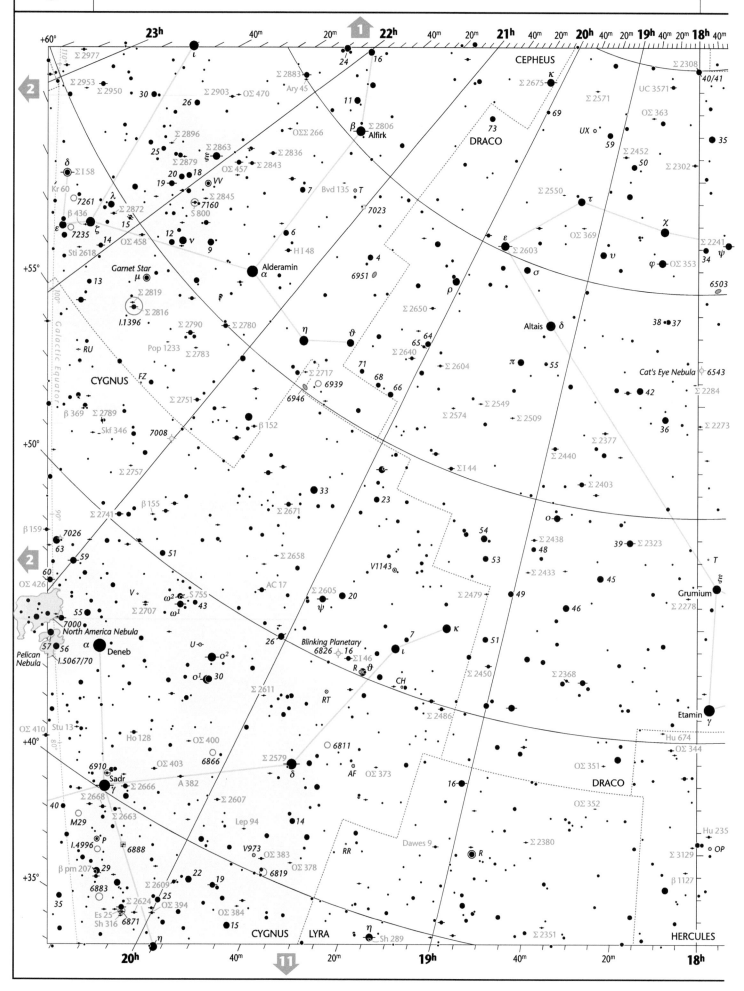

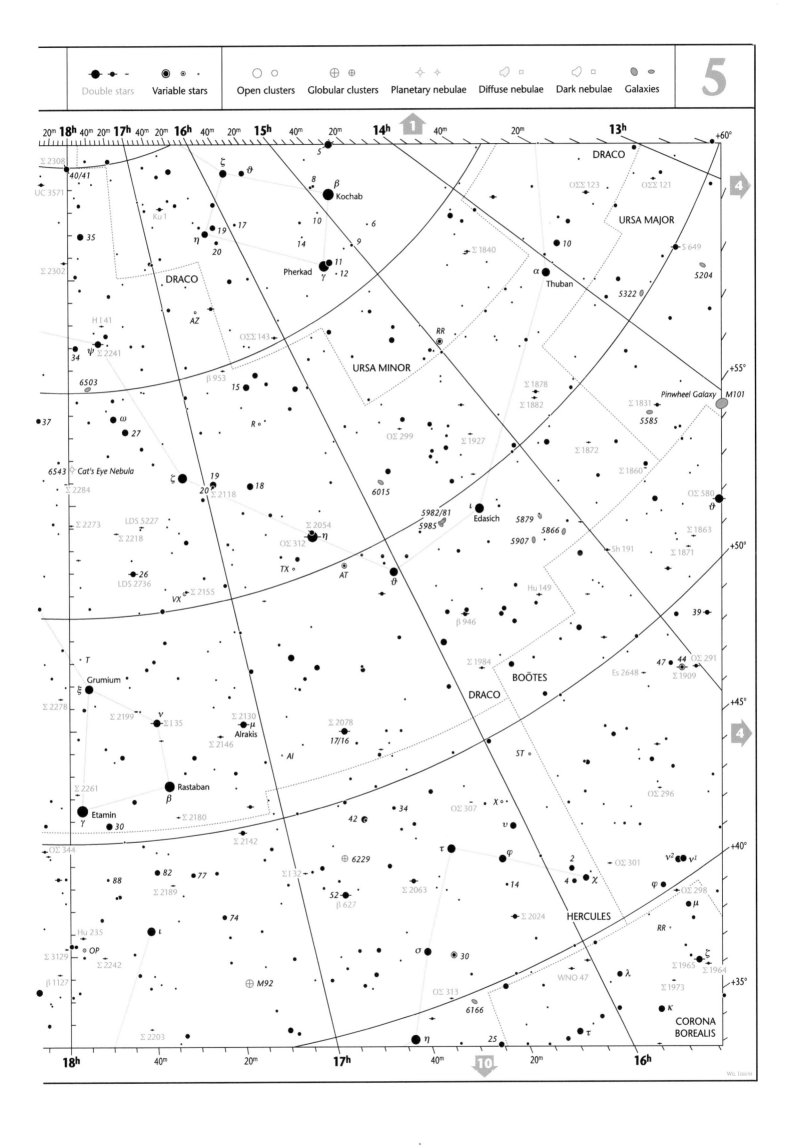

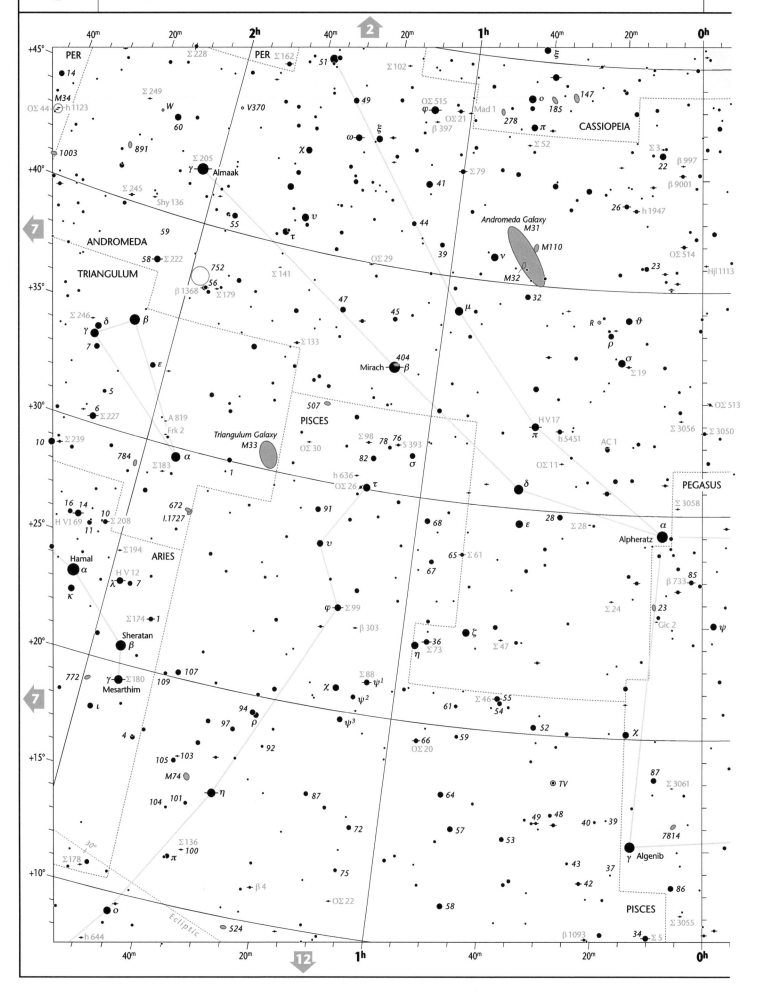

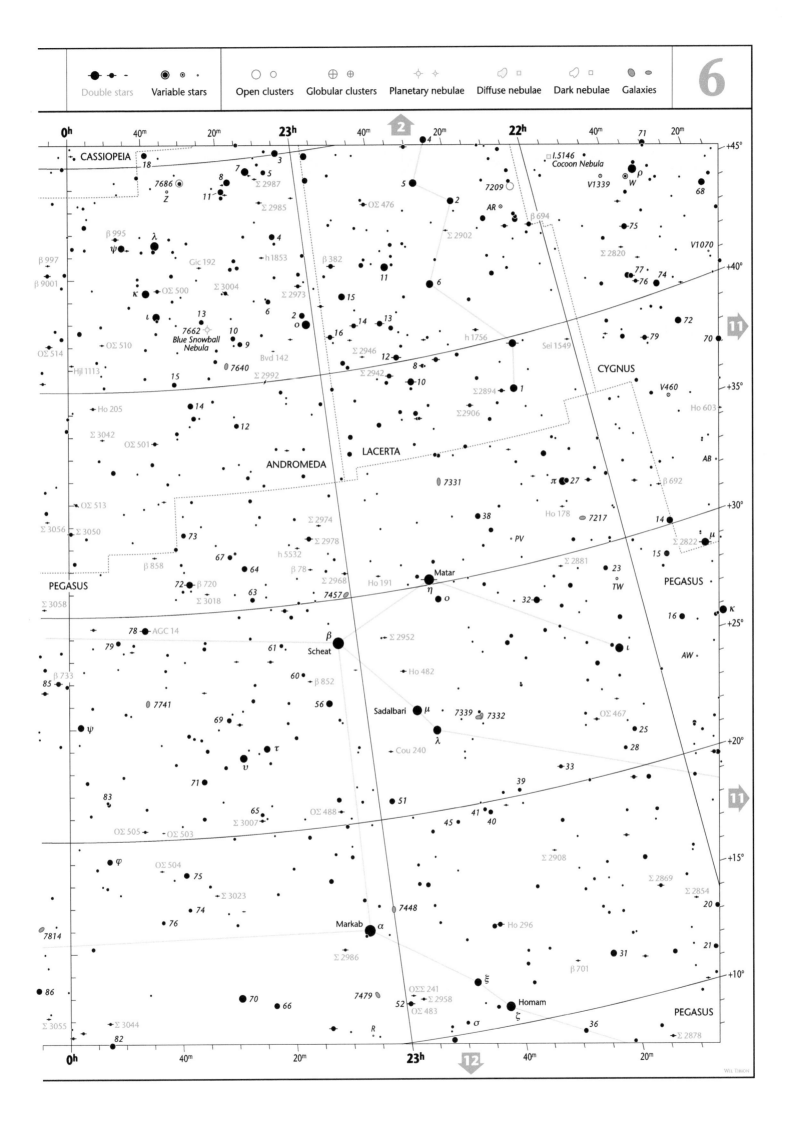

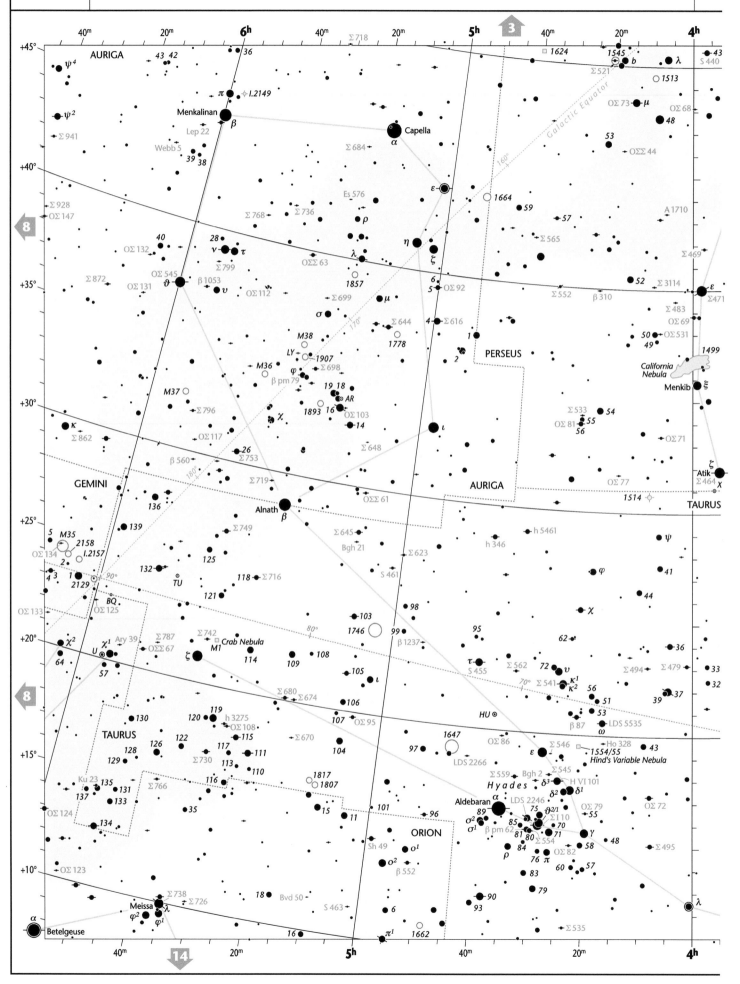

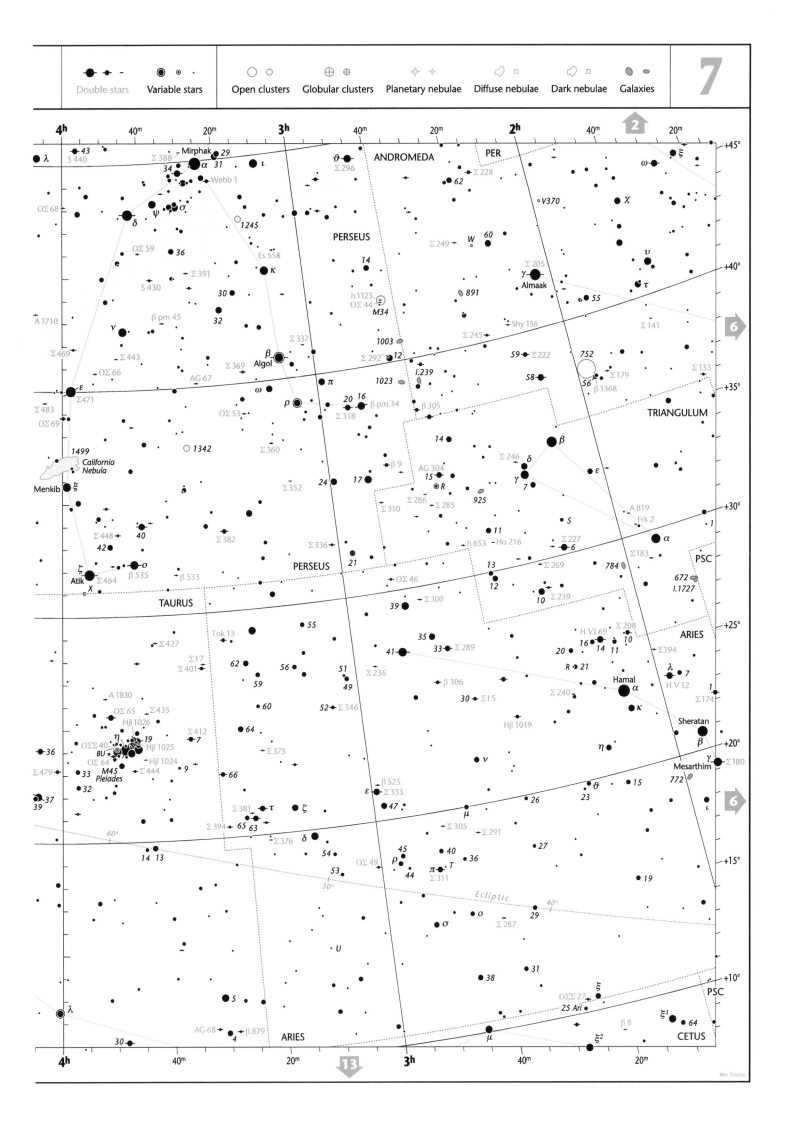

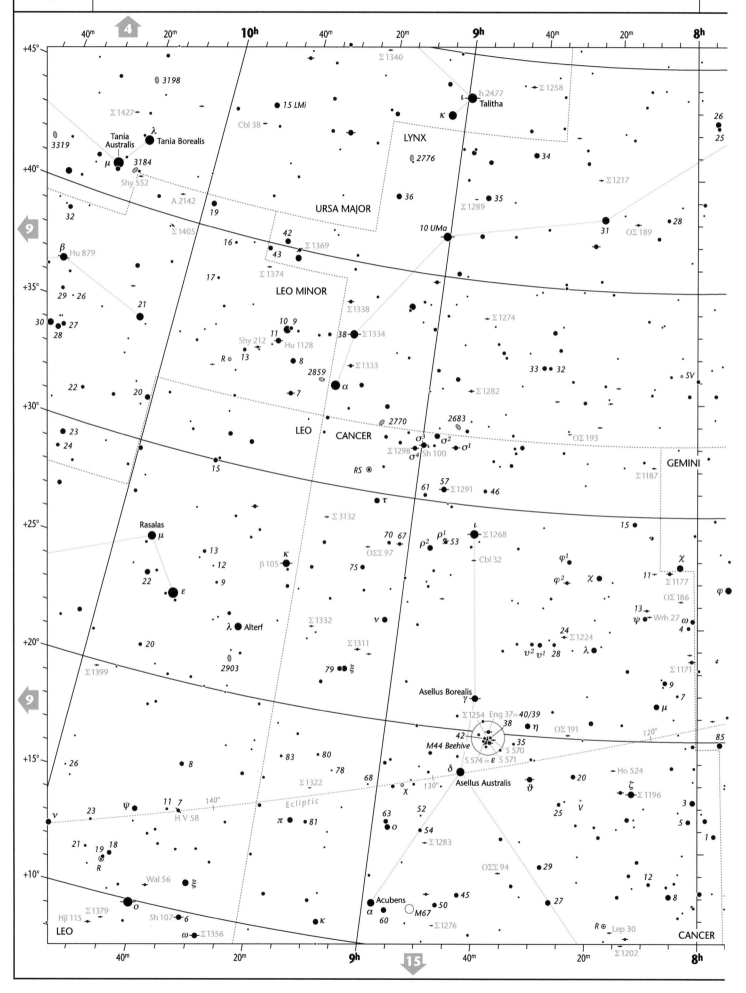

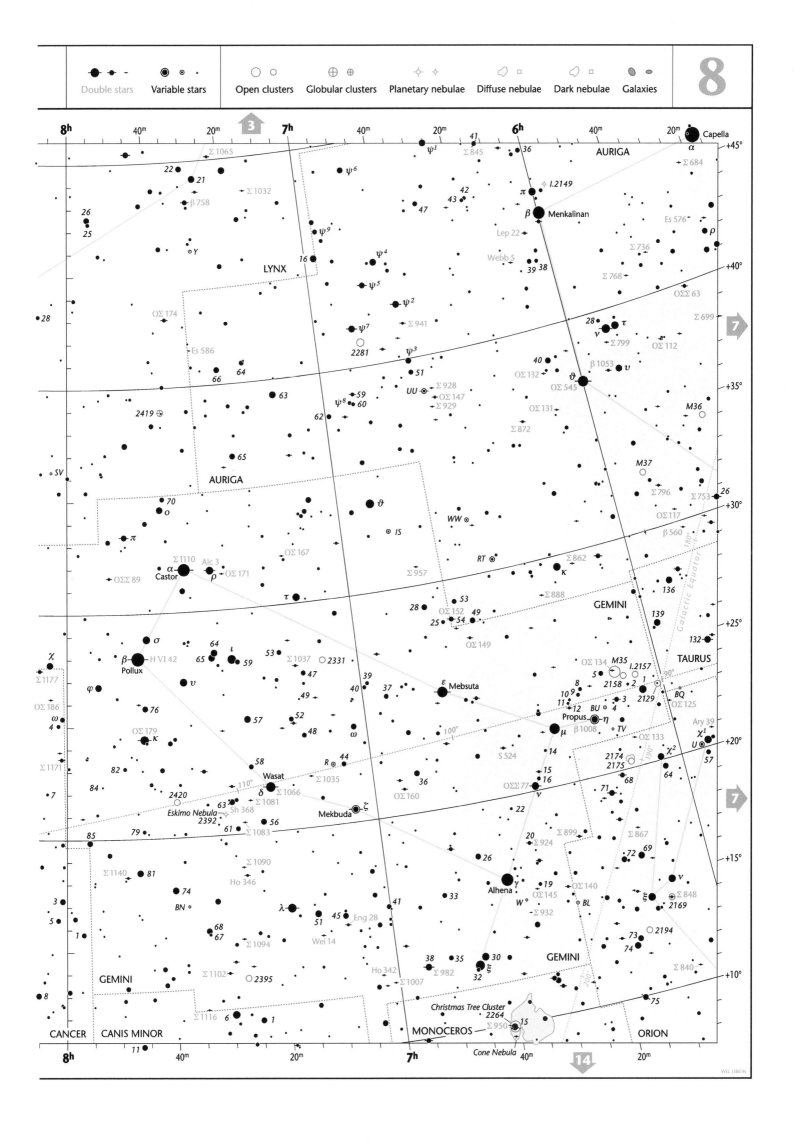

Double stars · Variable stars · Open clusters · Globular clusters · Planetary nebulae · Diffuse nebulae · Dark nebulae · Galaxies

Capella
AURIGA
α
Σ 684
Σ 1065
22
21
Σ 1032
β 758
26
25
Y
ψ¹
Σ 845
36
42
43
47
ψ⁶
I.2149
π
β   Menkalinan
Lep 22
39 38
Es 576
ρ
Σ 736
LYNX
ψ⁹
16
ψ⁴
ψ⁵
ψ²
Σ 941
ψ⁷
2281
Webb 5
28
ν
τ
Σ 799
OΣ 112
28
Σ 768
OΣΣ 63
Σ 699
OΣ 174
Es 586
66   64
ψ³
51
40
β 1053
ϑ
υ
OΣ 132
OΣ 545
M36
63
ψ⁸ 59
60
UU
Σ 928
Σ 147
Σ 929
OΣ 131
Σ 872
62
M37
Σ 796
Σ 753
26
2419
65
AURIGA
Σ 699
OΣ 117
β 560
70
o
ϑ
WW
IS
70
π
OΣ 167
RT
Σ 862
κ
Σ 957
Σ 888
GEMINI
136
139
Σ 1110
Alc 3
α
Castor
ρ
OΣ 171
OΣΣ 89
53
28
OΣ 152
49
25
54
OΣ 149
132
TAURUS
τ
χ
Σ 1177
σ
β   H VI 42
Pollux
64
65
ι
59
53
Σ 1037
2331
47
39
37
ε   Mebsuta
OΣ 134
M35
I.2157
5
2158
8
9
10 9
11
2 1
2129
90°
BQ
OΣ 125
OΣ 186
ω
4
υ
76
OΣ 179
κ
49
52
48
ω
12
BU
4
3
Propus
η
TV
β 1008
μ
Ary 39
χ¹
U
57
82
58
R
44
36
S 524
14
15
16
2174
2175
68
χ²
64
Σ 1171
84
7
2420
63
Sh 368
Eskimo Nebula
2392
δ   Wasat
Σ 1066
Σ 1081
56
Σ 1035
OΣ 160
OΣΣ 77
ν
22
β 867
71
Σ 899
85
79
61
Σ 1083
20
Σ 924
72
69
Σ 1140
81
Ho 346
Σ 1090
26
γ   Alhena
19
OΣ 145
OΣ 140
BL
ν
Σ 848
2169
74
BN
λ
51
45
Eng 28
Wei 14
33
W
Σ 932
73
74
2194
3
5
1
68
67
Σ 1094
38
Σ 982
35
32
30
ξ
GEMINI
Σ 840
8
Σ 1102
2395
Ho 342
Σ 1007
Christmas Tree Cluster
2264
15
75
CANCER   CANIS MINOR
Σ 1116
6
1
MONOCEROS
Σ 950
Cone Nebula
ORION

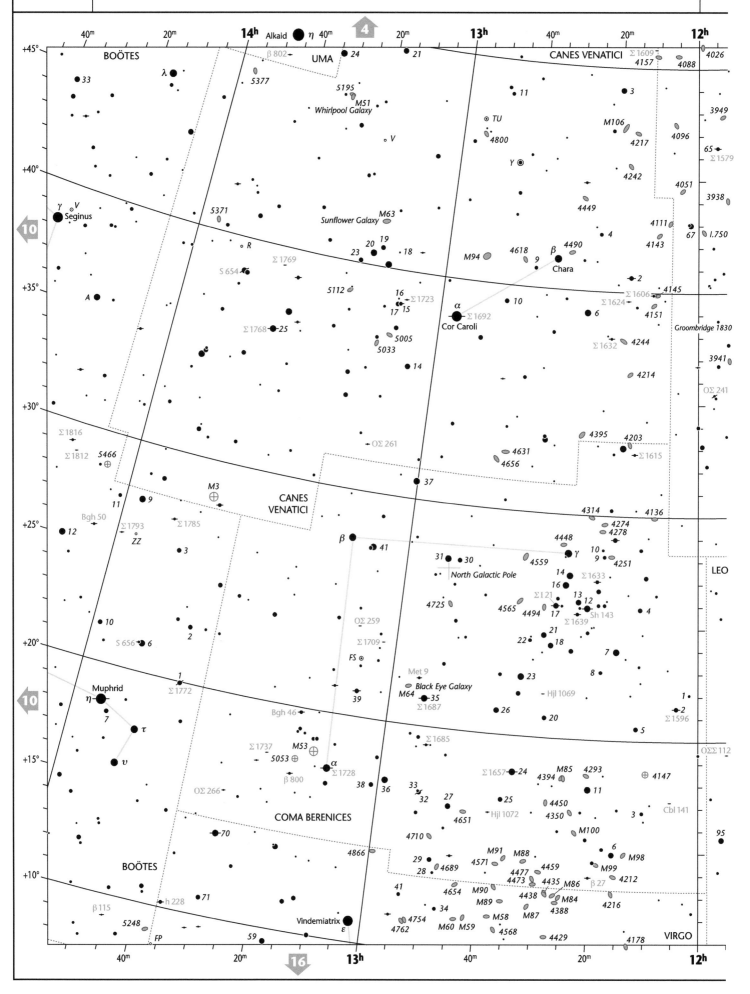

Magnitudes

● ● ● ● ● ● ● ● ● ● ● ● ● ● · ·
brighter than 0.0 0.0–0.5 0.5–1.0 1.0–1.5 1.5–2.0 2.0–2.5 2.5–3.0 3.0–3.5 3.5–4.0 4.0–4.5 4.5–5.0 5.0–5.5 5.5–6.0 6.0–6.5 6.5–7.0 7.0–7.5 fainter

BOÖTES

UMA

14ʰ Alkaid η 40ᵐ   20ᵐ   13ʰ

CANES VENATICI

β 802

24

21

Σ 1609
4157

Σ 1609
4088

4026

33

λ

5377

5195

M51
Whirlpool Galaxy

11

3

Σ 1609

3949

M106
4217

4096

65
Σ 1579

V

TU
4800

γ

4242

4449

4111
4143

4051

4490

3938

3941

γ V
Seginus

A

5371

Sunflower Galaxy

M63

20 19
23   18

M94   4618   9
β   Chara

4   4051

67 I.750

4145

5005
5033

5112

16
17 15

Σ 1723

α   Σ 1692
Cor Caroli

10

6

4244

4214

4151

Groombridge 1830

R

Σ 1769

S 654

Σ 1768 25

14

37

Σ 1606

Σ 1624

Σ 1632

O Σ 241

Σ 1816
Σ 1812

5466

CANES
VENATICI

4631
4656

4395   4203
Σ 1615

M3

11   9

Bgh 50

Σ 1793
ZZ

Σ 1785

3

β

41

31   30

4314   4136

4448
4274
4278

γ   10
9   4251

4559

12

North Galactic Pole

14   Σ 1633
16
Σ 121   13 12
4565   4494   17   Sh 143   4
Σ 1639

LEO

10

S 656 6

2

O Σ 259

Σ 1709

4725

21
22   18

23

7
8

1
Σ 1772

Muphrid
η
7
τ

FS

Met 9

Black Eye Galaxy
M64   35
Σ 1687

26

20

Hjl 1069

1
2
Σ 1596

υ

Σ 1737

M53
5053

β 800

α
Σ 1728

38   36

Σ 1685

Σ 1657 24   4394   M85 4293
4147

11

O Σ 266

33
32   27

25
Hjl 1072

4450
4350

3
Cbl 141

O ΣΣ 112

COMA BERENICES

4651

M100

70

29
28   4689

M91   M88
4571   4477 4459
4473 4435 M86
4438   M84
M89   4388

M99

6   M98

M90

4212
β 27

4216

95

BOÖTES

41
4654

34

M58 M87

β 115

h 228

71

Vindemiatrix
ε

4754
4762

M60 M59

M84

4429

4178

5248

FP

59

VIRGO

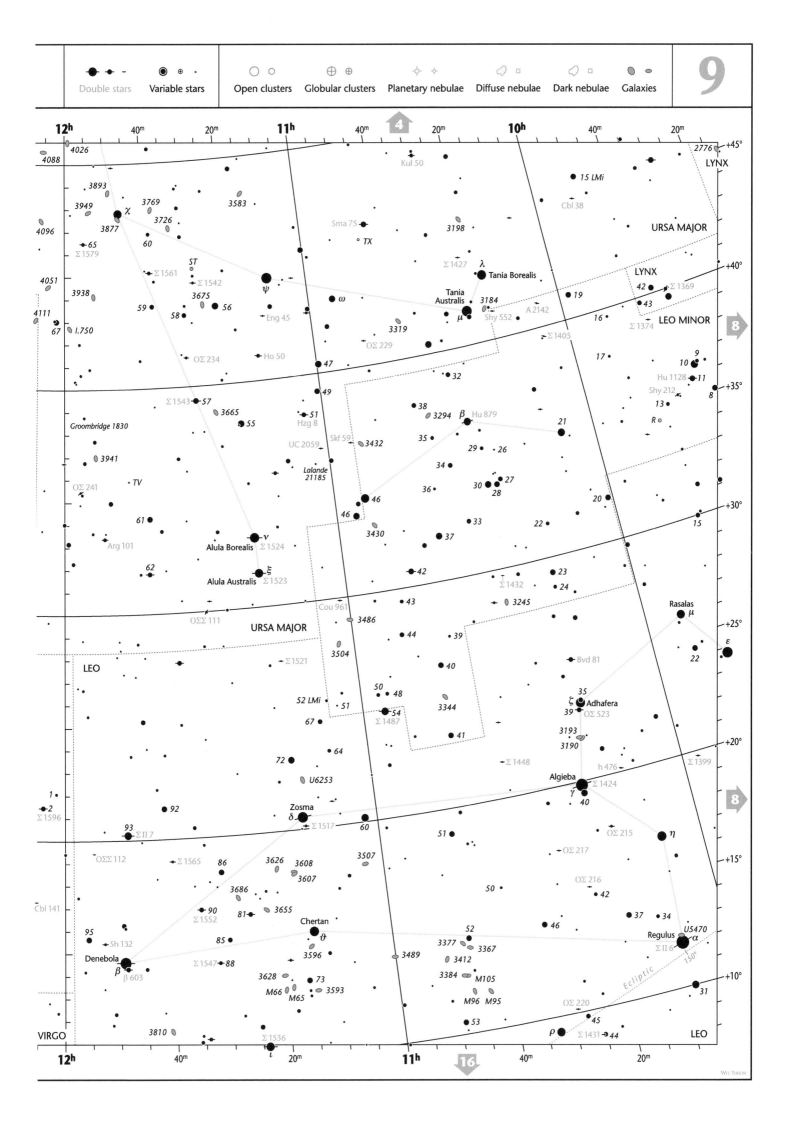

Magnitudes

brighter than 0.0 | 0.0–0.5 | 0.5–1.0 | 1.0–1.5 | 1.5–2.0 | 2.0–2.5 | 2.5–3.0 | 3.0–3.5 | 3.5–4.0 | 4.0–4.5 | 4.5–5.0 | 5.0–5.5 | 5.5–6.0 | 6.0–6.5 | 6.5–7.0 | 7.0–7.5 | fainter

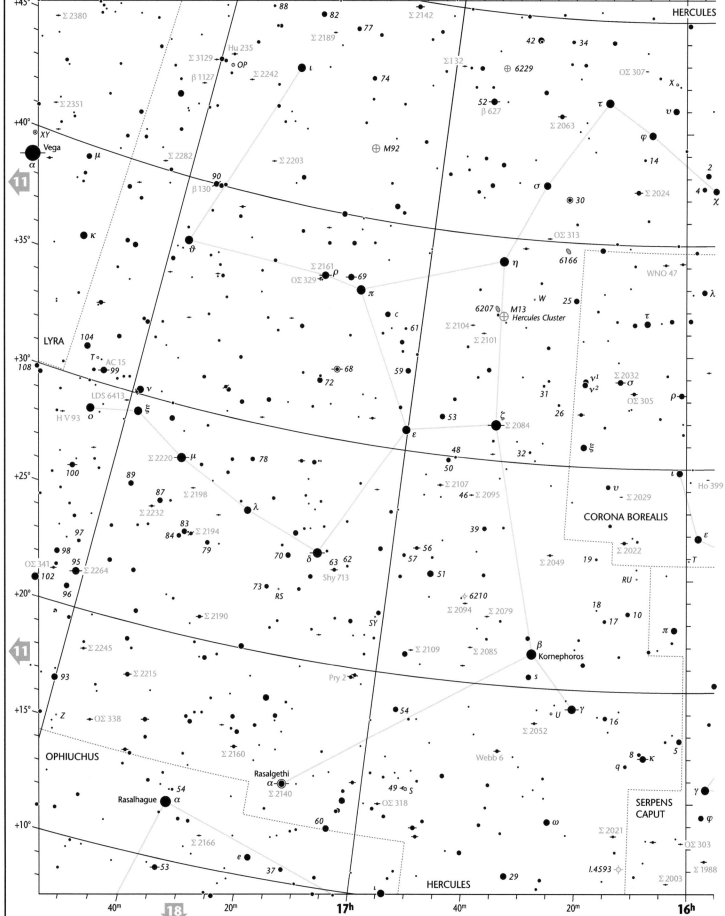

HERCULES

LYRA

OPHIUCHUS

CORONA BOREALIS

SERPENS CAPUT

HERCULES

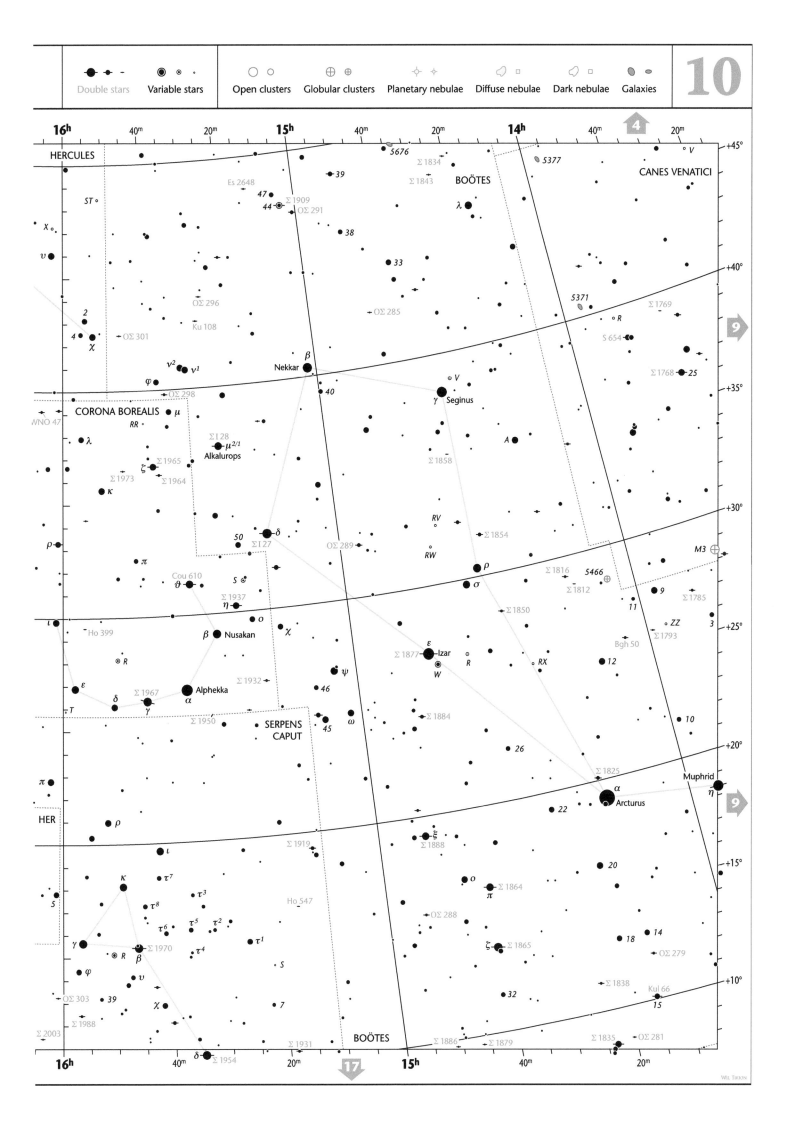

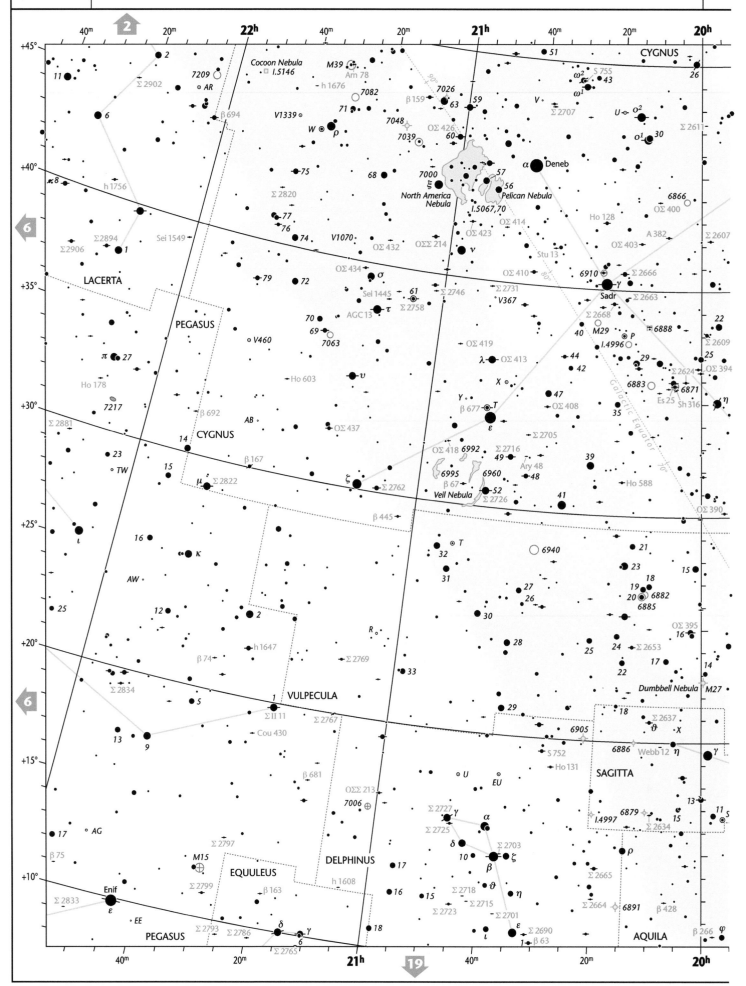

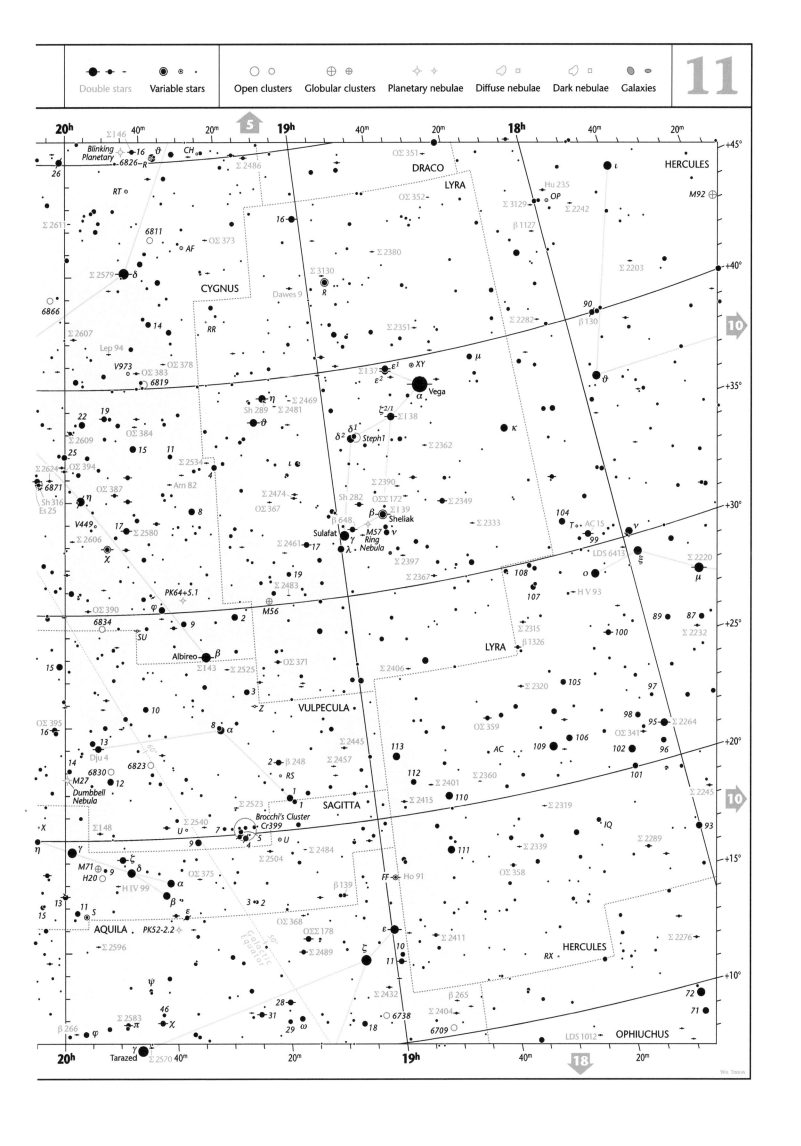

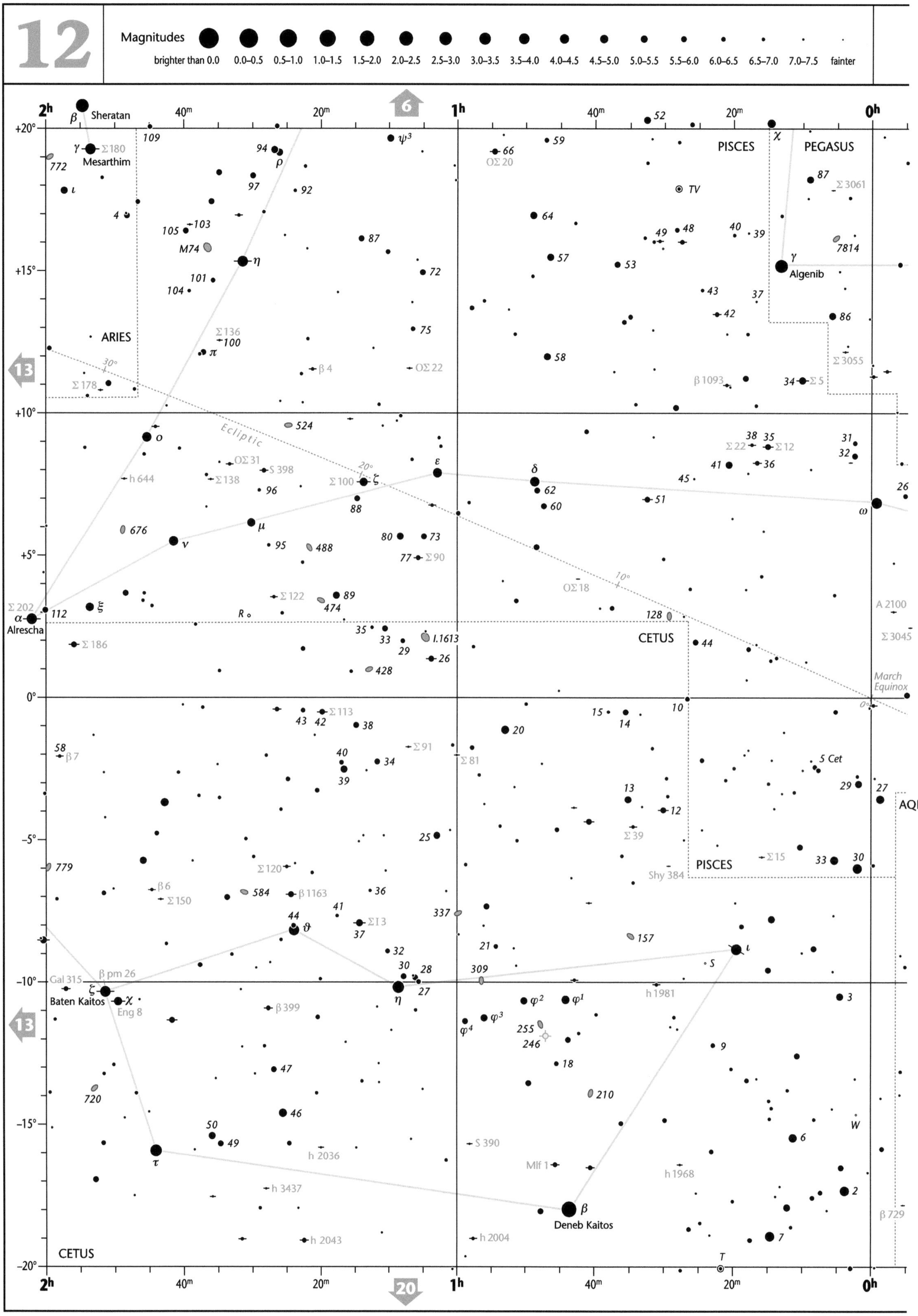
12
Magnitudes
brighter than 0.0
0.0–0.5
0.5–1.0
1.0–1.5
1.5–2.0
2.0–2.5
2.5–3.0
3.0–3.5
3.5–4.0
4.0–4.5
4.5–5.0
5.0–5.5
5.5–6.0
6.0–6.5
6.5–7.0
7.0–7.5
fainter
2h
40m
20m
6
1h
40m
20m
0h
+20°
+15°
+10°
+5°
0°
-5°
-10°
-15°
-20°
β Sheratan
γ Σ180
Mesarthim
772
ι
109
94
ρ
97
92
ψ3
66
OΣ20
59
52
χ
PISCES
PEGASUS
87
Σ3061
TV
4
105
103
M74
η
101
104
87
72
64
57
49
48
40
39
53
7814
γ
Algenib
43
37
42
86
ARIES
Σ136
100
π
75
58
13
30°
Σ178
β4
OΣ22
β1093
34 Σ5
Σ3055
524
Ecliptic
ο
OΣ31
S398
h644
Σ138
96
20°
Σ100
ζ
ε
δ
62
60
88
38
35
Σ22
Σ12
41
36
31
32
45
51
26
ω
676
ν
μ
95
488
80
73
77 Σ90
10°
OΣ18
Σ202
ξ
Σ122
89
474
α
Alrescha
112
R
128
A 2100
CETUS
35
33
29
I.1613
26
44
Σ186
Σ3045
428
March Equinox
0°
10
15
14
43
42
Σ113
38
20
58
β7
40
39
34
Σ91
Σ81
5 Cet
29
27
AQR
13
12
25
Σ39
779
Σ120
Σ15
33
30
PISCES
Shy 384
β6
Σ150
584
β1163
36
41
Σ13
37
337
157
44
ϑ
21
32
30
28
27
309
S
ι
Gal 315
β pm 26
ζ
Baten Kaitos
χ
Eng 8
β399
φ2
φ1
φ3
φ4
h1981
3
13
255
246
18
9
47
720
210
46
50
49
h2036
S390
6
W
τ
h3437
Mlf 1
h1968
2
β
Deneb Kaitos
β729
h2043
h2004
7
CETUS
T
20
2h
40m
20m
1h
40m
20m
0h

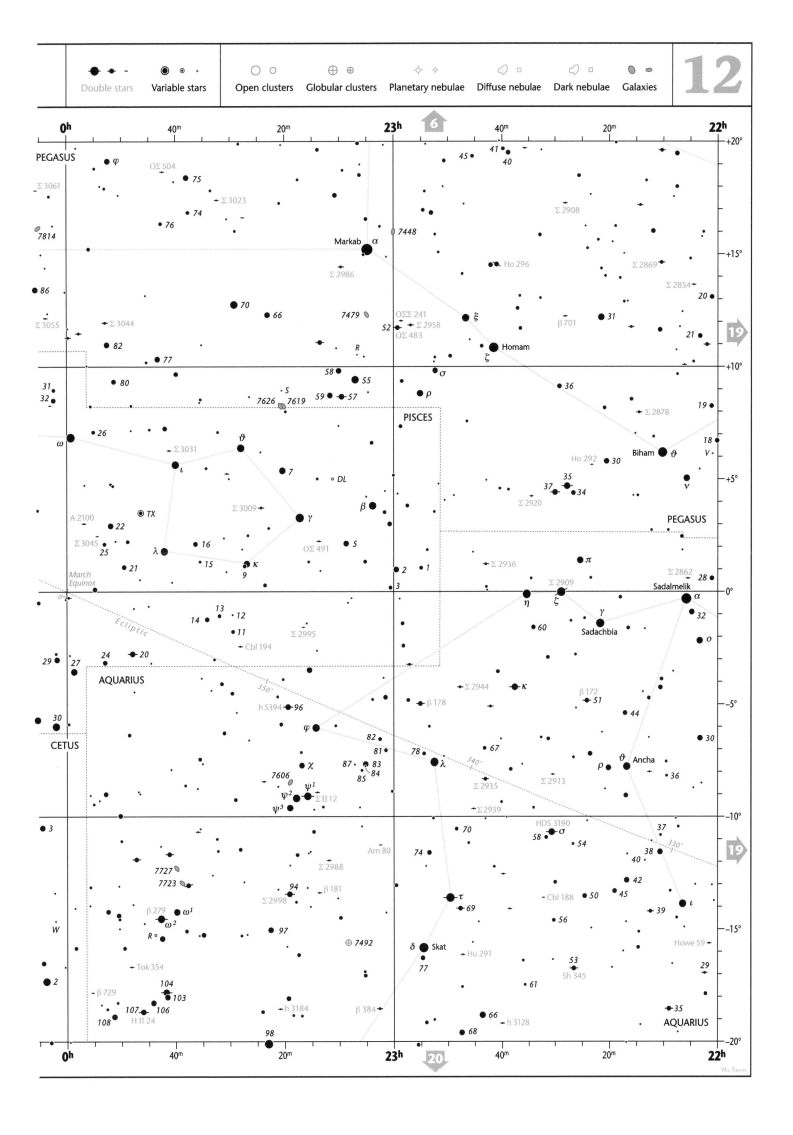

Double stars   Variable stars   Open clusters   Globular clusters   Planetary nebulae   Diffuse nebulae   Dark nebulae   Galaxies

Magnitudes
brighter than 0.0   0.0–0.5   0.5–1.0   1.0–1.5   1.5–2.0   2.0–2.5   2.5–3.0   3.0–3.5   3.5–4.0   4.0–4.5   4.5–5.0   5.0–5.5   5.5–6.0   6.0–6.5   6.5–7.0   7.0–7.5   fainter

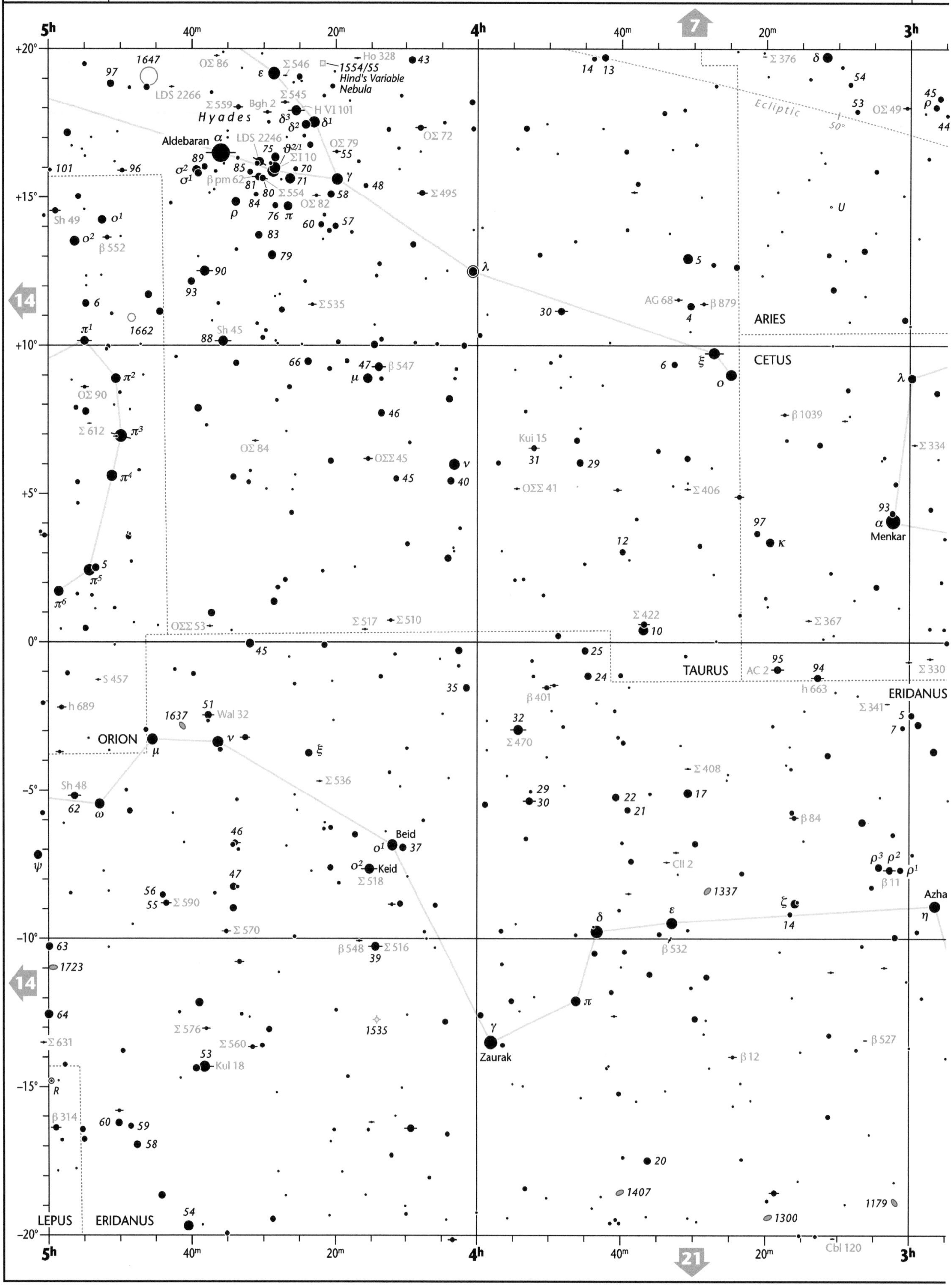

5ʰ   40ᵐ   20ᵐ   4ʰ   40ᵐ   7   20ᵐ   3ʰ

+20°
1647
97
OΣ 86   ΟΣ 376   δ
Σ 546   Ho 328   43   14 13   54
ε   1554/55   53   ΟΣ 49
LDS 2266   Σ 545   Hind's Variable   45
Σ 559   Nebula   ρ
Bgh 2   H VI 101   44
Hyades   δ³
δ²   δ¹
Aldebaran α   ΟΣ 79   U
LDS 2246   75   θ²/¹   55
89   Σ 110   70
σ²   85   71   ΟΣ 72
σ¹   β pm 62   81   γ   5
101   96   80   Σ 554   58   48   AG 68   β 879
84   ΟΣ 82   57   4
Sh 49   o¹   76 π   60   30   ξ   ARIES
o²   83   6   o   CETUS
β 552   79   Σ 535   6

14   90   Σ 495   λ
93
π¹   88   Sh 45
1662

+10°
π²   66   47 β 547   β 1039   λ
ΟΣ 90   μ   Σ 334
Σ 612   π³   46
ΟΣ 84   Kui 15
π⁴   ΟΣΣ 45   31   29   Σ 406
ν   97
45   40   12   κ   93
+5°   ΟΣΣ 41   α
π⁵   5   Menkar
π⁶   Σ 367
Σ 422   Σ 330
ΟΣΣ 53   Σ 517 Σ 510   10   AC 2   95 94
0°   TAURUS   h 663   ERIDANUS
45   25
35   24   Σ 341   5
S 457   β 401   7
h 689   51   32   ΟΣ 408   β 84
1637   Wal 32   Σ 470   ρ³ ρ²
ORION   ν   29   22   17   ρ¹
μ   ξ   30   21   β 11
Sh 48   Σ 536   Cll 2   Azha
62   ω   1337   ζ   η
–5°   46   Beid   14
ψ   47   o¹   37   δ   ε
o² Keid   β 532
56   Σ 518   π
55   Σ 590   γ   β 12
63   Σ 570   Zaurak   β 527
–10°   β 548 Σ 516
1723   39
64   1535
–15°   Σ 631   Σ 576   20
R   53   1407   1179
β 314   60   59   Kul 18   1300
58
LEPUS   ERIDANUS   54   Cbl 120
–20°
5ʰ   40ᵐ   4ʰ   21   20ᵐ   3ʰ

14

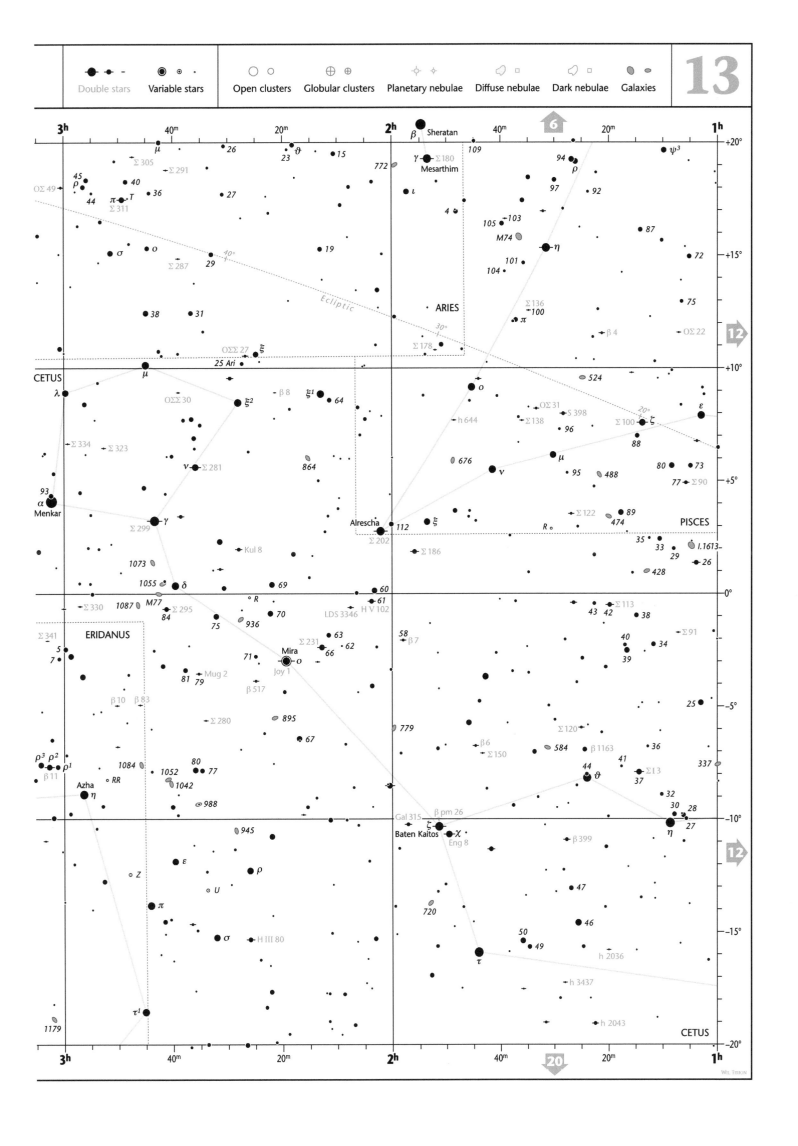

Double stars Variable stars Open clusters Globular clusters Planetary nebulae Diffuse nebulae Dark nebulae Galaxies

3ʰ   40ᵐ   20ᵐ   2ʰ   40ᵐ   20ᵐ   1ʰ

6 ↑

β Sheratan   +20°

μ   26
23 ϑ   15   772   γ Σ180   109   94 ρ   ψ³
Σ305   Mesarthim   97   92
45 ρ Σ291   ι
 OΣ49   40   36   27   4   105 103   87
44 π T   M74
Σ311   19   η
σ o   40°   101   72
Σ287 29   104   +15°
Ecliptic   Σ136 100
38 31   30°   π   75
Σ178   β4 OΣ22

ξ   +10° ↦ **12**
OΣΣ27
25 Ari   ARIES   524
μ   o
CETUS   OΣΣ30   β8 ξ¹ 64   h644 OΣ31 S398 ζ ε
λ   ξ²   Σ138 96   88
Σ334 Σ323   ν Σ281   864   676 ν μ 95 488   80 73   +5°
93   864   77 Σ90
α   γ   Alrescha 112 ξ   Σ122 89
Menkar   Σ299   Kul 8   R   474   35
1073   Σ202   33 29 I.1613
1055 δ   69   60   Σ186   428   26   0°
Σ330 1087 M77 Σ295   R   61   43 42 Σ113 38
ERIDANUS   84   75 936   70 LDS 3346 H V 102   40 Σ91
Σ341   63   58 β7   39 34
5   71 Mira Σ231 66 62   25
7   81 79 o   β517   779   36   -5°
β10 β83   Mug 2   895   β6   584 β1163 41 Σ13 337
ρ³ ρ² ρ¹   1084 80   67   Σ150   44 ϑ 37
β11   1052 77   32
Azha RR 1042   Gal 315 ζ   30 28
η   988   β pm 26   Baten Kaitos χ   η 27 -10°
945   Eng 8   β399

↦ **12**
ε   ρ
Z   720   47
U   50 49 46
π   τ   h 2036
σ H III 80   h 3437
τ¹
1179   h 2043   CETUS   -20°

3ʰ   40ᵐ   20ᵐ   2ʰ   40ᵐ   20ᵐ   1ʰ

20 ↓

WIL TIRION

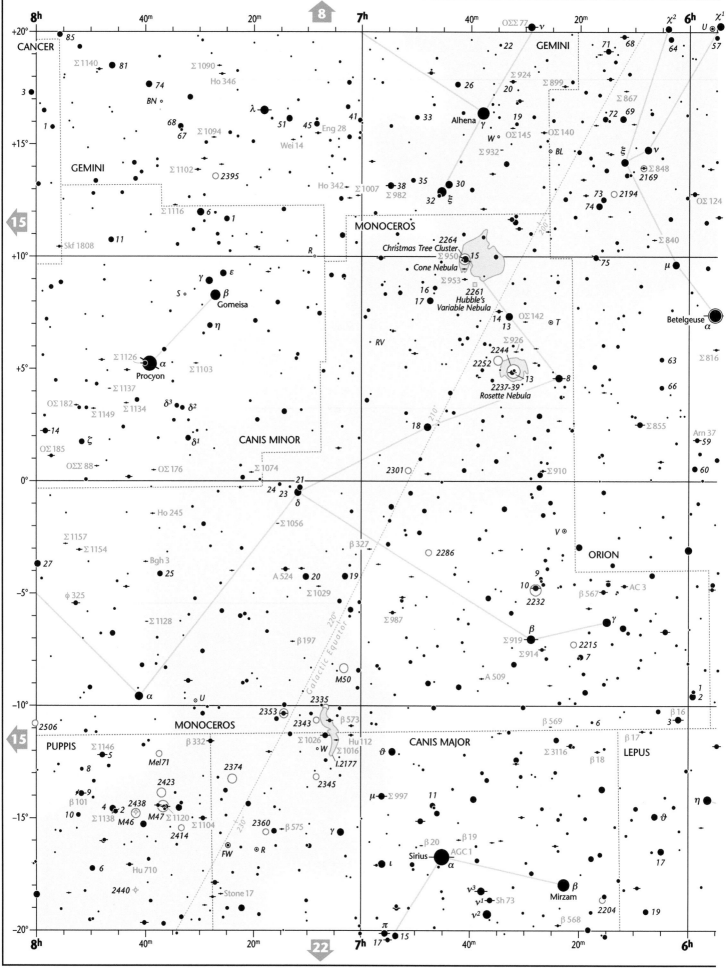

Magnitudes

brighter than 0.0   0.0–0.5   0.5–1.0   1.0–1.5   1.5–2.0   2.0–2.5   2.5–3.0   3.0–3.5   3.5–4.0   4.0–4.5   4.5–5.0   5.0–5.5   5.5–6.0   6.0–6.5   6.5–7.0   7.0–7.5   fainter

8h   40m   20m   8   7h   40m   20m   χ²   6h   χ¹

+20°

CANCER

85

Σ 1140   81

OΣΣ 77   ν   GEMINI

22   71   68

64   57   U

3

74

Σ 1090   Ho 346

26   Σ 924   Σ 899

20

Σ 867

72   69

BN

33   Alhena   γ   19

λ

51   45   Eng 28   41

1

68   67   Σ 1094

W   OΣ 145 → OΣ 140

ξ   ν

+15°

GEMINI

Wei 14

Σ 932

BL

Σ 848   2169

OΣ 124

11

Σ 1102   2395

6   1

38   35   30

73   2194

74

Skf 1808

Ho 342   Σ 1007

32   ξ

Σ 982

Σ 840

+10°

γ   ε

MONOCEROS

2264

Christmas Tree Cluster

Σ 950   15

75

μ

β

S

Cone Nebula

Σ 953

Gomeisa

η

16

17

2261

Hubble's

Variable Nebula

+5°

Σ 1126   α

Procyon

Σ 1103

Σ 1137

δ³   δ²

OΣ 182   Σ 1134

Σ 1149

14   ζ

OΣ 185

OΣΣ 88   OΣ 176   δ¹

CANIS MINOR

Σ 1074

14   13   OΣ 142   T

Σ 926

2244

2252

2237-39

Rosette Nebula

RV

18

63   Σ 816

66

13

8

Σ 855

Arn 37

59

60

0°

21

24   23   δ

2301

Σ 910

Betelgeuse

α

V

-5°

Σ 1157

Σ 1154

27

φ 325

Bgh 3

25

Ho 245

Σ 1056

β 327

A 524   20   19

Σ 1029

2286

9

10

2232

β 567   AC 3

ORION

1

2

Σ 1128

Σ 987

γ

β

Σ 919

Σ 914

2215

7

β 197

-10°

2506

PUPPIS

Σ 1146

5

8

β 101

9

4   2438

10   2   M47   Σ 1120

Σ 1138   M46

Mel71

2423

α

U

M50

2335

2353

2343

MONOCEROS

β 332

2374

Galactic Equator

2440

Σ 1104

2414

FW   R

Hu 710

Stone 17

β 573

Σ 1026

W   Σ 1016

Hu 112   I.2177

2345

2360   β 575

ϑ

μ   Σ 997

11

β 20   β 19

Sirius   AGC 1

α

ι

ν³

ν¹   Sh 73

ν²

CANIS MAJOR

Σ 3116

β 569   6

3

β 17

LEPRUS

β 18

η

ϑ

17

Mirzam

β

2204

β 568

19

π

17   15

8h   40m   20m   22   7h   40m   20m   6h

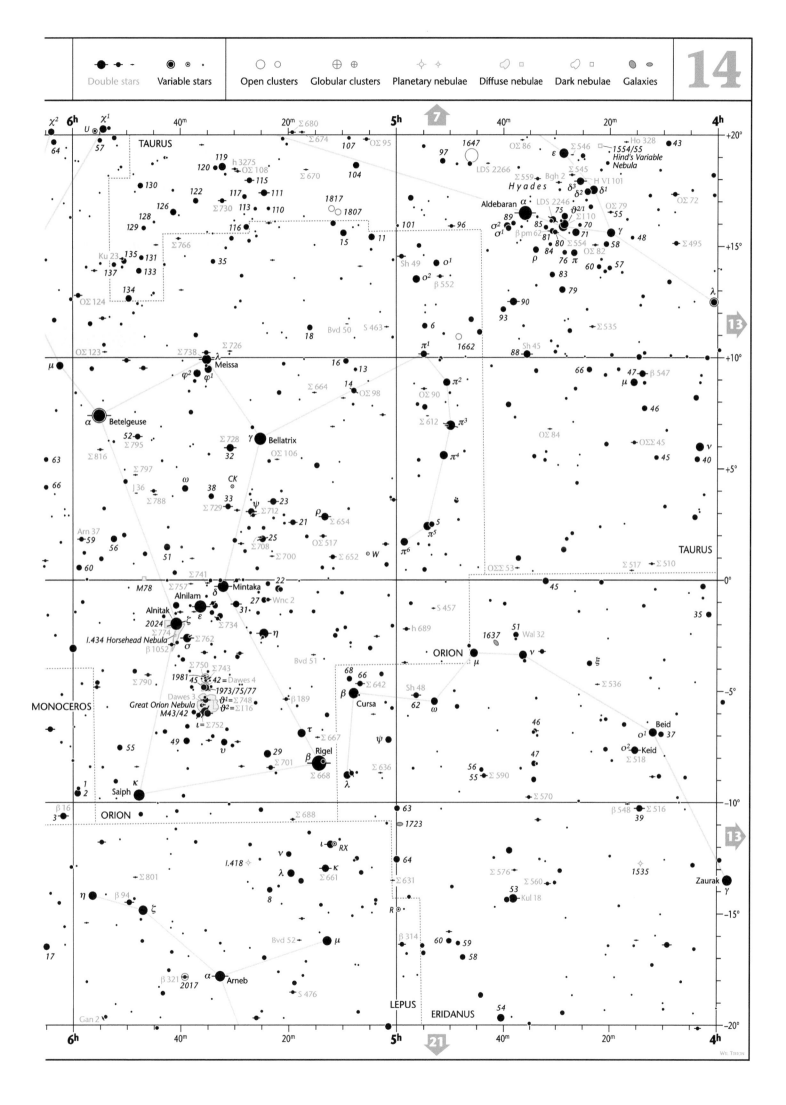

Magnitudes

brighter than 0.0   0.0–0.5   0.5–1.0   1.0–1.5   1.5–2.0   2.0–2.5   2.5–3.0   3.0–3.5   3.5–4.0   4.0–4.5   4.5–5.0   5.0–5.5   5.5–6.0   6.0–6.5   6.5–7.0   7.0–7.5   fainter

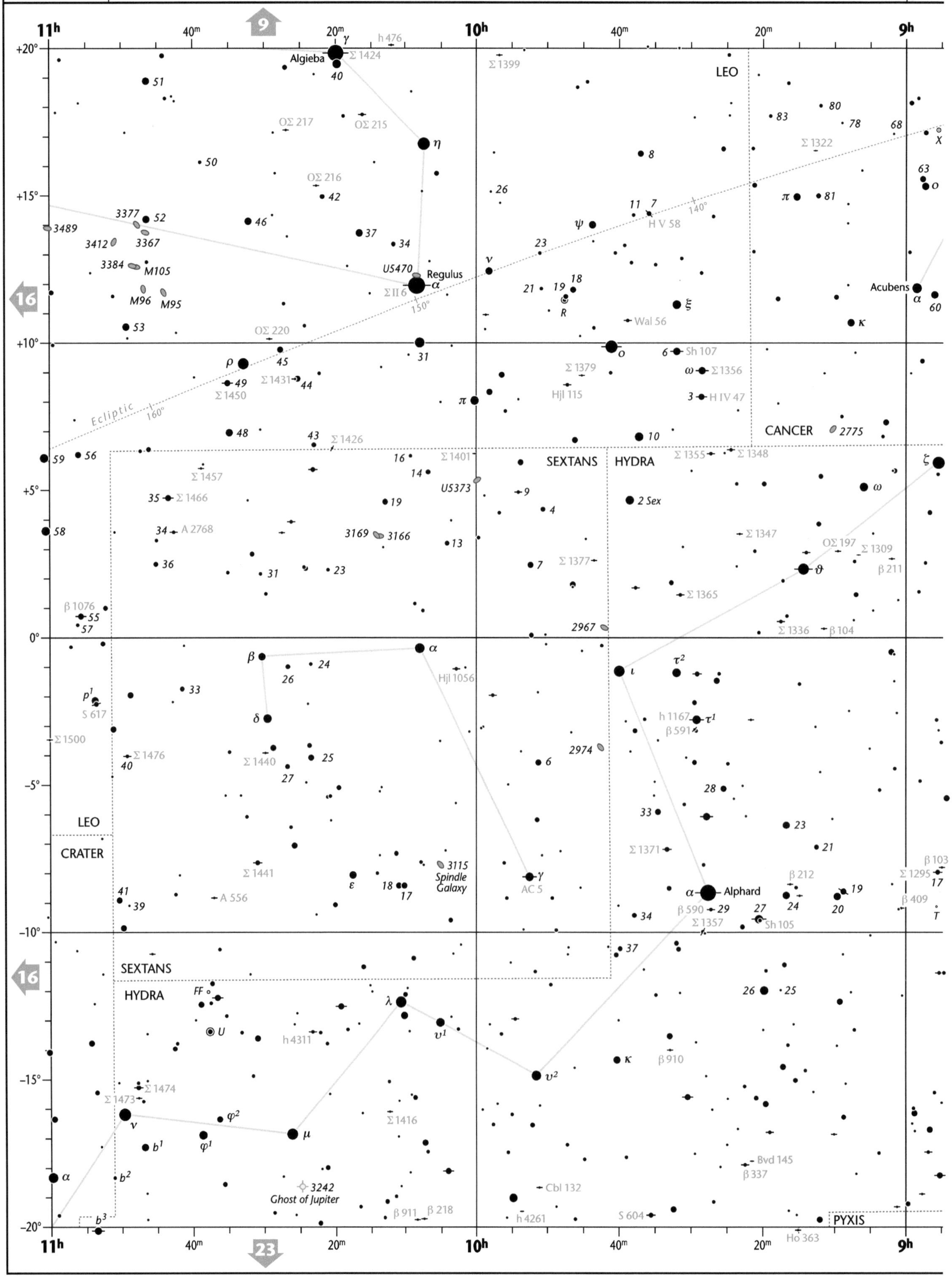

11^h 40^m 9   20^m 10^h 40^m 20^m 9^h

+20°

γ h 476
Algieba Σ 1424
40
51
Σ 1399
LEO
OΣ 217   OΣ 215
80
83
78   68
Σ 1322
χ
50
η
8
63
OΣ 216
o
42
11   7
π 81
+15°
26
H V 58
140°
3377   52
46
ψ
3489
37
23
3412   3367
34
ν
Acubens
3384   M105
U5470   Regulus
21   19   18
α 60
M96   M95
α
ξ
κ
150°
R
53
ΣII 6
Wal 56
+10°
OΣ 220
31
o
6   Sh 107
ρ 45
Σ 1379
ω Σ 1356
49   44
3   H IV 47
Σ 1450 Σ 1431
Hjl 115
48
π
CANCER   2775
43   Σ 1426
10
Ecliptic   160°
16   Σ 1401
SEXTANS   HYDRA
59   56
14
Σ 1355 Σ 1348
ζ
Σ 1457
U5373
9
+5°
35   Σ 1466
4
2 Sex
ω
34   A 2768
19
Σ 1347
58
3169   3166
OΣ 197 Σ 1309
36
13
7   Σ 1377
ϑ β 211
31   23
Σ 1365
β 1076
Σ 1336 β 104
55
2967
57
0°
β α
ι τ^2
24
p^1 33
26
Hjl 1056
h 1167 τ^1
S 617
δ
β 591
Σ 1500
25
2974
6
28
Σ 1476
27
33
23
40
Σ 1440
21
–5°
Σ 1371
LEO
β 212 β 103
CRATER
3115   Spindle Galaxy
19   Σ 1295   17
Σ 1441
α Alphard   24   20 β 409
41   ε 18   17
γ
β 590   29   27   T
39   A 556
AC 5
34   Sh 105
Σ 1357
–10°
16
37
SEXTANS
HYDRA
FF
λ
26   25
U
υ^1
h 4311
κ β 910
–15°
Σ 1474
υ^2
Σ 1473
φ^2
ν
Σ 1416
b^1 φ^1
μ
Bvd 145
b^2
β 337
α
3242
Cbl 132
PYXIS
Ghost of Jupiter
b^3 β 911 β 218
h 4261   S 604
Ho 363
–20°

11^h 40^m 23   10^h 40^m 20^m 9^h

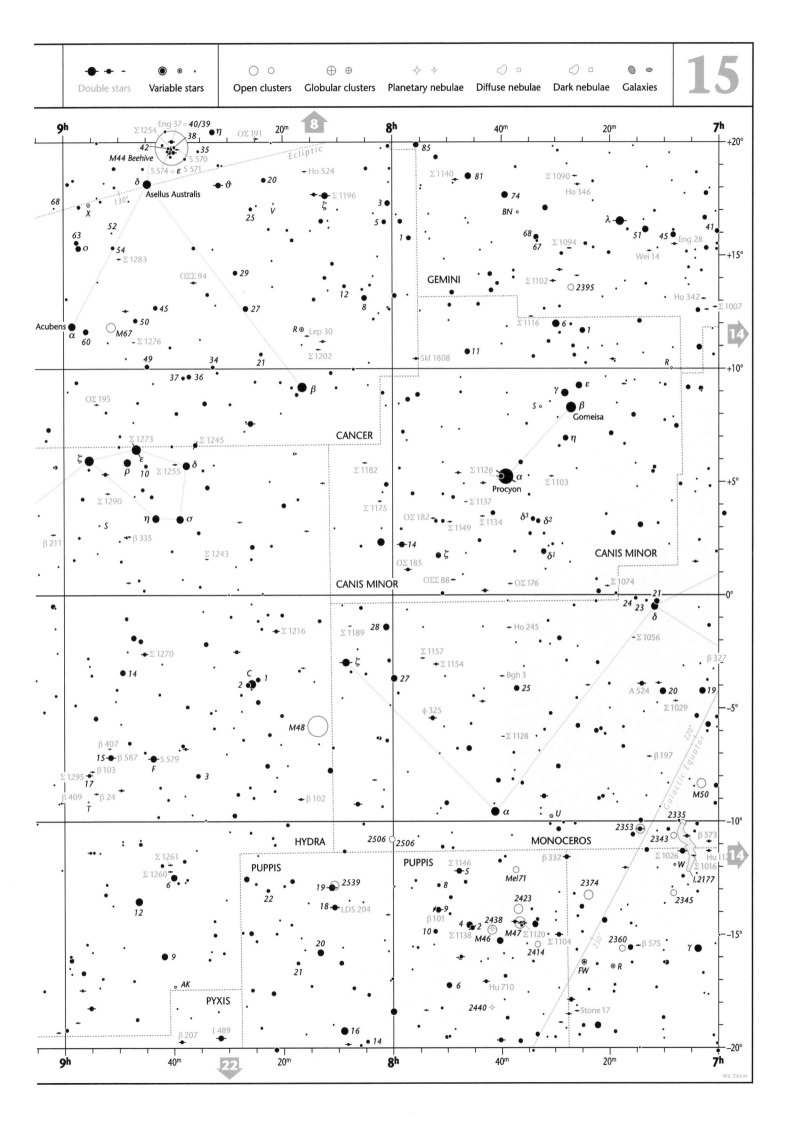

Double stars Variable stars Open clusters Globular clusters Planetary nebulae Diffuse nebulae Dark nebulae Galaxies

9h · 8 · 8h · 7h +20°

Σ1254 Eng 37 = 40/39 η ΟΣ 191 20m 85
38 η Σ1140 81 Σ1090
42 35 74 Ho 346
M44 Beehive S 570 BN
S 574 = ε S 571 68 Σ1094 λ 51 45 Eng 28 41
δ ϑ 20 67 Wei 14
Asellus Australis V GEMINI Σ1102 2395 Ho 342 +15°
68 130° 25 Σ1196 ζ 3 Σ1116 6 1 Σ1007
X 5
63 52 29 12 8 1 Skf 1808 11 R +10°
o 45 27 R Lep 30
Acubens 50 M67 Σ1202 β γ ε
α 60 Σ1276 34 ΟΣ 195 S β
49 37 36 21 CANCER Gomeisa
Σ1273 Σ1245 η
ζ ε δ Σ1182 α +5°
ρ 10 Σ1255 Σ1126 Procyon Σ1103
Σ1290 η σ Σ1175 Σ1137
S β 335 ΟΣ 182 Σ1149 Σ1134 δ³ δ²
β 211 Σ1243 14 δ¹ CANIS MINOR
ζ ΟΣ 185
CANIS MINOR ΟΣΣ 88 ΟΣ 176 Σ1074 21 0°
24 23 δ
Σ1216 28 Σ1189 Ho 245 Σ1056
Σ1270 ζ Σ1157 β 327
14 27 Σ1154 Bgh 3 A 524 20 19
C 1 25 Σ1029
2 φ 325 -5°
M48 Σ1128 β 197
β 407 220°
15 β 587 S 579 β 197
β 103 F 3 M50
Σ1295 17 α U 2335
β 409 β 24 β 102 -10°
T 2353 2343 β 573
HYDRA 2506 2506 MONOCEROS 2335 Hu 112 14
Σ1261 PUPPIS PUPPIS Σ1146 β 332 Σ1026 W Σ1016
Σ1260 19 2539 5 I.2177
6 22 8 Mel71 2374 2345
12 18 LDS 204 9 2423
20 β 101 4 2438 2 M47 Σ1120 2360 β 575
9 10 Σ1138 M46 2414 γ
21 6 Hu 710 FW R
AK 2440 Stone 17
PYXIS β 207 I 489 16 14

9h 40m 20m 8h 40m 20m 7h -20°

22

WIL TIRION

Magnitudes

brighter than 0.0 | 0.0–0.5 | 0.5–1.0 | 1.0–1.5 | 1.5–2.0 | 2.0–2.5 | 2.5–3.0 | 3.0–3.5 | 3.5–4.0 | 4.0–4.5 | 4.5–5.0 | 5.0–5.5 | 5.5–6.0 | 6.0–6.5 | 6.5–7.0 | 7.0–7.5 | fainter

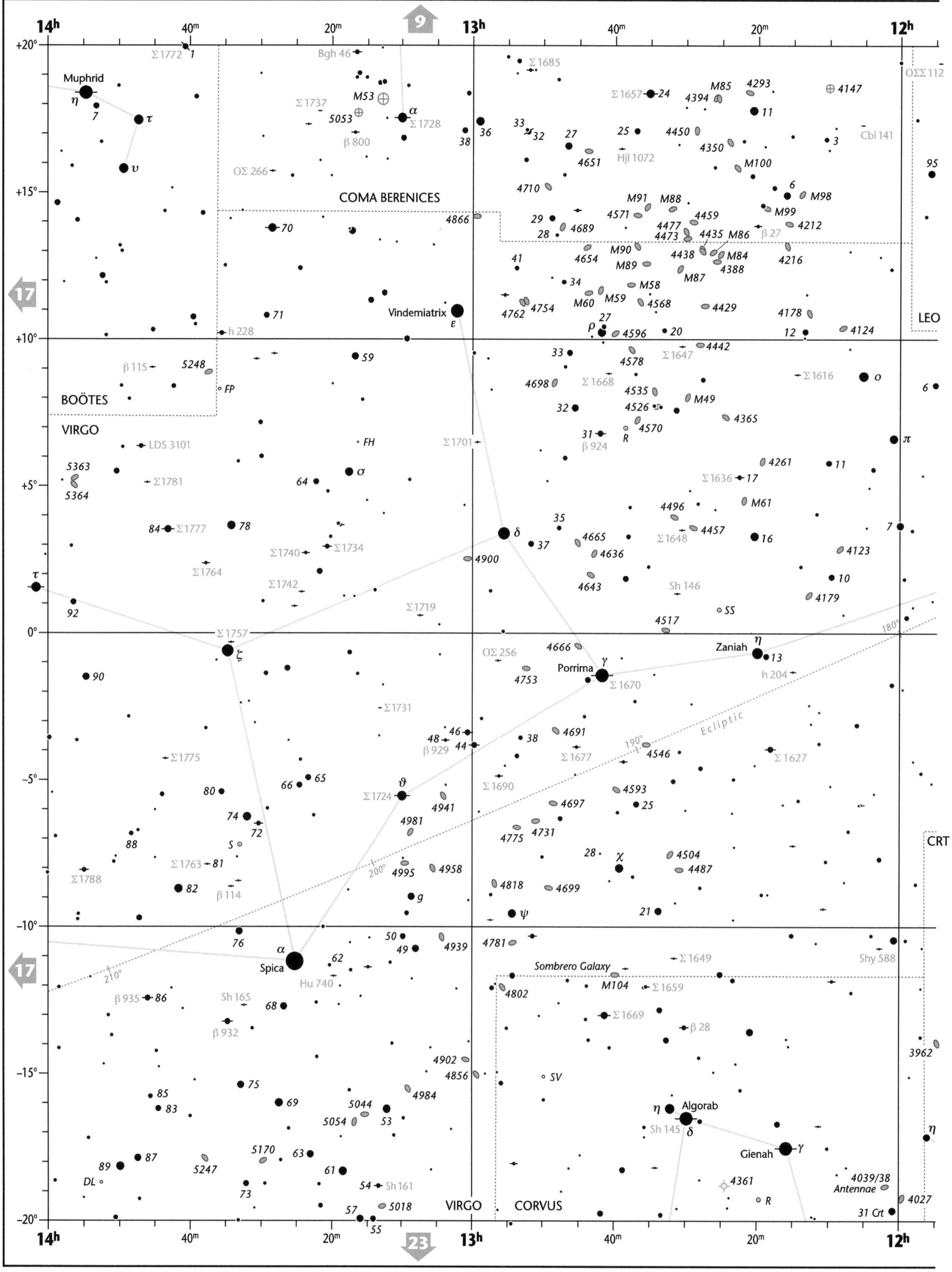

14ʰ 40ᵐ 20ᵐ **9** 13ʰ 40ᵐ 20ᵐ 12ʰ

+20°

Σ1772 1 OΣΣ112

Muphrid
η
7
τ
Σ1737 M53 ⊕ α Σ1685 24 M85 4293 ⊕ 4147
5053 ⊕ Σ1728 4394 4147
β 800 36 33 32 27 25 4450 4350 11 3 Cbl 141 95
υ 38 4651 M100
OΣ 266 4710 6 M98
COMA BERENICES M91 M88 M99 4212
+15° 4866 29 4571 4459 4477 4435 M86 β 27
70 28 4689 4473 4438 M84 4216
M90 M89 4388
41 4654 M87
34 M58 4178
71 M60 M59 4568 4429
Vindemiatrix 4762 4754 12 4124
h 228 ε ρ 27
+10° β 115 5248 33 4596 20
BOÖTES FP 59 4698 Σ1668 4578 Σ1647 4442 Σ1616 o
VIRGO 4535 M49 6
LDS 3101 32 4526 M49 4365
5363 Σ1781 FH 31 4570 π
5364 64 β 924 R 4261 11
+5° σ Σ1636 17 M61 7
84 Σ1777 78 4496 4457 16
Σ1740 Σ1734 35 4665 Σ1648 10
Σ1764 37 4636 4123
Σ1742 4900 4643 Sh 146 4179
Σ1719 4517 SS
τ Σ1757 OΣ 256 4666 γ η Zaniah
0° ξ 4753 Porrima Σ1670 13 h 204 180°
92 90 Σ1731 38 4691 4546 Σ1627
48 46 β 929 44 Σ1677 190° Ecliptic
Σ1775 66 65 4593
-5° 80 ϑ 4941 25 CRT
74 Σ1724 4981 4697
72 S 4775 4731 χ 4504
88 Σ1763 81 4995 4958 28 4487
Σ1788 82 β 114 g 4818 4699 21
ψ 3962
-10° 76 50 4781 Sombrero Galaxy Σ1649 Shy 588
α 62 49 4939 4802 M104 Σ1659
Spica Hu 740 Σ1669
β 935 86 Sh 165 68 β 28
β 932 SV
4902 η Algorab
85 75 4856 δ
83 69 5044 Sh 145
87 5054 53 Gienah γ
89 5170 63 4361 4039/38
DL 5247 61 R Antennae η
73 54 Sh 161 31 Crt 4027
57 5018 VIRGO CORVUS
55
-20°

14ʰ 40ᵐ 20ᵐ **23** 13ʰ 40ᵐ 20ᵐ 12ʰ

LEO

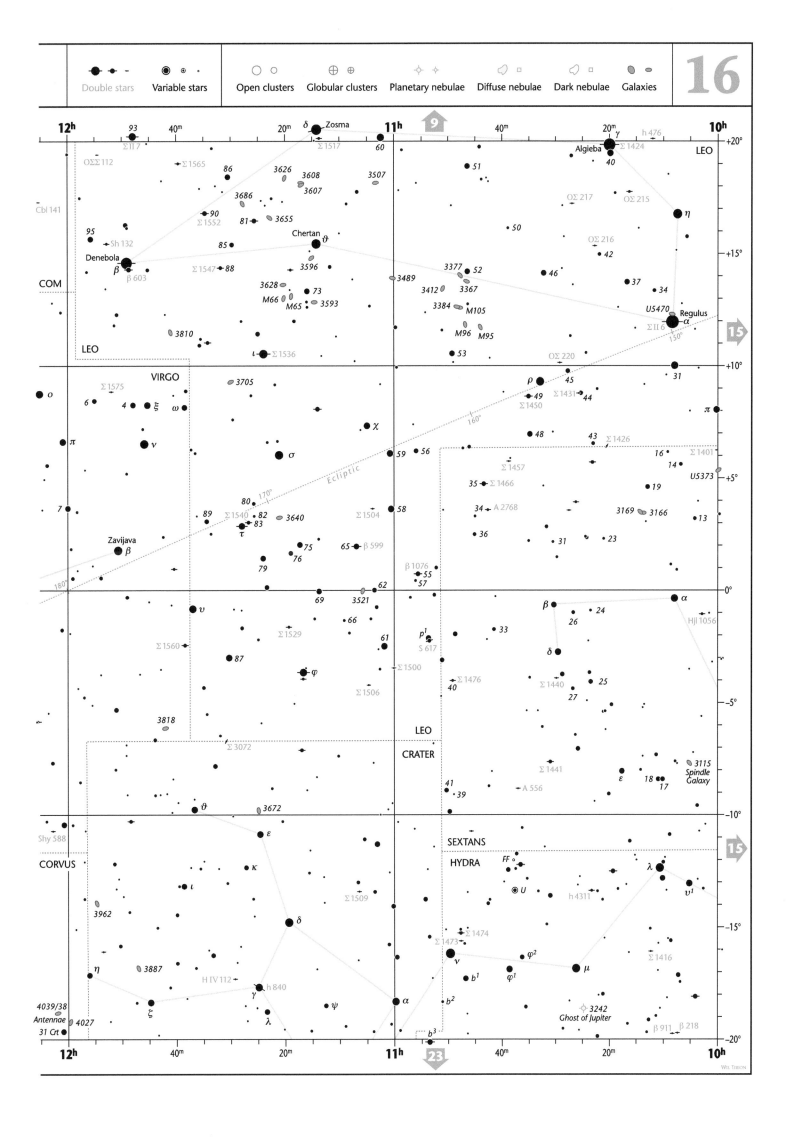

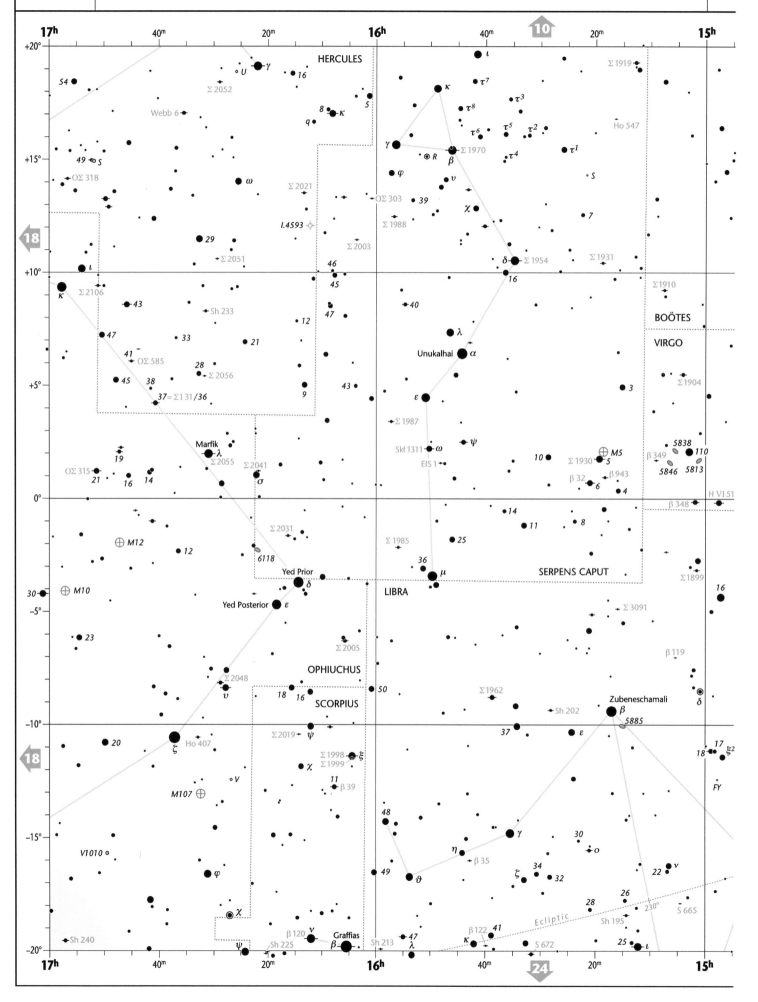

Magnitudes

brighter than 0.0 0.0–0.5 0.5–1.0 1.0–1.5 1.5–2.0 2.0–2.5 2.5–3.0 3.0–3.5 3.5–4.0 4.0–4.5 4.5–5.0 5.0–5.5 5.5–6.0 6.0–6.5 6.5–7.0 7.0–7.5 fainter

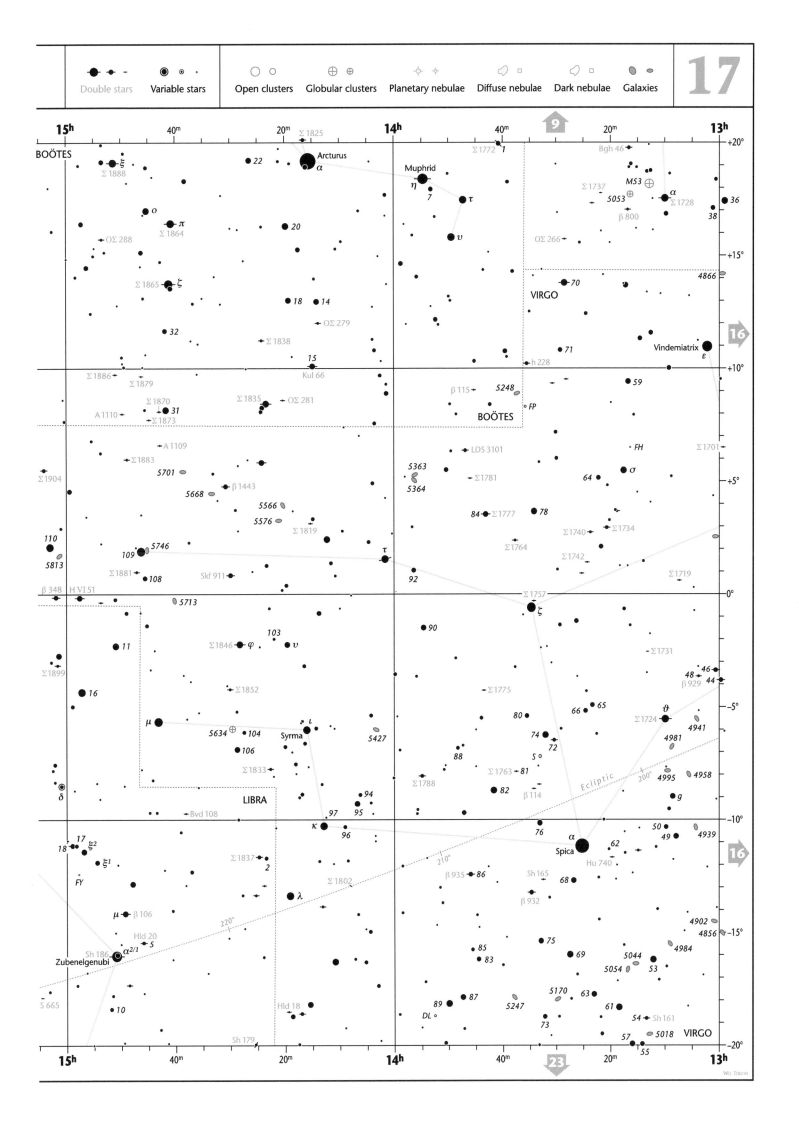

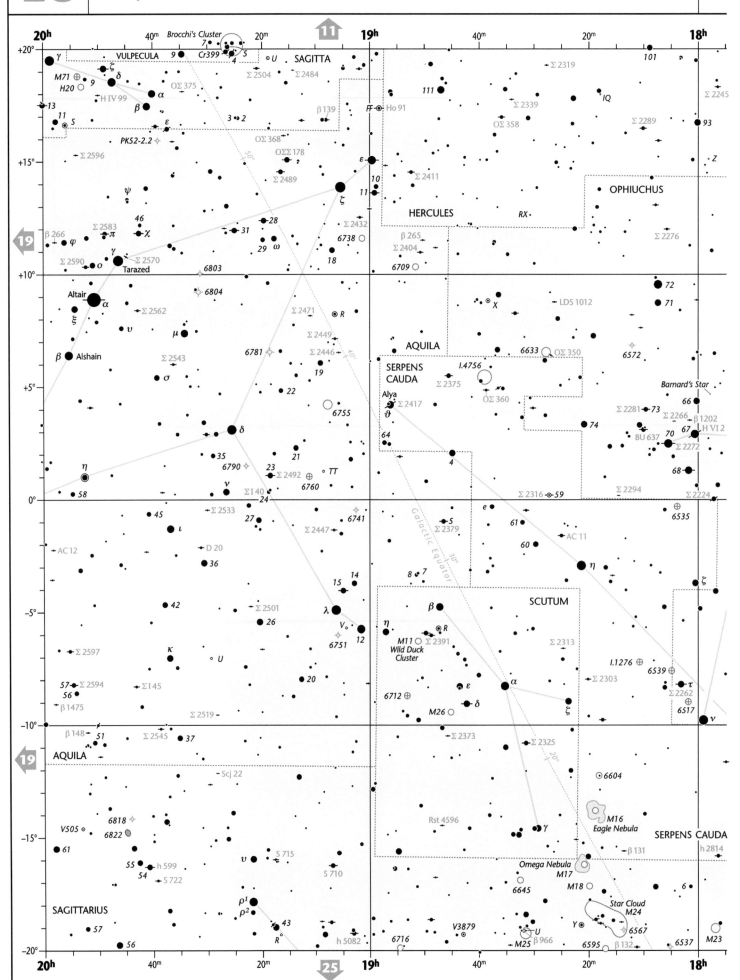

Magnitudes

brighter than 0.0   0.0–0.5   0.5–1.0   1.0–1.5   1.5–2.0   2.0–2.5   2.5–3.0   3.0–3.5   3.5–4.0   4.0–4.5   4.5–5.0   5.0–5.5   5.5–6.0   6.0–6.5   6.5–7.0   7.0–7.5   fainter

20ʰ   40ᵐ   *Brocchi's Cluster*   20ᵐ   11   19ʰ   40ᵐ   20ᵐ   18ʰ

+20°

γ

VULPECULA   9   Cr399   4   5   U   SAGITTA   101

M71   ζ   δ   Σ2504   Σ2484   Σ2319   111   IQ   Σ2245

H20   9   α   OΣ375   Σ2339   Σ2289   93

13   H IV 99   β   υ   3   2   β139   FF   Ho 91   OΣ358

11   S   ε   OΣ368   Z

PKS2-2.2   OΣΣ178   ε   HERCULES   RX   Σ2276

+15°   Σ2596   Σ2489   10   Σ2411

ψ   11   OPHIUCHUS

46   28   ζ

β266   φ   π   χ   31   Σ2432   β265   Σ2404

19   Σ2583   29   ω   6738   6709

Σ2590   o   γ   Σ2570   18

Tarazed   6803   72

+10°   Altair   6804   χ   LDS 1012   71

ξ   α   Σ2562   Σ2471   R   AQUILA

υ   μ   Σ2449   6633   OΣ350   6572

β   Alshain   6781   Σ2446   SERPENS   I.4756   Barnard's Star

Σ2543   19   CAUDA   66

+5°   σ   22   Σ2375   OΣ360   Σ2281   73   Σ2266   β1202

6755   Alya   Σ2417   H VI 2

δ   ϑ   74   BU 637   70   67

35   64   Σ2272

η   6790   21   4   68

23   Σ2492   TT   Σ2294   Σ2224

58   ν   6760   Σ2316   59

ΣI40   24   6535

0°   Σ2533   27   6741   e   Galactic Equator

45   5   61   AC 11

AC 12   ι   Σ2447   Σ2379   60

D 20   η   ζ

36   14   8   7   SCUTUM

15   β   I.1276   6539

–5°   42   Σ2501   λ   η   R   Σ2313   τ

26   V   M11   Σ2391   Σ2262

κ   U   6751   12   Wild Duck   Σ2303   6517

     Cluster

57   Σ2594   20   ε   α

56   ΣI45   6712   δ   ζ

β1475   M26

–10°   Σ2519   ν

β148   51   Σ2545   37   Σ2373   Σ2325

19   AQUILA   Scj 22   6604

6818   M16

V505   6822   Rst 4596   γ   Eagle Nebula   SERPENS CAUDA

–15°   61   β131   h 2814

55   h 599   υ   S 715   Omega Nebula   M17

54   S 722   S 710   6645   M18   6

     Star Cloud

ρ¹   M24

SAGITTARIUS   ρ²   V3879   γ   6567

57   43   U   6595   6537

56   R   h 5082   6716   M25   β966   β132   M23

–20°

20ʰ   40ᵐ   20ᵐ   25   19ʰ   40ᵐ   20ᵐ   18ʰ

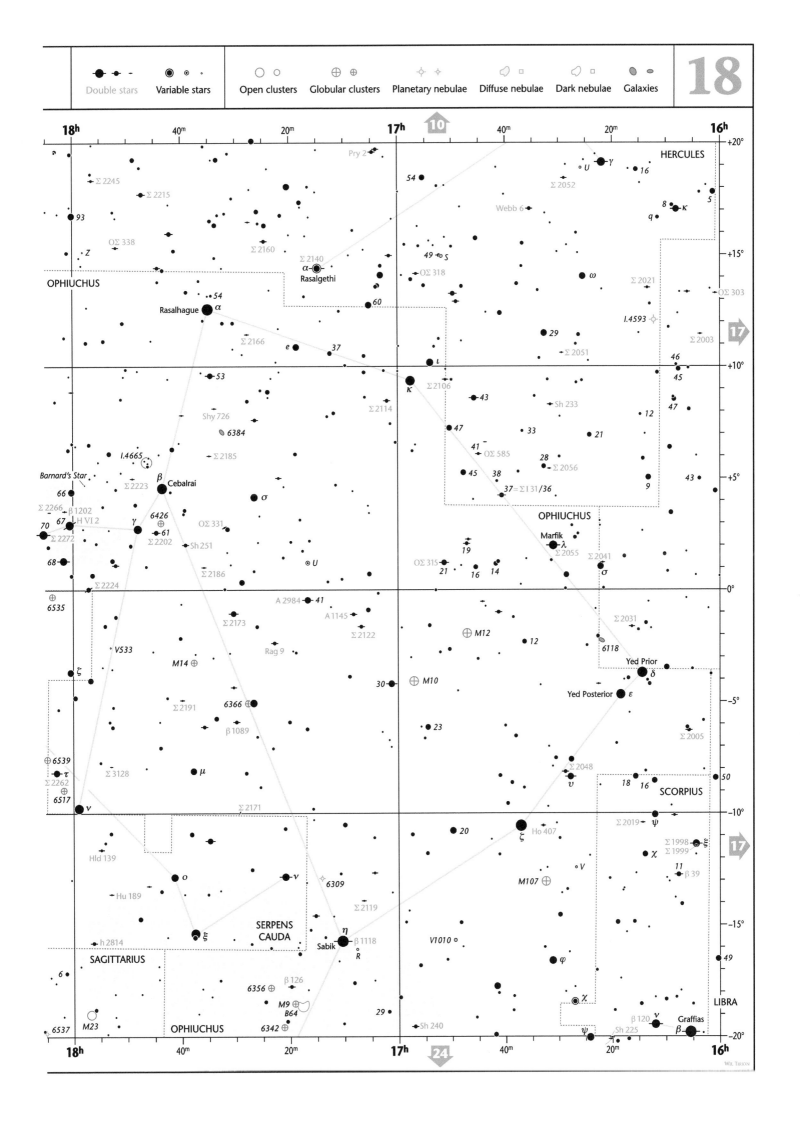

Magnitudes

brighter than 0.0   0.0–0.5   0.5–1.0   1.0–1.5   1.5–2.0   2.0–2.5   2.5–3.0   3.0–3.5   3.5–4.0   4.0–4.5   4.5–5.0   5.0–5.5   5.5–6.0   6.0–6.5   6.5–7.0   7.0–7.5   fainter

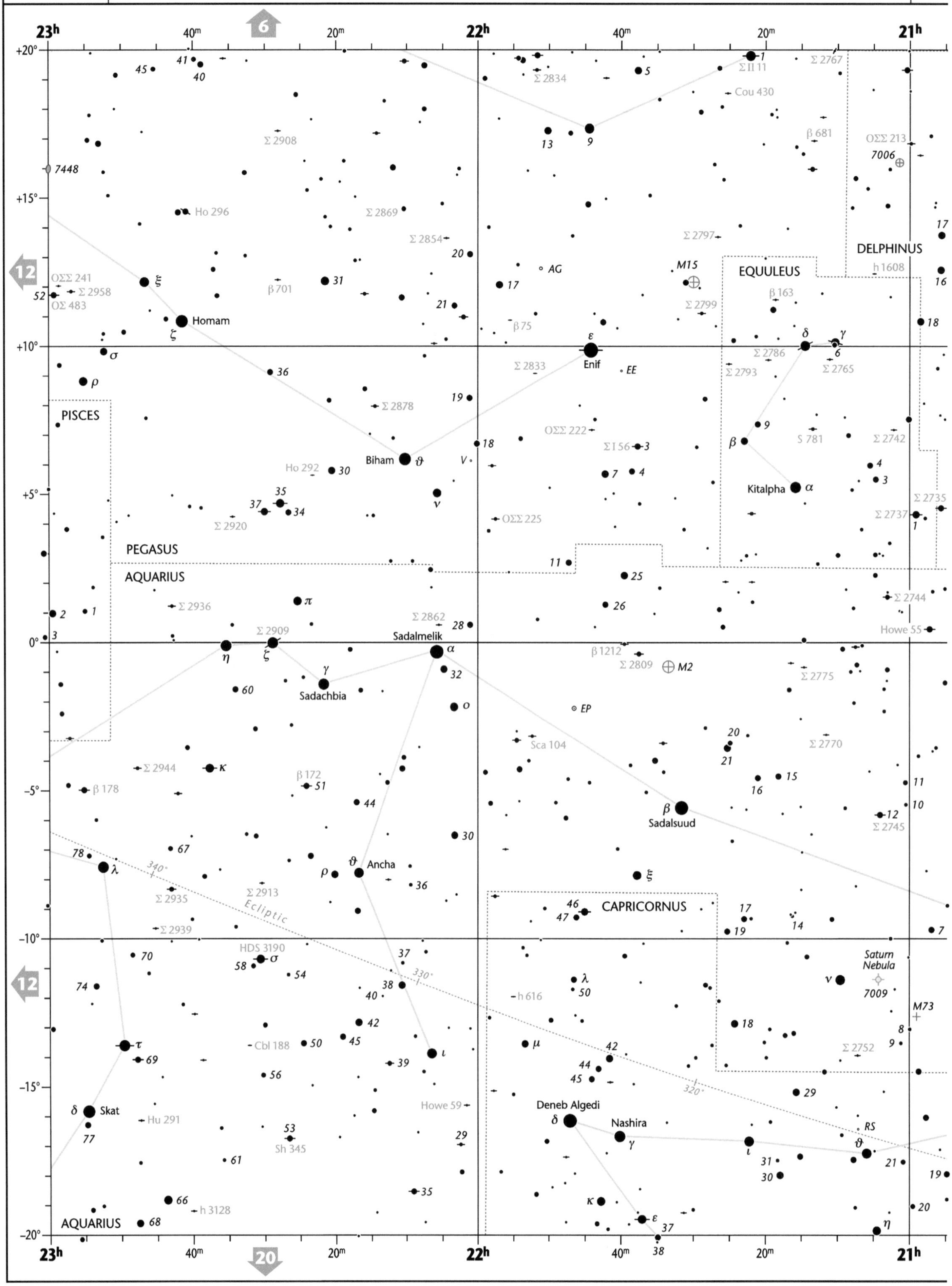

23h   40m   6   20m   22h   40m   20m   21h

+20°

45   41   40   Σ 2834   5   Σ II 11   1   Σ 2767

Cou 430

Σ 2908   13   9   β 681   OΣΣ 213   7006

+15°

7448   Ho 296   Σ 2869   Σ 2797   DELPHINUS   h 1608   17

Σ 2854   20   AG   M15   EQUULEUS   16

12

OΣΣ 241   ξ   31   17   Σ 2799   β 163   18

52   Σ 2958   β 701   21   β 75   δ   γ

OΣ 483   Homam   ε   Σ 2786   6

ζ   Enif   Σ 2793   Σ 2765

+10°   σ   36   Σ 2833   EE   β   9   S 781   Σ 2742

ρ   19   4

PISCES   Σ 2878   18   OΣΣ 222   Kitalpha   α   3

Biham   ϑ   V   Σ I 56   3   Σ 2735

Ho 292   30   7   4

+5°   35   ν   Σ 2737

37   34   1

Σ 2920   OΣΣ 225

PEGASUS   11

AQUARIUS   25   Σ 2744

2   1   π   26   Howe 55

3   Σ 2936   Σ 2909   28   β 1212

0°   η   ζ   Sadalmelik   α   Σ 2809   M2   Σ 2775

60   γ   32

Sadachbia   o   EP

κ   20   Σ 2770

Σ 2944   β 172   Sca 104   21

β 178   51   15

−5°   44   16   11

67   30   β   12   10

78   λ   340°   ρ   ϑ   Ancha   Sadalsuud   Σ 2745

Σ 2913   36   ξ

Σ 2935   Ecliptic   46   CAPRICORNUS   17   7

Σ 2939   47   19   14

−10°   37   h 616   Saturn Nebula

70   HDS 3190   λ   ν   7009

12   58   σ   54   38   330°   50   M73

74   40   18

42   μ   42   8

τ   Cbl 188   50   45   44   9   Σ 2752

69   39   ι   45   29

56   320°   RS

δ   Skat   Deneb Algedi   ϑ

Hu 291   53   δ   Nashira   ι   31   21

77   Sh 345   29   γ   30   19

61   Howe 59   20

66   h 3128   35   κ   η

AQUARIUS   68   ε   37

−20°   38

23h   40m   20   20m   22h   40m   20m   21h

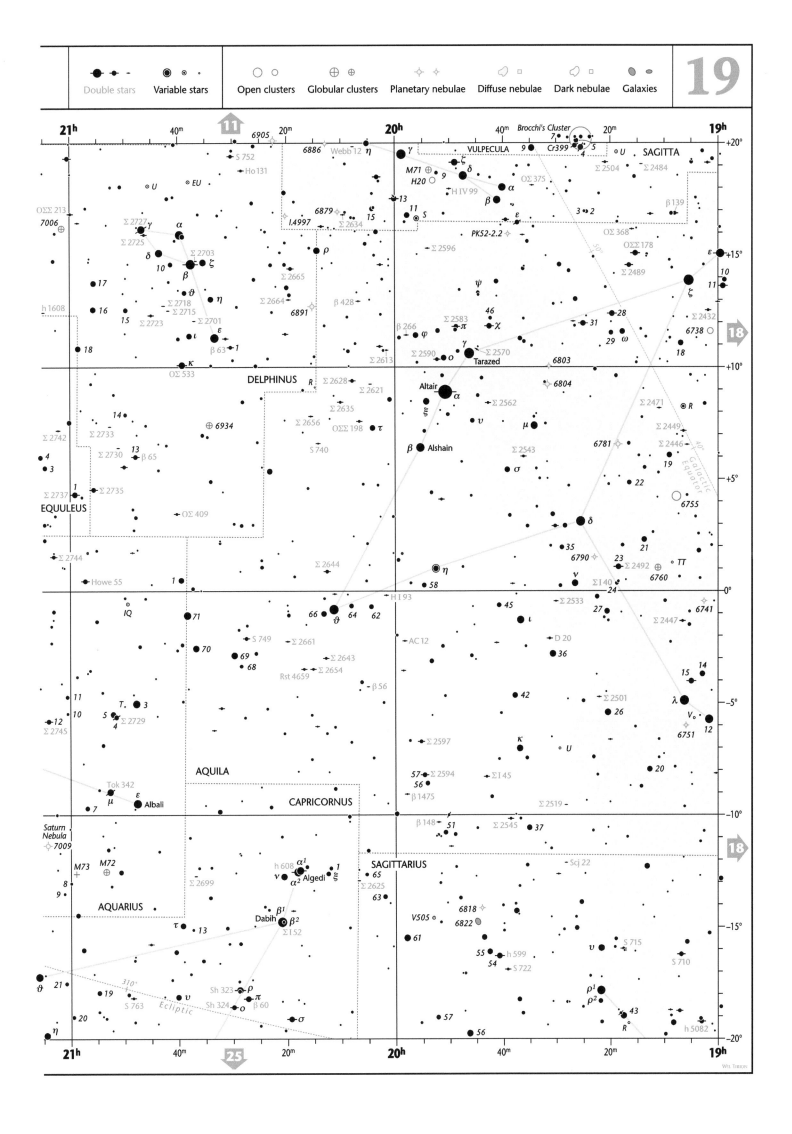

Magnitudes

brighter than 0.0   0.0–0.5   0.5–1.0   1.0–1.5   1.5–2.0   2.0–2.5   2.5–3.0   3.0–3.5   3.5–4.0   4.0–4.5   4.5–5.0   5.0–5.5   5.5–6.0   6.0–6.5   6.5–7.0   7.0–7.5   fainter

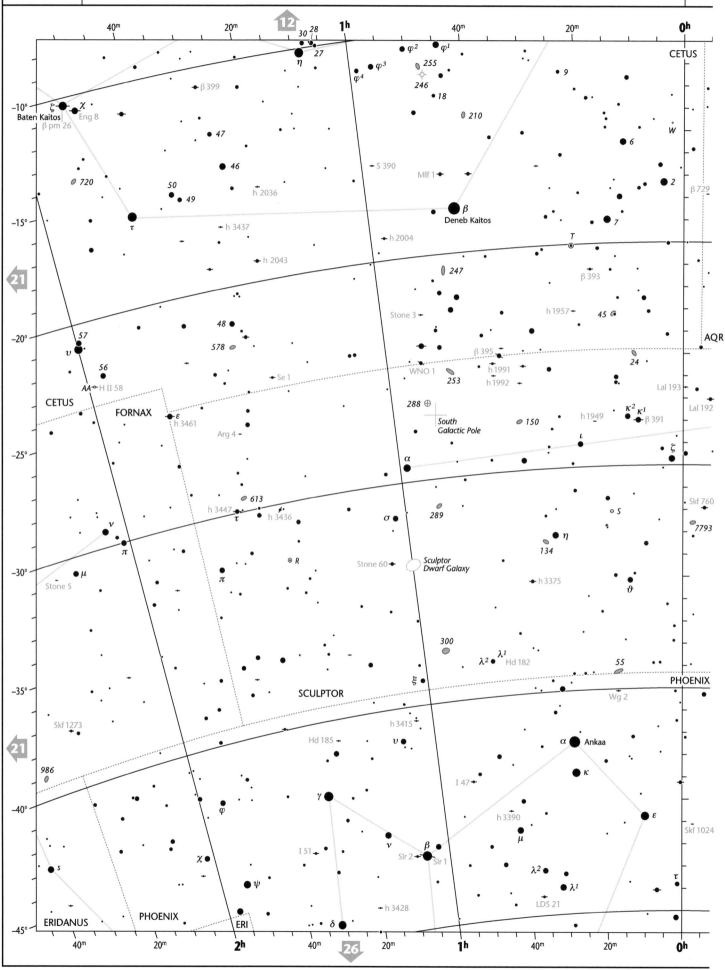

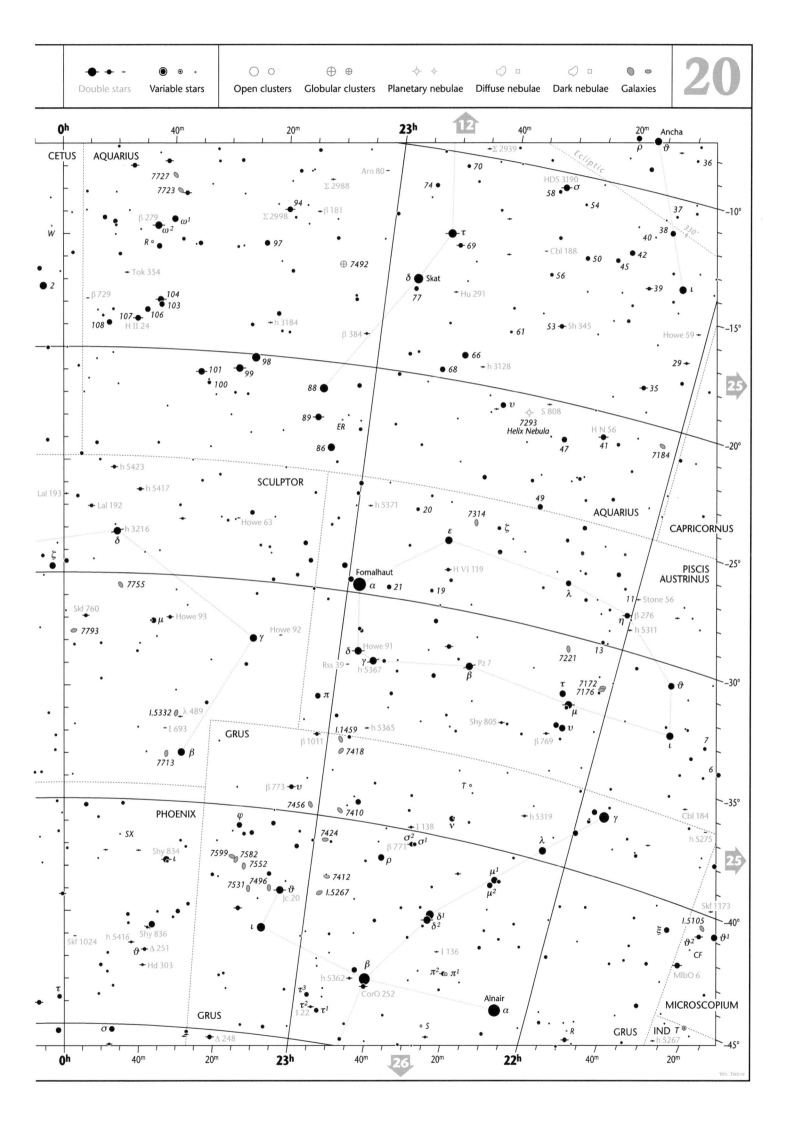

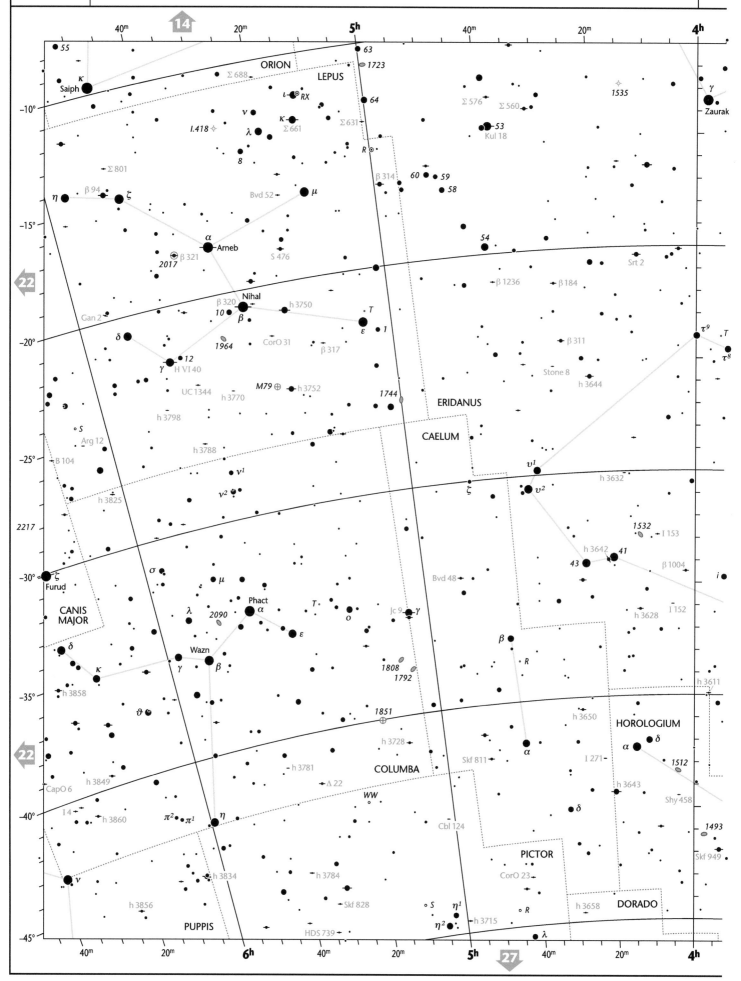

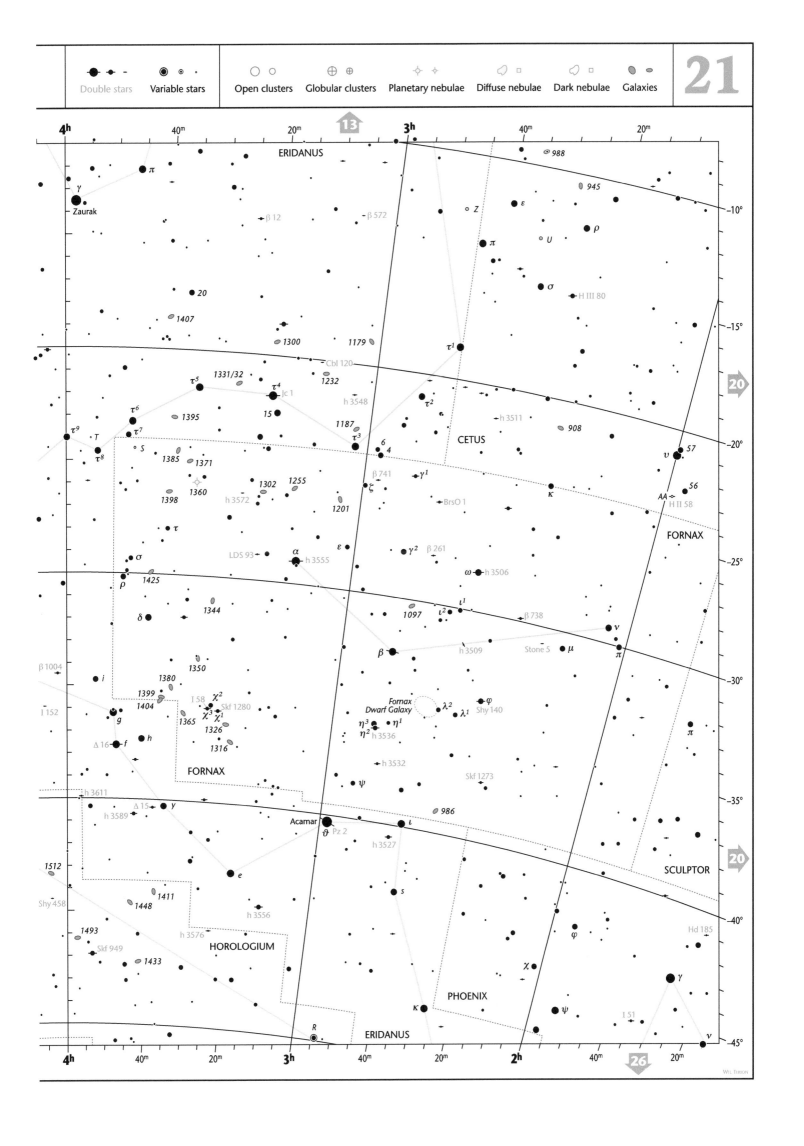

Magnitudes

brighter than 0.0 · 0.0–0.5 · 0.5–1.0 · 1.0–1.5 · 1.5–2.0 · 2.0–2.5 · 2.5–3.0 · 3.0–3.5 · 3.5–4.0 · 4.0–4.5 · 4.5–5.0 · 5.0–5.5 · 5.5–6.0 · 6.0–6.5 · 6.5–7.0 · 7.0–7.5 · fainter

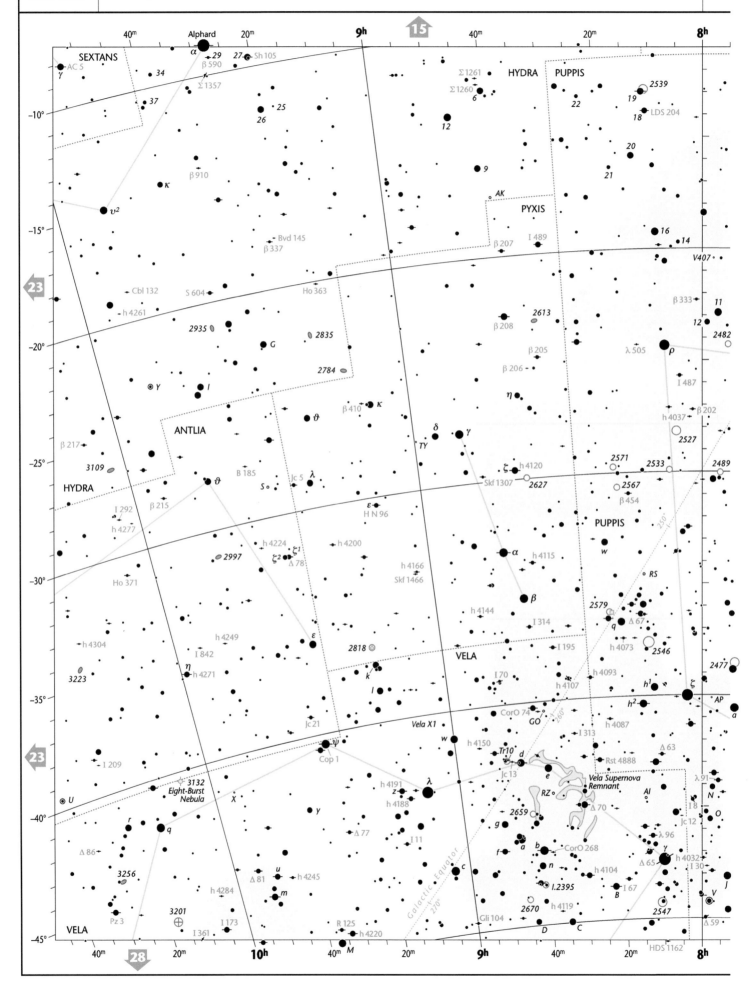

SEXTANS

Alphard

α

29

27

Sh 105

β 590

Σ 1357

γ   AC 5

34

37

25

26

κ

β 910

υ²

Σ 1261

Σ 1260

6

HYDRA   PUPPIS

22

2539

19

18   LDS 204

12

9

20

21

AK

PYXIS

16

14

V407

β 337

Bvd 145

I 489

β 207

11

12

2482

Cbl 132

S 604

Ho 363

β 208

2613

β 333

h 4261

2935

G

2835

2784

β 205

β 206

λ 505

I 487

β 202

h 4037

2527

Y

I

ANTLIA

β 410   κ

ϑ

δ   γ

TY

η

h 4120   2571

2533   2489

β 217

3109

HYDRA

ϑ

β 215

B 185

S

Jc 5   λ

ζ

Skf 1307

2627

2567

β 454

PUPPIS

w

I 292

h 4277

ε

H N 96

α   h 4115

2997

h 4224

ζ¹

ζ²   ξ

Δ 78

h 4200

h 4166

Skf 1466

RS

Ho 371

h 4249

ε

h 4144

β

2579

q   Δ 67

I 314

2546

h 4073

h 4304

I 842

2818

VELA

I 195

2477

3223

η

h 4271

k

l

I 70

h 4093

h¹

ξ

AP

a

CorO 74   h 4107   h²

GO   I 313

Δ 63

Jc 21

Vela X1   w   h 4150

Tr10   d

Jc 13   e

Rst 4888

λ 91

N

ψ

Cop 1

λ

z   h 4191

h 4188

RZ

Vela Supernova Remnant

AI

Δ 70

18

O

Jc 12

λ 96

3132

Eight-Burst Nebula

X

y

2659

g

f   a   b   CorO 268

γ   h 4032

Δ 65   I 30

U

r

q

Δ 86

Δ 77

I 11

c

n

h 4104

I 67

B

J

V

3256

u   h 4245

Δ 81

m   h 4284

I.2395

2670   h 4119

2547

Δ 59

3201   Pz 3

VELA

I 361   I 173

R 125   h 4220

Galactic Equator

Gli 104

D   C

HDS 1162

Double stars **Variable stars** **Open clusters** **Globular clusters** **Planetary nebulae** **Diffuse nebulae** **Dark nebulae** **Galaxies**

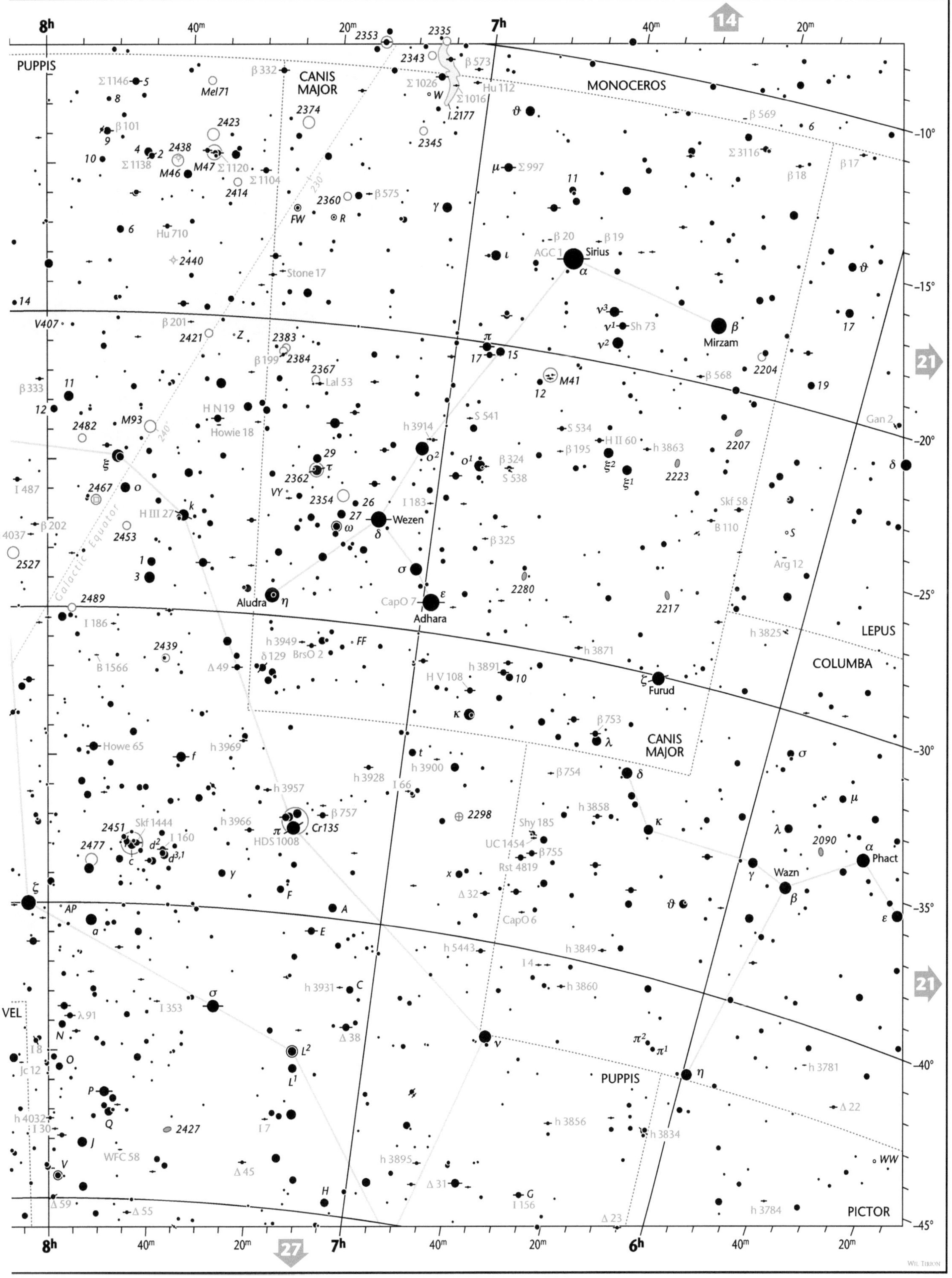

8h   40m   20m   7h   40m   20m   14

−10°
−15°
−20°
−25°
−30°
−35°
−40°
−45°

21
21

PUPPIS

CANIS MAJOR

MONOCEROS

Σ 1146   5
8
Mel 71
β 332
2374
2335
2343
β 573
Σ 1026
W
Hu 112
ϑ
β 569
6
β 101
9
2423
2438
M47 Σ 1120
2345
ι.2177
Σ 1016
β 3116
β 17
4 2
M46   Σ 1104
β 575
μ Σ 997
11
β 18
10 Σ 1138
2414
2360
FW   R
γ
β 20   β 19
6 Hu 710
2440
AGC 1   Sirius
α
ϑ
14 Stone 17
ν3
ν1 Sh 73
β
17
V407 β 201
π
17 15
ν2
Mirzam
β 333 11 β 199
Z 2383
M41
β 568
2204
12 2421 2384
2367
12
19
β202 M93 Lal 53
H N19 Howie 18
S 541
S 534
Gan 2
2482 ξ
29 τ
h 3914
o2
H II 60
h 3863
2207
δ
I 487 2467 o
2362
VY 2354
o1
β 324
ξ2
2223
h 4037 β 202 H III 27 k
26
I 183 S 538 ξ1
Skf 58 S
2453 27 ω
Wezen
β 325
B 110
1 δ
σ
Arg 12
3 σ
CapO 7 ε
2280 h 3825
2489 Aludra η
Adhara
2217
LEPUS
I 186 h 3949
FF h 3871
COLUMBA
2439 δ129 BrsO 2
h 3891 10
B 1566 Δ 49 H V 108
ζ
Furud
Howe 65 h 3969 κ
β 753
f h 3900 λ β 754
h 3957 t σ
h 3928 I 66 δ μ
Skf 1444 h 3966 β 757 h 3858
κ λ
2451 π Cr 135 2298 2090
2477 I 160 HDS 1008 Shy 185 α Phact
d2 c d3,1 UC 1454 β 755 γ Wazn
γ Rst 4819 x β ε
ξ F Δ 32 ϑ
AP A CapO 6
a E h 5443
VEL h 3849 I 4
C h 3860
λ 91 h 3931
I 353 σ Δ 38 π2 π1
N ν η
O L2 PUPPIS
I8 L1 h 3781
Jc 12 h 3856 h 3834
P
h 4032 Q 2427 I 7 Δ 22
I 30 WFC 58 h 3895 G
J Δ 45 WW
V H I 156 h 3784
Δ 59 Δ 55 Δ 31 Δ 23 PICTOR

8h   40m   20m   27   7h   40m   20m   6h   40m   20m

WIL TIRION

Magnitudes

brighter than 0.0 | 0.0–0.5 | 0.5–1.0 | 1.0–1.5 | 1.5–2.0 | 2.0–2.5 | 2.5–3.0 | 3.0–3.5 | 3.5–4.0 | 4.0–4.5 | 4.5–5.0 | 5.0–5.5 | 5.5–6.0 | 6.0–6.5 | 6.5–7.0 | 7.0–7.5 | fainter

VIRGO

82
76
β 114
50
49
4939
4781
Σ 1649
Ecliptic
α Spica
62
Hu 740
Sombrero Galaxy
M104
Σ 1659
-10°
210°
68
Σ 1669
β 28
η
86
Sh 165
β 932
3962
β 935
η Algorab
δ Sh 145
SV
53
5044
γ Gienah
75
69
5054
85
4361
4039/38
83
R
Antennae
63
61
31 Crt
4027
5170
54
Sh 161
-15°
5247
87
73
5018
ζ
ε
89
57 55
β 341
DL
β 610
5068
CORVUS
β 920
3
β 921
α Alchiba
PK303+40.1
β
6
-20°
γ
ψ
VIRGO
R
M68
HYDRA
5085
Stone 28
h 4556
4105
48 47
5061
4106
UC 2694
5078
h 4505
-25°
β 938
H N 69
5101
π
CENTAURUS
Δ 116
50
δ 164
r
β
W
Mug 3
HYDRA
h 4478
M83
Howe 72
Skf 1958
x² x¹
V335
-30°
5253
p
Jc 17
4
H N 51
β 1197
Slr 10
Howe 74
3
1
LDS 436
h 4500
h 4608
T
I.4696
2
λ 184
λ 143
Howe 94
5102
u
γ
Howe 28
z
ι h 4578
l
I.3370
Menkent ϑ
n
-35°
V1002
d
ψ
Skf 1128
4696
I 895
Howe 95
I 233
I 80
5367
WFC 145
ν
Centaurus A
χ
λ 180
Skf 900
h 4491
φ
μ
5128
I 423
R 218
Rmk 14 D
Pol 9
UY
5483
-40°
η
h 4672
5530
I 917
5643
v¹
I 83
LUPUS
I.4406
Omega
f
CapO 13
γ τ
E
v²
Centauri
5139 ω
w
h 4539
τ¹
4945
e
τ²
Slr 18
ι
ξ¹
σ
δ
CENTAURUS
ζ
Pol 3
CapO 61
4976
-45°
h 4600
ξ²
VIRGO

CRT

CENTAURUS

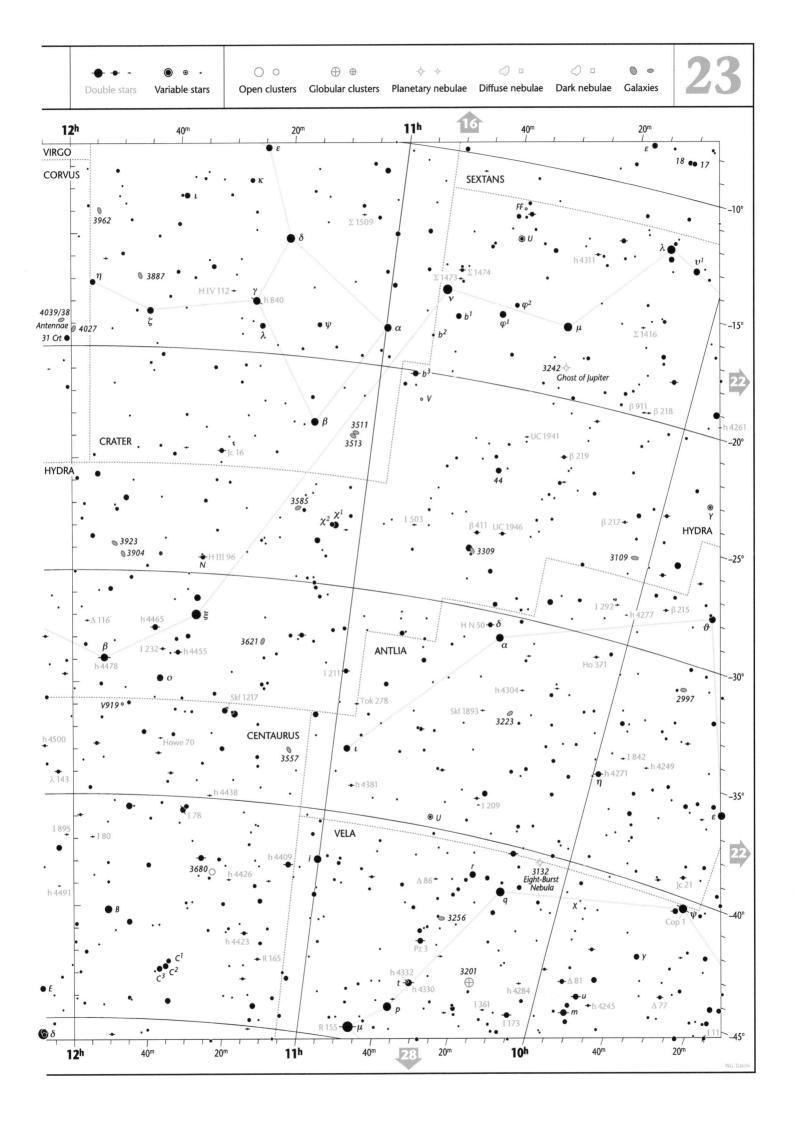

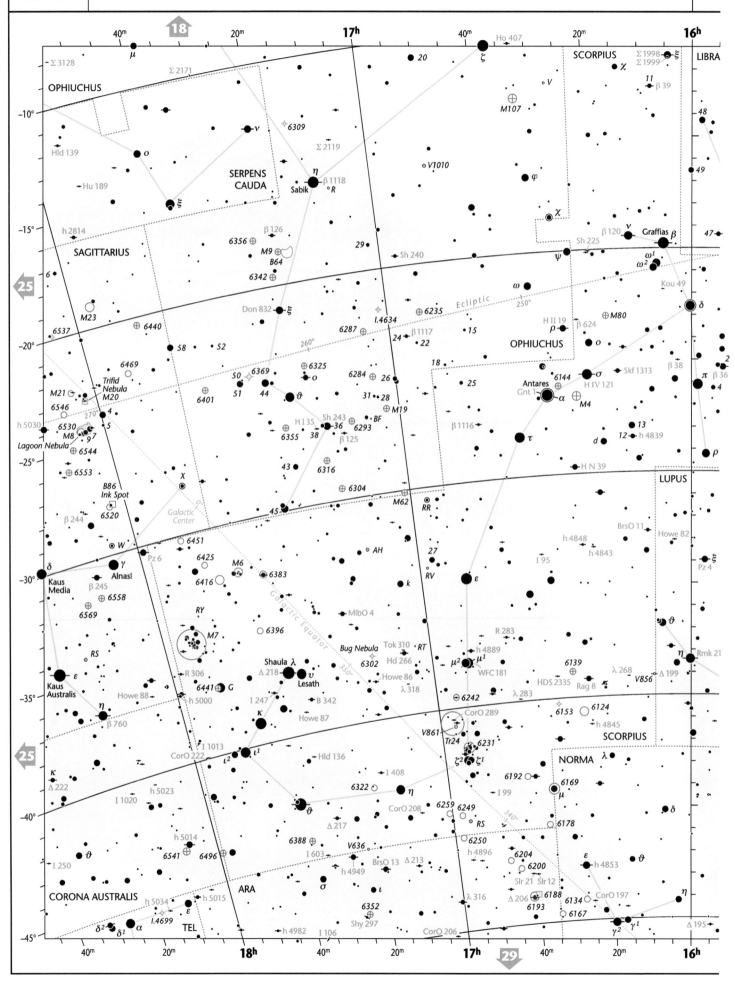

Magnitudes

brighter than 0.0 | 0.0–0.5 | 0.5–1.0 | 1.0–1.5 | 1.5–2.0 | 2.0–2.5 | 2.5–3.0 | 3.0–3.5 | 3.5–4.0 | 4.0–4.5 | 4.5–5.0 | 5.0–5.5 | 5.5–6.0 | 6.0–6.5 | 6.5–7.0 | 7.0–7.5 | fainter

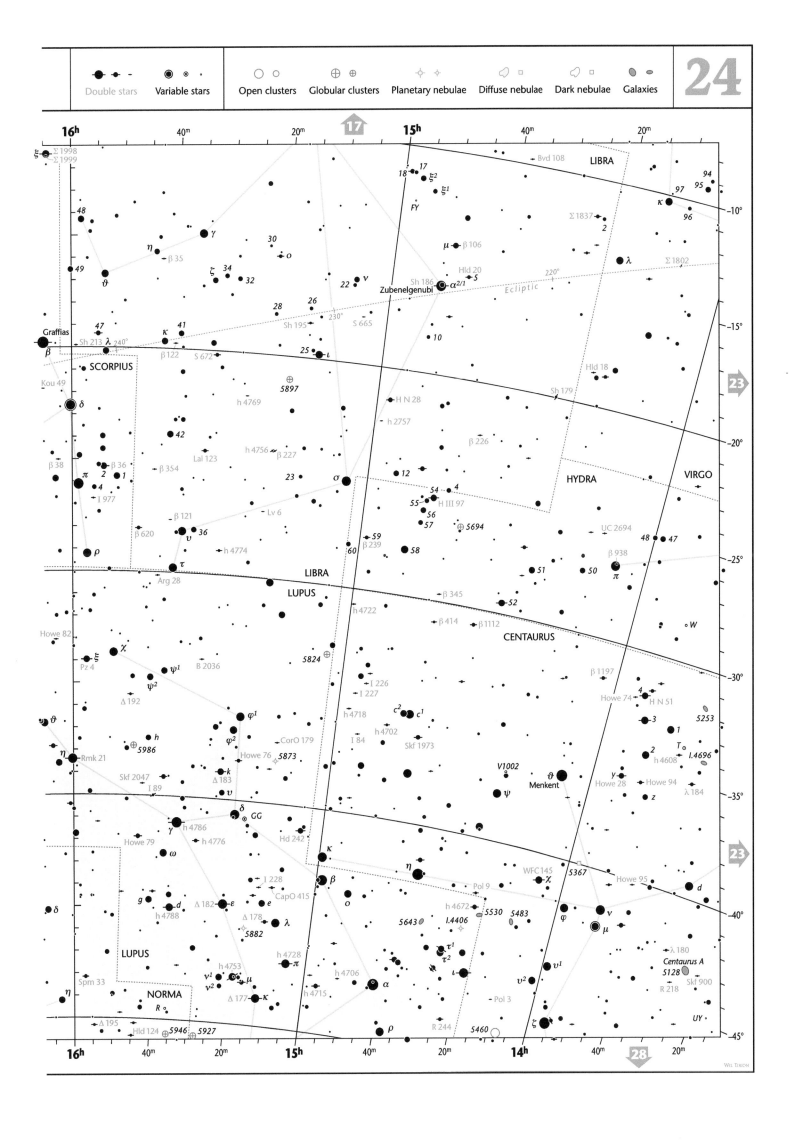

Magnitudes

brighter than 0.0 0.0–0.5 0.5–1.0 1.0–1.5 1.5–2.0 2.0–2.5 2.5–3.0 3.0–3.5 3.5–4.0 4.0–4.5 4.5–5.0 5.0–5.5 5.5–6.0 6.0–6.5 6.5–7.0 7.0–7.5 fainter

40ᵐ 20ᵐ 21ʰ 19 40ᵐ 20ᵐ 20ʰ

AQUARIUS 17 14 AQUARIUS CAPRICORNUS AQUILA
19 Saturn SAGITTARIUS
46 Nebula h 608 α¹ 1
47 ν 7009 M73 M72 ν α² Algedi ξ Σ 2625 65
Σ 2699 63
–10° 8 9 β¹ 61
λ 18 τ 13 Dabih β²
50 Σ 2752 ΣI 52
h 616 29 ρ β 60
42 Ecliptic 19 υ Sh 323 π
μ 320° 310° Sh 324 o σ
44 RS S 763
45 ϑ 21 20
Deneb Algedi ι 31 300°
δ Nashira 30 η β 674 RT
Howe 59 γ κ ε 26 β 974 17 4 M75
29 37 27 χ h 2975
33 φ
35 h 3003 6907
36 ζ 60 ω
7184 ψ 62 59
M30 24 β 153
41 ω h 5226 T h 5188 h 5168 RR
CAPRICORNUS Skf 2097 6923
AQUARIUS 8 δ Stone 64
Bvd 37 β 251 2925
Stone 56 5 γ ϑ²
11 ε β ϑ¹
β 276 η Skf 2098 α h 5178
λ h 5311 13 ϑ 7 R 321 RT
7221 6 Hd 294
7172 ι PISCIS AUSTRINUS ζ UC 4365
7176 Cbl 184 Lacaille Jc 18 U Δ 230
τ 8760 ζ η h 5228 RU
μ h S275 Skf 1173 Gli 259 ι
υ I.5105 ϑ¹ κ¹
β 769 ϑ² κ²
Shy 805 γ CF Δ 236 h 5211 Rss 36 β 763
ξ ι
I 138 MlbO 6 ν SAGITTARIUS
σ² σ¹ MICROSCOPIUM λ 415
β 771 μ¹ ζ I 41
μ² T Skf 1168 I 256
δ¹ h 5261 I 1429 α
δ² I 130
I 136 R 7049
π² π¹ Alnair Skf 376
α 7144 GRUS INDUS TELESCOPIUM I 17

40ᵐ 20ᵐ 22ʰ 40ᵐ 20ᵐ 21ʰ 40ᵐ 20ᵐ 20ʰ

26

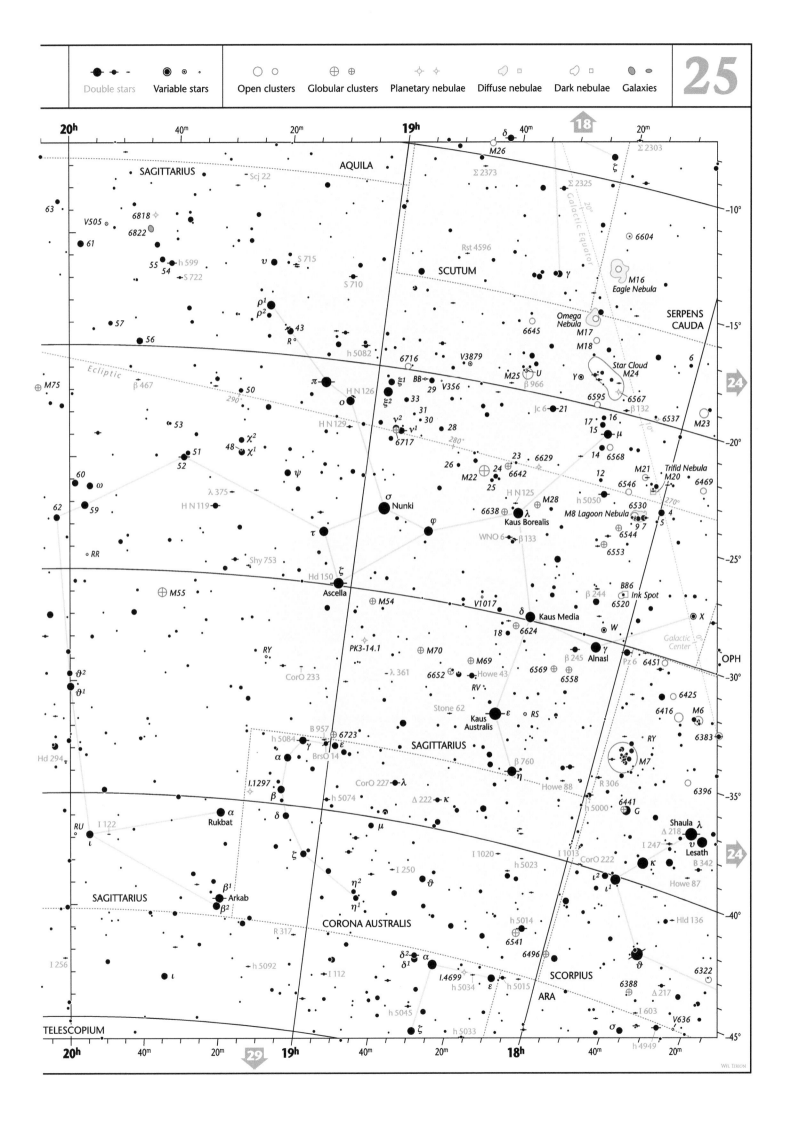

Double stars   Variable stars   Open clusters   Globular clusters   Planetary nebulae   Diffuse nebulae   Dark nebulae   Galaxies

20ʰ   40ᵐ   20ᵐ   19ʰ   40ᵐ   18   20ᵐ

SAGITTARIUS   AQUILA   δ
                                     M26   Σ 2303
                                     ζ
                                                  Σ 2373
63   Σ 2325   −10°
V505   6818   Rst 4596   ⊙ 6604
6822   SCUTUM
61
55 h 599   υ   S 715
54   S 722   γ   M16
                     ρ¹   Eagle Nebula
57   ρ²   Omega   SERPENS
56   γ 43   6645   Nebula   CAUDA
             R°   M17   −15°
Ecliptic   h 5082   M18   6
             β 467   6716   V3879   Star Cloud
50   ξ¹   BB   M25   U   M24
π   H N 126   ξ²   29   V356   γ   6567
                     o   33   6595   β 132
53   ν²   31   30   28   17   16   6537
290°   H N 129   ν¹   15   μ   M23
48 χ²   6717   26   23   6629   14   6568   −20°
   χ¹   24   12   M21   Trifid Nebula
51   ψ   M22   6642   6546   M20
52   25   H N 125   6469
60 ω   λ 375   σ   6638   λ   M8 Lagoon Nebula
62   59   H N 119   Nunki   φ   M28   h 5050   6530   9 7   4
                                              Kaus Borealis   6544   5
         τ   φ   WNO 6   β 133   6553
RR   Shy 753   −25°
                                   ζ   V1017   B86
M55   Hd 150   Ascella   M54   β 244   Ink Spot   X
                                              6520
                              δ   Kaus Media   W   Galactic
                        RY   18   6624   Center   OPH
PK3-14.1   M70   γ   Alnasl   −30°
CorO 233   λ 361   M69   β 245   Pz 6   6451
                                   6652   Howe 43   6569   6558   6425
ϑ²   RV°   6416   M6
ϑ¹   Stone 62   Kaus   RS   6383
            B 957   6723   Australis   ε   RY
h 5084   γ   SAGITTARIUS   M7
α   BrsO 14   β 760   6396
I.1297   CorO 227   λ   η   Howe 88
Hd 294   β   h 5074   Δ 222   κ   −35°
RU   I 122   α   δ   μ   6441   G
ι   Rukbat   ζ   Shaula   λ
                                                          Δ 218
                                   η²   ϑ   I 247   υ   Lesath
β¹   η¹   I 1020   h 5023   CorO 222   B 342
β²   Arkab   I 250   I 1013   ι²   Howe 87
SAGITTARIUS   CORONA AUSTRALIS   h 5014   ι¹   Hld 136
                                              h 5000   6496
I 256   h 5092   R 317   δ²   α   6541   ϑ   6322
         ι   δ¹   I.4699   SCORPIUS   6388   Δ 217
TELESCOPIUM   I 112   ε   h 5015   ARA   I 603
                                   h 5045   h 5034   V636
                                   ζ   h 5033   σ   h 4949

20ʰ   40ᵐ   29   19ʰ   40ᵐ   18ʰ   20ᵐ

WIL TIRION

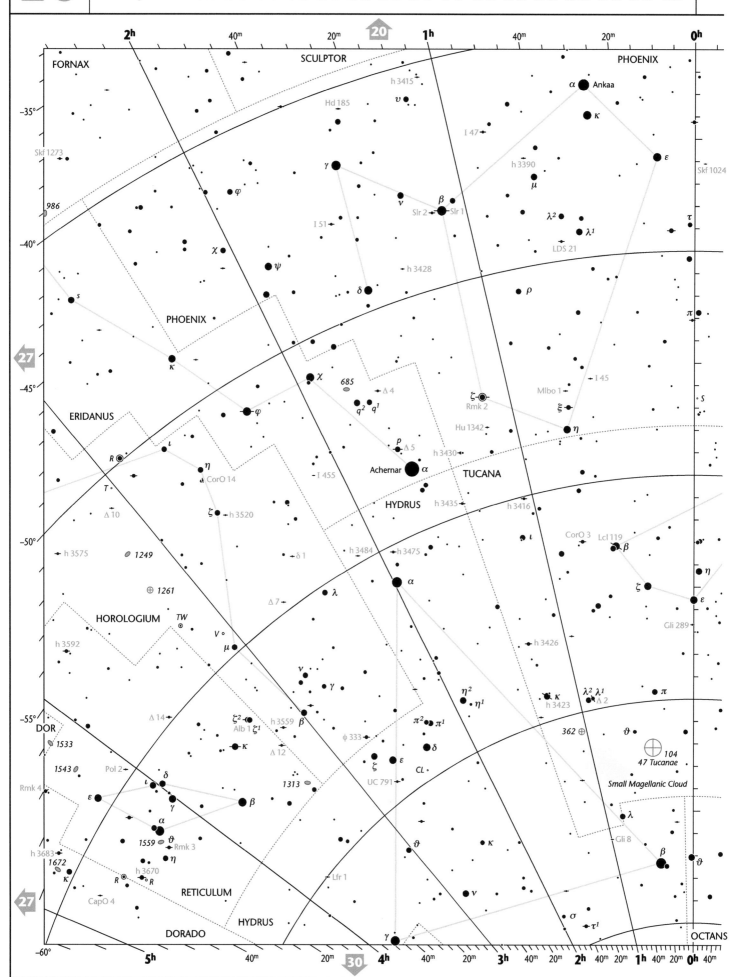

Magnitudes

brighter than 0.0 | 0.0–0.5 | 0.5–1.0 | 1.0–1.5 | 1.5–2.0 | 2.0–2.5 | 2.5–3.0 | 3.0–3.5 | 3.5–4.0 | 4.0–4.5 | 4.5–5.0 | 5.0–5.5 | 5.5–6.0 | 6.0–6.5 | 6.5–7.0 | 7.0–7.5 | fainter

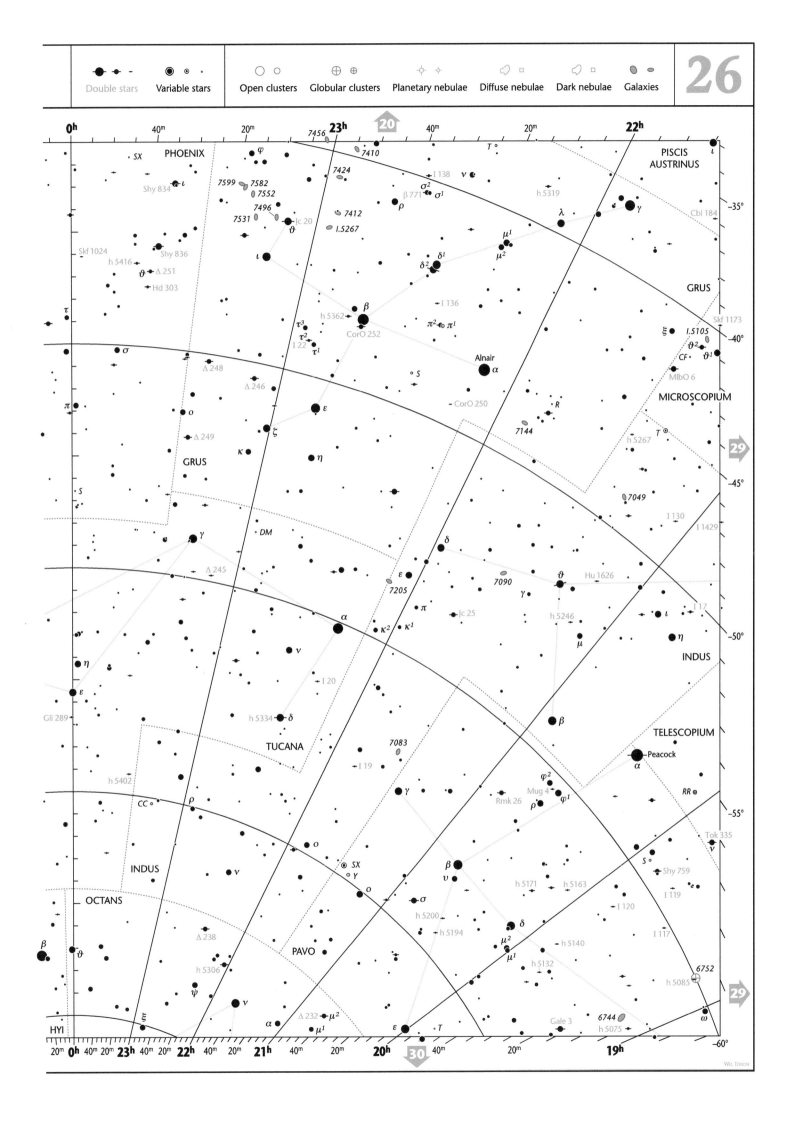

Double stars Variable stars Open clusters Globular clusters Planetary nebulae Diffuse nebulae Dark nebulae Galaxies

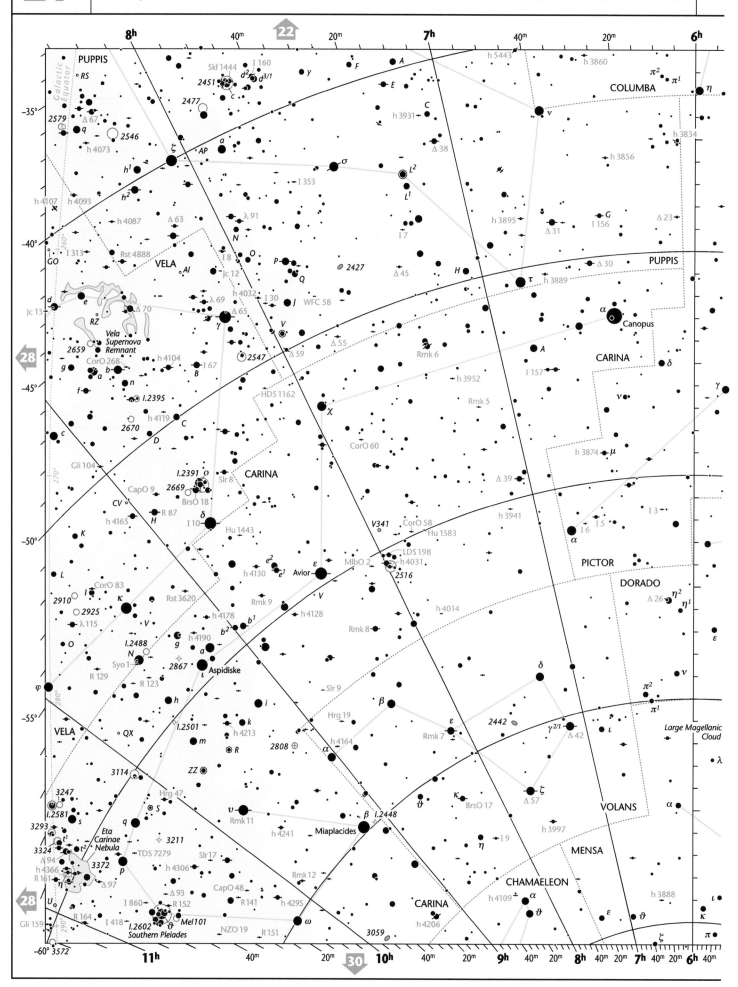

Magnitudes

● ● ● ● ● ● ● ● ● ● · · · · ·

brighter than 0.0 0.0–0.5 0.5–1.0 1.0–1.5 1.5–2.0 2.0–2.5 2.5–3.0 3.0–3.5 3.5–4.0 4.0–4.5 4.5–5.0 5.0–5.5 5.5–6.0 6.0–6.5 6.5–7.0 7.0–7.5 fainter

PUPPIS

COLUMBA

VELA

CARINA

Canopus

PUPPIS

Vela
Supernova
Remnant

CARINA

PICTOR

DORADO

Avior

Aspidiske

VOLANS

Large Magellanic
Cloud

MENSA

Miaplacides

VELA

Eta
Carinae
Nebula

Southern Pleiades

CHAMAELEON

CARINA

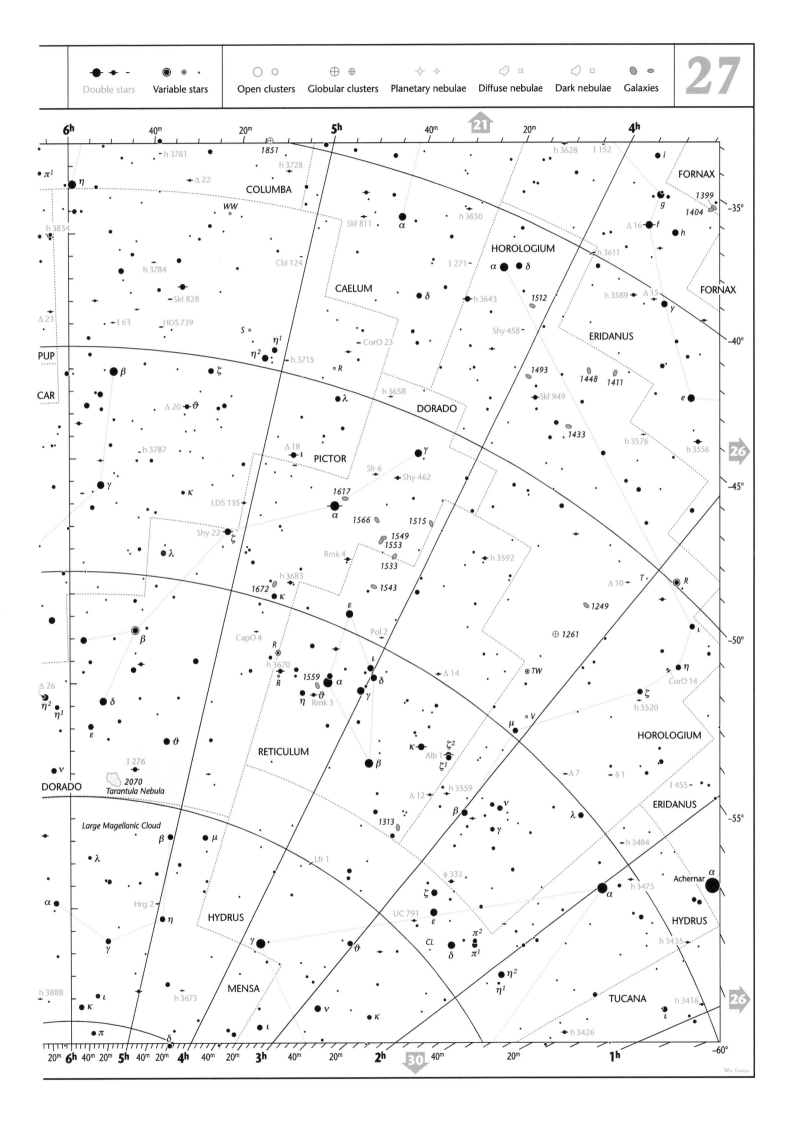

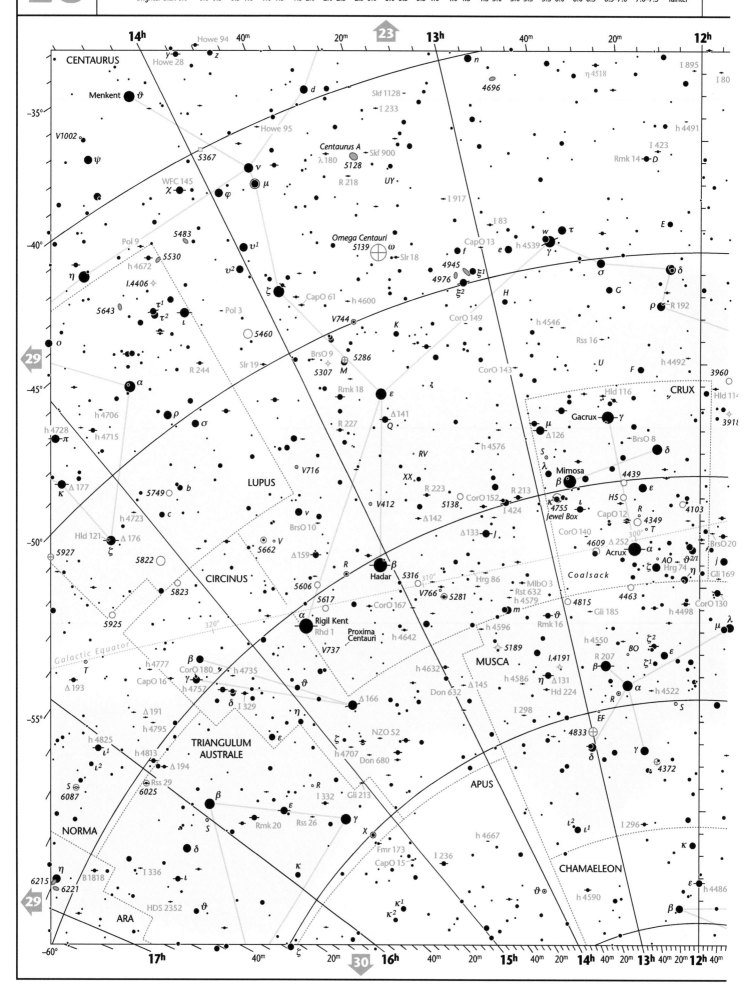

Magnitudes

brighter than 0.0 0.0–0.5 0.5–1.0 1.0–1.5 1.5–2.0 2.0–2.5 2.5–3.0 3.0–3.5 3.5–4.0 4.0–4.5 4.5–5.0 5.0–5.5 5.5–6.0 6.0–6.5 6.5–7.0 7.0–7.5 fainter

CENTAURUS

Menkent ϑ

V1002
ψ

Howe 94 40ᵐ 20ᵐ 23 13ʰ 40ᵐ 20ᵐ 12ʰ
γ z n η 4518 I 895 I 80
Howe 28 d 4696 h 4491
−35° Howe 95 Skf 1128 I 233
5367 Centaurus A Skf 900 I 423
λ 180 5128 Rmk 14 D
WFC 145 ν R 218 UY
χ μ I 917
φ I 83 E
5483 Omega Centauri CapO 13 w τ
Pol 9 5530 5139 ω f h 4539 γ
h 4672 υ¹ Slr 18 e
I.4406 4945 ξ¹ σ G
η 5643 υ² CapO 61 h 4600 4976 ξ² H ρ R 192
τ¹ ζ CorO 149 δ
−40° τ² ι V744 K h 4546 CRUX
o Pol 3 Rss 16 h 4492 3960
5460 BrsO 9 5286 CorO 143 U F Hld 116 Hld 114
R 244 Slr 19 5307 M 3918
α Rmk 18 ε Gacrux γ
−45° Δ 141 μ BrsO 8
h 4706 Q Δ 126 δ
ρ R 227 RV S
h 4728 σ h 4576 λ Mimosa 4439
π h 4715 V716 XX β ε
κ Δ 177 LUPUS R 223 CorO 152 R 213 HS R
5749 b 5138 I 424 4755 CapO 12 4349 4103
h 4723 c ν V412 Δ 142 Jewel Box T
BrsO 10 Δ 133 J CorO 140 4609 Δ 252 BrsO 20
Hld 121 Δ 176 V Acrux α AO ϑ²/¹ j
−50° ζ 5662 Δ 159 R β ζ Hrg 74 η Gli 169
5927 5822 CIRCINUS Hadar 5316 310° Hrg 86 MlbO 3 Coalsack 4463 CorO 130
5925 5823 5606 V766 5281 Rst 632 4815 h 4498
5617 CorO 167 h 4579 m ϑ Gli 185
α Rigil Kent Proxima h 4642 Rmk 16 h 4550 ζ²
Rhd 1 Centauri I.4191 BO ε
Galactic Equator V737 5189 R 207 ζ¹
T h 4777 MUSCA β α
Δ 193 β CorO 180 h 4735 ϑ h 4632 η Δ 131 R h 4522
CapO 16 γ h 4757 η Don 632 Δ 145 Hd 224 S
δ I 329 ζ EF
Δ 191 Δ 166 I 298 4833 γ
−55° h 4795 ε NZO 52 δ 4372
h 4825 η APUS
TRIANGULUM ζ CHAMAELEON
ι¹ h 4813 AUSTRALE Don 680 κ
ι² Δ 194 h 4707 ι² ι¹ I 296
S Rss 29 β R I 332 Gli 213
6087 6025 ε κ
NORMA Rmk 20 Rss 26 γ h 4667
η B 1818 I 336 δ X h 4590 κ
6215 ι S Fmr 173 I 236 ε h 4486
6221 HDS 2352 ϑ CapO 15 ϑ β
29 ARA κ¹
κ²

17ʰ 40ᵐ ζ 20ᵐ 30 16ʰ 40ᵐ 20ᵐ 15ʰ 40ᵐ 20ᵐ 14ʰ 40ᵐ 20ᵐ 13ʰ 40ᵐ 20ᵐ 12ʰ 40ᵐ
−60°

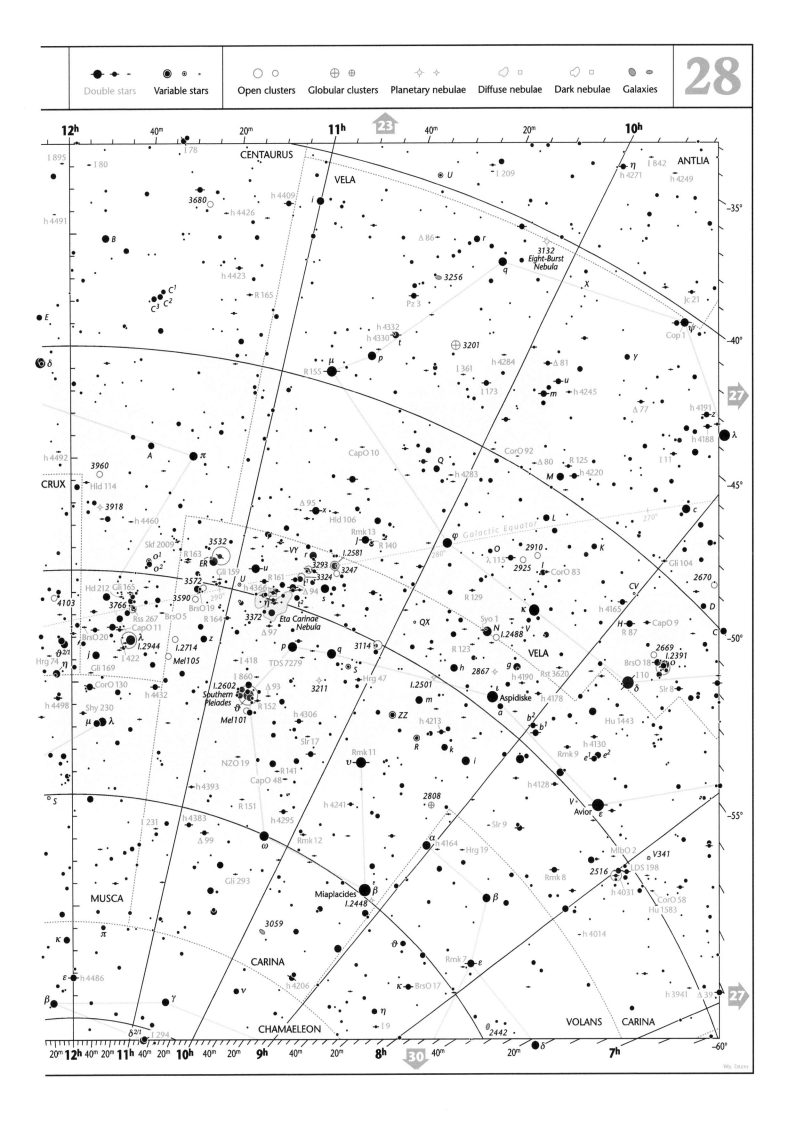

Double stars   Variable stars   Open clusters   Globular clusters   Planetary nebulae   Diffuse nebulae   Dark nebulae   Galaxies

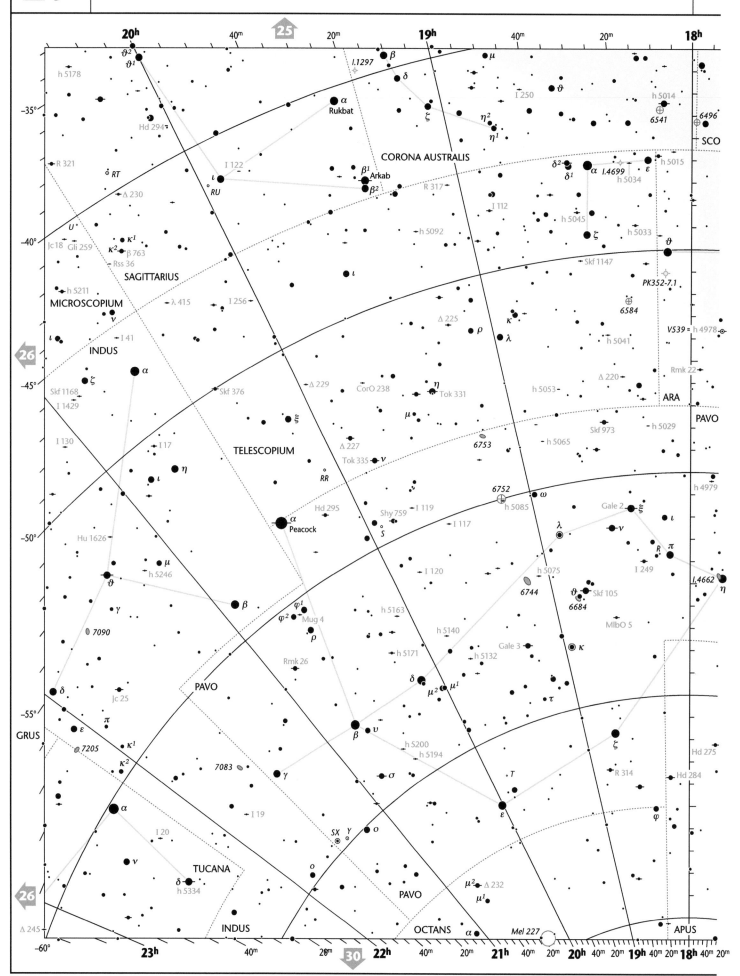

Magnitudes

brighter than 0.0   0.0–0.5   0.5–1.0   1.0–1.5   1.5–2.0   2.0–2.5   2.5–3.0   3.0–3.5   3.5–4.0   4.0–4.5   4.5–5.0   5.0–5.5   5.5–6.0   6.0–6.5   6.5–7.0   7.0–7.5   fainter

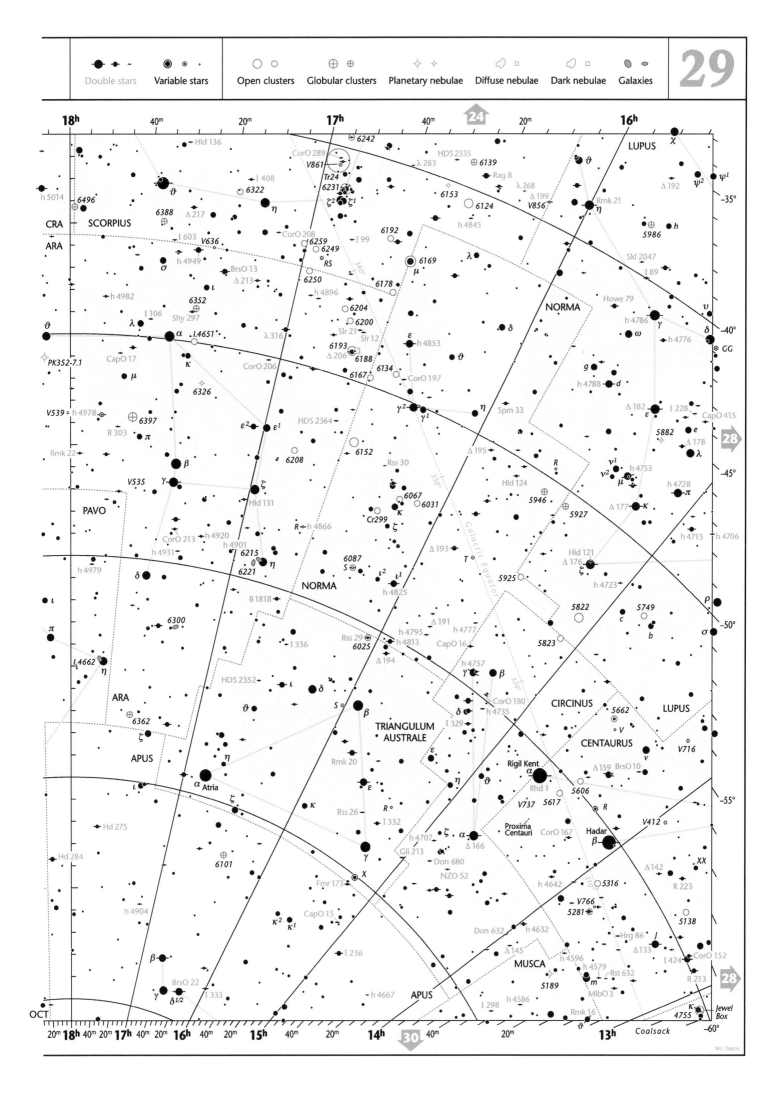

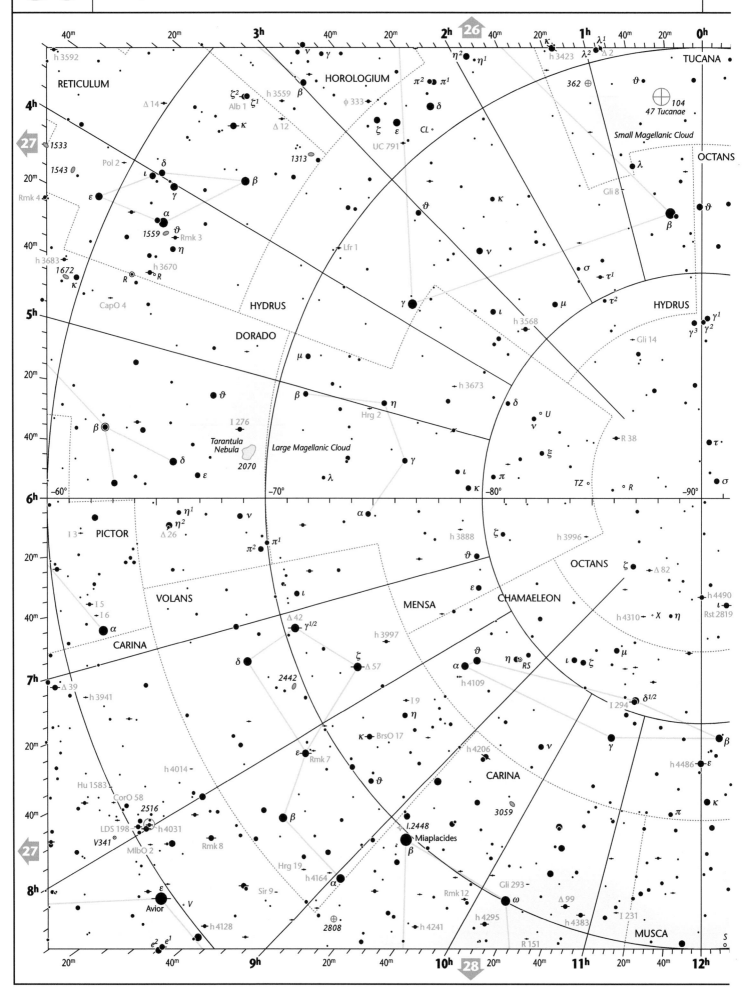

Magnitudes

brighter than 0.0 | 0.0–0.5 | 0.5–1.0 | 1.0–1.5 | 1.5–2.0 | 2.0–2.5 | 2.5–3.0 | 3.0–3.5 | 3.5–4.0 | 4.0–4.5 | 4.5–5.0 | 5.0–5.5 | 5.5–6.0 | 6.0–6.5 | 6.5–7.0 | 7.0–7.5 | fainter

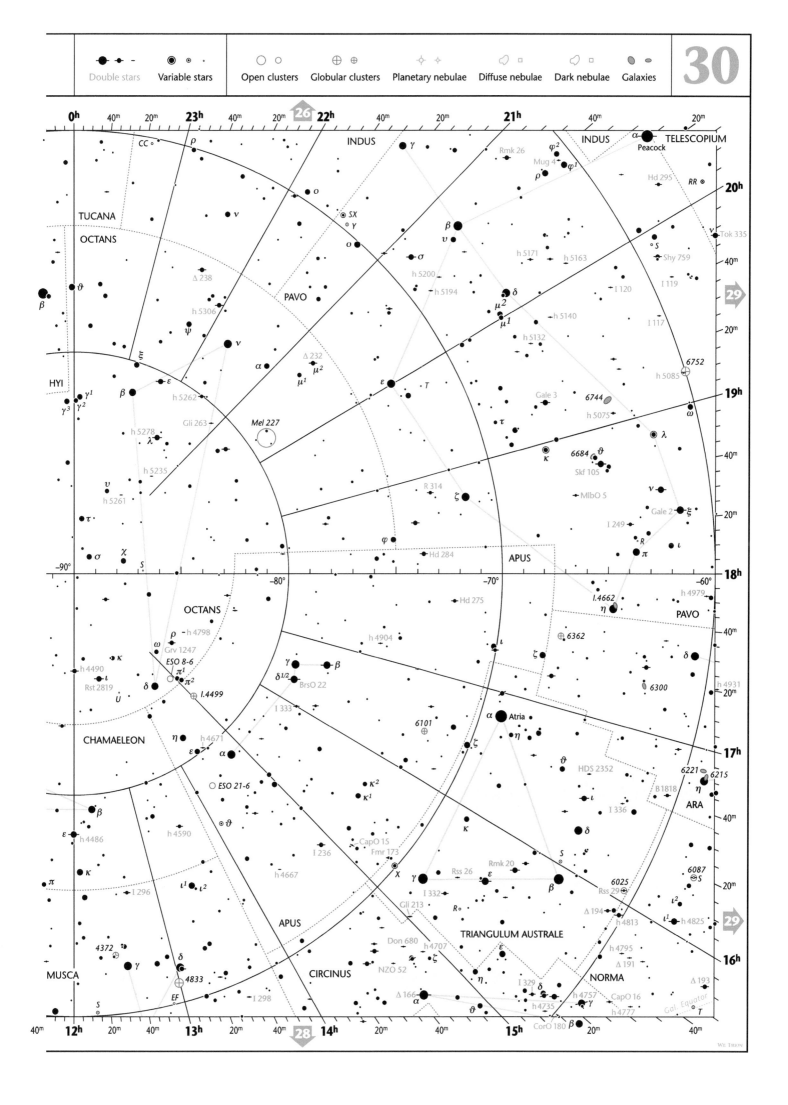

Appendix A: The target list

Two criteria were used to select the following 2,500 "high probability" double stars from the 110,950 unique systems in the January, 2015 *Washington Double Star Catalog*.

The first is visual limits to magnitude and resolution. With few exceptions, targets are only included down to a combined visual magnitude of 7.75 and secondary stars down to magnitude 13.5; separations are not less than the Abbe resolution limit of $R_o = 0.5''$ feasible with a 250 mm or larger objective. Over 90% of the pairs can be resolved with a 150 mm aperture.

The second is evidence of a physical bond. This includes a projected separation less than 5,000 AU, a divergence between proper motion vectors less than 30% of the larger proper motion, a note in WDS that the system is physical based on CPM and/or parallax data, or an orbital solution (weighted by the quality rating assigned in the *6th Catalog of Orbits of Visual Binary Stars*). Negative indicators included a projected separation greater than 50,000 AU, a CPM deviation greater than 80%, a WDS flag indicating a "bogus binary" or optical pair, or a linear solution (a straight line path rather than curved orbital motion). Every record in WDS was evaluated against these criteria and the highest scoring systems and components (within the visual limits) were selected. These systems were further vetted using sky survey images at SIMBAD, observing notes and other references. Although this procedure cannot ensure every target is a physical system, it does reveal more accurately the true variety in physical double and multiple stars outside the traditional and ambiguous criterion of "looking like" a double star.

To conserve space, data are listed on a single line. This includes the Bayer, Flamsteed or Gould (southern hemisphere) designation, the catalog ID (as it appears in the charts); the component letter codes; the celestial coordinates for epoch J2000; magnitudes, position angle (θ) and separation (ρ) for the listed components; distance from the Sun (in parsecs);

spectral types (with alternate or dual types, most special codes and giant subtypes omitted for simplicity), and the Henry Draper (HD) and Smithsonian (SAO) catalog numbers. Distances estimated by spectroscopic parallax are flagged with an asterisk (*).

Again to conserve space, remarks are given primarily for "showpiece" (★) stars, systems frequently observed (200 or more positional measures) or "neglected" (not measured since 2000), high CPM systems (annual proper motion of $0.5''$ or more), local systems (within 25 parsecs of the Sun), and systems with a calculated orbit. Period (P) and average orbit radius (r) are indicated for binaries in the *6th Orbits*, with eccentricity and year of next periastron or apastron given for high quality orbits. If the period is less than 300 years, the orbit diagram and ephemeris were examined to determine how the system will change visually. If there is no orbital solution, the projected separation with the Couteau correction for foreshortening (see Appendix B) provides a visual scale. "MSC" system masses are from A. Tokovinin's *Multiple Star Catalog*, other mass estimates are from WDS notes. Star color for a primary/secondary pair is noted using the letter codes on p. 15; a dash indicates a lack of color. The remark line ends with the epoch in parentheses.

The "Component" column enumerates the members of the system as these are described in WDS, MSC and SIMBAD, in the order the parameters are listed. Many lines have no component codes at all, which indicates the pair is listed as binary in WDS ("A B" is understood). The presence of binary component codes ("A B," "A C," etc.) indicates that the pair was extracted from a WDS multiple system comprising low probability or optical components. Pairs that cannot be resolved with a 300 mm aperture ($\rho < 0.4''$ or a large Δm) are grouped by parentheses with no space or comma between the component codes; the magnitude, θ and ρ describe the pair as a single component. To make plain the relative separations in "1+2" triples, brackets are used to indicate measures made within the binary: a component list of "A [B C]" with separations of "ρ [ρ]" means the first separation is measured from A to B, but the second is from B to C. This system of formatting the component codes explicitly departs from the notation (described on p. 10) used in WDS and other double star catalogs and respected in the target list remarks. It is designed to indicate both the likely visual appearance and probable dynamic structure of the double star, so far as the most recent data describe it.

Label	Catalog ID	Components	Coordinates (J2000)	Mag.	θ	ρ	Dist.	Spectral type(s)	HD, SAO No.
	Andromeda		And						*Chart 6*
LN	Σ 2973		23 02.8 +44 04	6.4, 10.1	39	7.4″	330	B2V	217811, 52626
KZ	Σ 2985	A (BaBb)	23 10.0 +47 58	7.2, 8.0	256	16″	25.0	G2IV	218739, 52754
			Local solar type 1+(2) spectroscopic triple; dominant in faint field; B is a BY Dra type variable (a low mass star with irregular surface brightness) and double line spectroscopic binary (P = 3.0 d). AB: *P* = ~6,300 y, ps = 540 AU. (2012)						
	Σ 2987		23 10.4 +49 01	7.4, 10.4	150	4.4″	45.4	G1V	218790, 52759
	Bvd 142	A [Ba Bb]	23 10.5 +41 19	7.8, [10.7, 11.4]	165, [89]	80″, [0.6″]	63	F5V K0V	218805, 52762
	Σ 2992	A B	23 13.1 +40 00	7.7, 9.6	284	14″	230	A7III	219127, 73090
	Σ 3004		23 20.7 +44 07	6.3, 10.1	178	13″	75	A5Vn	220105, 52927
	Gic 192	A C	23 26.7 +45 20	7.5, 10.9	354	57″	36.3	G0	220821, 53025
	OΣ 500	A B	23 37.5 +44 26	6.1, 7.4	12	0.5″	250	B8V	222109, 53202
			High mass binary; sparse field, faint double 5′ s.p. AB: *P* = 351 y, orbit *r* =100 AU, closing. (2012)						
	OΣ 501		23 40.0 +37 39	6.5, 10.6	162	15″	91	F2IV	222399, 73422
	β 995	(AaAb) B	23 47.6 +46 50	6.1, 8.7	250	0.8″	320	B3IV	223229, 53374
	OΣ 510	A B	23 51.6 +42 05	7.9, 8.4	119	0.6″	220	A6V	223672, 53427
			A type binary; dark field. AB: *P* = 1,523 y, orbit *r* = 175 AU, widening. (2012)						
	Σ 3042		23 51.9 +37 53	7.6, 7.8	86	5.7″	73	F5V	223718, 73565
	Ho 205	A B C	23 54.1 +39 17	6.7, 12.8, 11.7	181, 250	4.8″, 93″	181	F8	223971, 73597
	OΣ 513		23 58.4 +35 01	6.8, 9.3	17	3.3″	125	A3	224492, 73640
	Hjl 1113	A B C	23 59.2 +41 12	7.8, 10.2, 8.2	188, 192	4.3″, 115″	44.3	G0	224602, 53549
★	Σ 3050	A B	23 59.5 +33 43	6.5, 6.7	338	2.4″	28.9	F8V	224635, 73656
			Solar type binary; pretty Y/– color, dark field. AB: *P* = 717 y, orbit *r* = 110 AU, widening. (596 measures; 2013)						
	OΣ 514	(AaAb) B	00 04.6 +42 06	6.2, 9.7	170	5.2″	390	B9III	225218, 36037
			Distant (2)+1 triple, visual binary. MSC 8.1 M$_\odot$. Aa,Ab: *P* = 70 y, widening; AB: ps = 2740 AU. (2002)						
	Σ 3056	A B D	00 04.7 +34 16	7.7, 8.1, 10.6	144, 238	0.7″, 95″	177	G8III	225220, 53617
	β 997		00 05.0 +45 40	7.6, 9.4	337	3.9″	71	F8IV	225291, —
	β 9001	A C	00 05.2 +45 14	6.7, 10.6	236	22″	139	A2V	3, 36042
	Σ 3	(AaAb) B	00 10.1 +46 23	7.8, 9.1	80	5.1″	150	A4V	556, 36118
★	OΣ 2	A B (CaCb)	00 13.4 +26 59	6.8, 7.7, 10.4	159, 224	0.4″, 18″	131	F8V	895, 73823
			Solar type 2+(2) quadruple, visual triple. MSC 5.8 M$_\odot$. AB: *P* = 422 y, orbit *r* = 85 AU, *e* = 0.72, widening. (2012)						
	h 1947	A B	00 16.4 +43 36	6.2, 9.8	78	9.7″	107	A2V	1185, 36221
	Σ 19		00 16.7 +36 38	7.1, 9.5	141	2.4″	126	A2V	1223, 53772
	Σ 24		00 18.5 +26 08	7.8, 8.4	247	5.1″	155	A2	1429, 73883
	AC 1		00 20.9 +32 59	7.3, 8.3	287	1.9″	68	F5V	1641, 53827
	Σ 28	A B	00 23.9 +29 30	8.3, 8.6	224	33″	85	F7V F8V	1942, 73956
	OΣ 11		00 30.7 +32 08	7.6, 7.7	317	3.3′	159	F2	2688, 53947
	h 5451		00 31.4 +33 35	6.0, 9.3	86	55″	116	K1III	2767, 53956
29 pi	H V 17	(AaAbAc) B	00 36.9 +33 43	4.4, 7.1	174	38″	183	B5V	3369, 54033
			High mass (2+1)+1 spectroscopic quadruple, visual double. MSC gives 14.5 M$_{\#x2299;}$. Aa,Ab (Mkt 1): *P* = 144 d, *e* = 0.54; AB: ps = 9,400 AU. (2012)						
	Σ 47	A B	00 40.3 +24 03	7.3, 8.8	206	16″	132	A4III	3743, 74185
V355	Σ 52	(AaAb) B	00 44.2 +46 14	7.9, 9.0	3	1.4″	118	F6V	4134, 36611
★ 65 i	Σ 61		00 49.9 +27 43	6.3, 6.3	115	4.5″	89	F5III	4758, 74295
			Matched CPM double with F giant; often measured but no orbit yet. AB: ps = 540 AU. (250 measures; 2014)						
★ 36	Σ 73	A B	00 55.0 +23 38	6.1, 6.5	330	1.0″	38.0	G6IV K6IV	5286, 74359
			Solar type subgiant binary, dark field. AB: *P* = 168 y, orbit *r* = 37 AU, *e* = 0.31, closing. (706 measures; 2013)						
	Σ 79	(AaAb) (BaBb)	01 00.1 +44 43	6.0, 6.8	194	7.9″	119	B9.5V A2V	5789, 36833
	Mad 1		01 00.6 +47 19	7.7, 9.1	2	0.8″	140*	A2	5842, 36840

Label	Catalog ID	Components	Coordinates (J2000)	Mag.	θ	ρ	Dist.	Spectral type(s)	HD, SAO No.
	OΣ 21		01 03.0 +47 23	6.8, 8.1	174	1.2″	96	A9IV	6114, 36875
	β 397	A B	01 07.8 +46 51	7.5, 10.3	142	8.7″	173	K0II	6645, —
★ 42 phi	OΣ 515	A B	01 09.5 +47 15	4.6, 5.6	118	0.5″	220	B6IV B9V	6811, 36972
		High mass binary; sparse field; system mass 6.5 M⊙. AB: *P* = 554 y, orbit *r* = 125 AU; decreasing θ. (226 measures; 2012)							
	Σ 102	A B C	01 17.8 +49 01	7.3, 9.2, 8.8	274, 224	0.5″, 10″	380	B9.5V	7710, 37087
	OΣ 29		01 18.9 +39 58	7.5, 11.7	266	20″	157	G5	7864, 54593
	Σ 133	A B	01 32.8 +35 51	6.8, 9.4	192	3.0″	136	K3III	9370, 54771
		Solar type double with K giant; the pairs CD (m.10, 5″) 20″ s. and EF (m.10, 33″) 3′ n. are apparently unrelated. (2009)							
	Σ 141		01 40.1 +38 58	8.3, 8.6	304	1.7″	108	F5	10156, 54884
		Tiny solar type double; YO/Y color, dark field. Double h 1087 (m.10, 13″) 9′ p. (2012)							
	Σ 179		01 53.2 +37 19	7.6, 8.1	160	3.5″	79	F2V	11430, 55058
	β 1368	A C	01 56.2 +37 15	5.8, 11.9	77	19″	97	K0III gM0	11749, 55107
★ 57 gam	Σ 205	A (BaBbC)	02 03.9 +42 20	2.3, 5.0	64	9.8″	120	K3II B8	12533, 37734
		Almach. High mass 1+(2+1) quadruple, visual double with K giant; striking Y/B color rivaling Albireo, typical of a K or M supergiant primary with OB or A component. MSC system mass 26.0 M⊙. AB discovered by C. Mayer (1777), C discovered by F.W. von Struve (1842), Ba,Bb by Morgan *et al.* (1978). AB: no orbit (353 measures), but MSC gives *P* = 6,600 y; ps = 1,590 AU. BC: *P* = 64 y, orbit *r* = 36 AU, *e* = 0.93; now at periastron, C (m.6) reappears s.f. around 2025 on its way to apastron 2047 (381 measures). Ba,Bb: *P* = 2.7 d. (2012)							
59	Σ 222		02 10.9 +39 02	6.1, 6.7	36	17″	140	B9V A1Vn	13294, 55330
		High mass, nearly matched double; rich field. h 1109 (m.10, 24″) 10′ n.p. with several unidentified pairs. (2013)							
	Shy 136	(AaAb) B	02 13.2 +40 30	7.4, 7.3	340	10.3′	26.2	G0V G0V	13531, 37868
		V451+V450 And. Local, solar type (2)+1 triple; visual wide matched double; both BY Dra type variables in rich field, southern pair of three stars visible at low power. From the Shaya & Olling 2011 catalog that identifies wide and comoving pairs by a statistical analysis of Hipparcos proper motion data. AB: ps = 16,500 AU. (2003)							
	Σ 228		02 14.0 +47 29	6.6, 7.2	294	0.8″	39.6	F2V F7V	13594, 37878
		Solar type binary; YW/– color, rich field. System mass 2.5 M⊙. AB: *P* = 144 y, orbit *r* = 35 AU, *e* = 0.27, periastron 2042. (331 measures; 2012)							
	Σ 245	(AaAb) B	02 18.6 +40 17	7.3, 8.0	293	11″	97	F3V F3V	14189, 37940
		Solar type (2)+1 triple; rich field. MSC gives 5.7 M⊙. Aa,Ab: *P* = 110 y, orbit *r* = 33 AU; AB: ps = 1440 AU. Neglected. (1999)							
	Σ 249		02 21.6 +44 36	7.2, 9.0	195	2.4″	168	A2	14477, 37971
	β 305	A B C	02 38.3 +37 44	6.2, 11.3, 11.4	262, 206	8.5″, 19″	82	F5IV	16327, 55729

Antlia Ant *Charts 22, 23*

Label	Catalog ID	Components	Coordinates (J2000)	Mag.	θ	ρ	Dist.	Spectral type(s)	HD, SAO No.
★ zet 1	Δ 78		09 30.8 −31 53	6.2, 6.8	212	8.1″	124	A1V A1V	82384, 200445
		Matched A type double; W or YW color, faint field with line of four stars 10′ f. Neglected. (1999)							
	Jc 21		09 33.1 −39 08	6.5, 9.2	205	56″	75	F0IV	82785, 200492
	B 185		09 36.0 −27 31	7.6, 9.7	201	3.2″	156	A4III	83196, 177722
	h 4224		09 36.1 −31 14	7.8, 8.2	117	7.5″	370	A4V	83232, 200538
	h 4249		09 48.8 −35 01	8.1, 8.1	122	4.2″	133	A9V	85100, 200758
	β 215		09 54.1 −28 00	7.2, 9.3	346	1.8″	430	B4V	85860, 178157
	I 842		09 54.5 −34 54	7.3, 11.1	28	3.5″	120	A3V	85963, 200855
eta	h 4271		09 58.9 −35 53	5.2, 11.3	318	31″	33.3	F1III	86629, 200926
	h 4277		10 01.9 −28 41	8.1, 8.8	33	22″	118	A3IIIm	87038, 178319
	I 292		10 04.3 −28 23	8.4, 7.8	306	0.7″	82	F6V	87416, 178366
	Ho 371		10 05.8 −30 54	6.7, 11.1	44	6.1″	158	G8III	87660, 201045
	h 4304		10 20.2 −33 08	7.6, 9.8	286	9.5″	126	A3III	89672, 201293
	I 209	A B	10 24.4 −38 35	8.4, 8.6	129	1.3″	147	F2IV	90256, 201362
	Skf 1893		10 27.8 −34 24	7.5, 12.8	80	33″	39.4	G2V	90712, 201414
del	H N 50		10 29.6 −30 36	5.6, 9.8	226	11″	133	B9.5V	90972, 201442

Label	Catalog ID	Components	Coordinates (J2000)	Mag.	θ	ρ	Dist.	Spectral type(s)	HD, SAO No.
	h 4381		10 54.6 −38 45	7.0, 8.5	42	26″	161	B8III	94565, 201886
	Tok 278		10 55.9 −35 07	7.5, 11.5	352	61″	53	G5V	94771, 201911

Apus   Aps   *Chart 30*

Label	Catalog ID	Components	Coordinates (J2000)	Mag.	θ	ρ	Dist.	Spectral type(s)	HD, SAO No.
	h 4667		14 22.6 −73 33	8.2, 8.6	138	2.3″	240	A0V	125292, 257145
	h 4671		14 29.4 −80 06	8.1, 8.7	125	4.2″	76	F5V	125856, 257158
	I 236		14 53.2 −73 11	5.9, 7.6	123	2.2″	101	G5III	130458, 257206
★	CapO 15		15 06.4 −72 10	7.2, 8.5	42	1.4″	250	B8V	132874, 257237
			High mass double; YW/W color, f. of 2 m.7 stars in faint field. AB: ps = 470 AU. Neglected. (1991)						
	Fmr 173		15 15.4 −70 32	6.9, 12.7	145	57″	188	G6IV+M3V	162157, 257754
	I 333		15 59.9 −78 02	6.9, 7.5	332	0.7″	88	F3IV	141846, 257341
★ del 1	BrsO 22	A B	16 20.3 −78 42	4.9, 5.4	10	103″	230	M5III K3III	145366, 257380
			Binary with M and K type giants, A is an irregular, long period, pulsating variable star; YO/Y color, dominant in faint rich field. AB: ps = 32,000 AU. (2000)						
	h 4904		17 10.3 −75 23	7.6, 9.1	187	6.8″	81	F3V	153880, 257465
	Hd 275		17 44.3 −72 13	6.9, 8.1	89	0.6″	43.4	F7IV F5V	159964, 257525
	Hd 284	(AaAb) B	18 12.6 −73 40	6.0, 9.1	270	2.2″	42.1	F5V	165259, 257571

Aquarius   Aqr   *Charts 12, 19, 20*

Label	Catalog ID	Components	Coordinates (J2000)	Mag.	θ	ρ	Dist.	Spectral type(s)	HD, SAO No.
4	Σ 2729	A B	20 51.4 −05 38	6.4, 7.4	29	0.8″	61	F5IV	198571, 144877
			Solar type binary; s.p. of two m.6 stars in sparse field. System mass 3.2 M⊙. AB: P = 194 y, orbit r = 52 AU, e = 0.49, periastron 2090. (314 measures; 2013)						
6 mu	Tok 342	(AaAb) B	20 52.7 −08 59	4.8, 9.9	116	8.4′	48.2	K3III	198743, 144895
	Howe 55	A B	20 57.2 +00 28	6.1, 11.8	71	27″	94	K2III	199442, 126396
	Σ 2744	A B	21 03.1 +01 32	6.8, 7.3	112	1.3″	71	F7IV	200375, 126491
			Solar type binary; poor quality orbit. AB: P = 1,532 y, orbit r = 180 AU. (368 measures; 2011)						
12	Σ 2745	(AaAb) B	21 04.1 −05 49	5.8, 7.5	196	2.5″	154	G4III	200497, 145065
	Σ 2752	A B	21 07.2 −13 55	7.3, 10.0	178	4.4″	17.6	K2V+M0V	200968, 164147
			Local, low mass double; sparse field, wide unidentified double 7′ s.f. AB: ps = 100 AU. (2012)						
	Σ 2770		21 11.6 −03 07	7.6, 10.8	247	7.5″	200	K0III	201719, 145185
	Σ 2775	(AB) C	21 14.7 −00 50	7.5, 10.4	178	21″	310	A1V	202260, 145230
			High mass, A type (2)+1 triple, visual binary; sparse field, small star group s.p. MSC system mass 7.3 M⊙. AB: P = 81 y, orbit r = 50 AU, e = 0.17, decreasing θ; periastron 2027. AC: ps = 8,800 AU. (2008)						
	Σ 2809	A B	21 37.6 −00 23	6.2, 9.4	162	31″	320	A2V	205765, 145533
24	β 1212	(AaAb) B	21 39.5 −00 03	6.9, 8.4	286	0.5″	44.0	F6V	206058, 145566
			Solar type (2)+1 spectroscopic triple; sparse field. MSC 2.5 M⊙. AB: P = 49 y, orbit r = 20 AU, e = 0.87, periastron 2020. (258 measures; 2009)						
	Sca 104		21 52.4 −03 10	6.6, 11.0	117	63″	184	A0	207888, 145716
	Howe 59	A B	22 01.5 −15 37	7.2, 10.3	270	8.9″	128	G8IV F0V	209154, 164819
	Σ 2862		22 07.1 +00 34	8.0, 8.4	95	2.5″	79	G0	209965, 127306
★ 41	H N 56	A B	22 14.3 −21 04	5.6, 6.7	112	5.2″	72	K0III+F2V	210960, 190986
			Striking close double with K giant; sparse field, optical pair CD (m.9, 12″) 3′ n.f. is unrelated. From the 1821 W. Herschel N catalog, discoveries he made while searching for nebulae or new planets. (2009)						
★ 51	β 172	A B C	22 24.1 −04 50	6.5, 6.6, 12.2	38, 343	0.4″, 53″	124	A0V	212404, 146067
			A type 2+1 visual triple. AB: P = 149 y, orbit r = 50 AU, e = 0.71, apastron 2062. (238 measures; 2011)						
	S 808	A B	22 25.8 −20 14	7.1, 8.0	154	6.8″	600	G5III	212600, 191131
★ 53	Sh 345	A B	22 26.6 −16 45	6.3, 6.4	59	1.3″	20.2	G0V G0V	212697, 165078
			Local, solar type matched binary; dark field. AB: P = 3,500y, orbit r = 300 AU, closing. (243 measures; 2013)						

Label	Catalog ID	Components	Coordinates (J2000)	Mag.	θ	ρ	Dist.	Spectral type(s)	HD, SAO No.
★ 55 zet 1,2	Σ 2909	(AaAb) B	22 28.8 −00 01	4.3, 4.5	165	2.3″	28.2	F3V F2IV	213051, 146107
			Gorgeous solar type (2)+1 triple, visual double; W/W color, cataloged by W. Herschel (1779). MSC system mass 3.4 M$_\odot$. Aa,Ab (Ebe 1): P = 26 y, e = 0.13, apastron 2016; AB: P = 487 y, orbit r = 95 AU, e = 0.34. (1,155 measures; 2013)							
	Σ 2913		22 30.5 −08 07	7.8, 8.6	329	8.6″	200	F0V	213293, 146130
	HDS 3190		22 30.6 −10 41	4.8, 8.5	10	3.7″	89	A0IV	213320, 165134
	Cbl 188		22 32.2 −13 36	7.8, 9.8	94	42″	115	G0V	213510, 165149
	h 3128		22 40.0 −19 12	7.3, 11.2	228	11″	61	F6V	214657, 165223
	Σ 2936		22 43.0 +01 13	7.0, 9.6	52	4.2″	145	A6III	215129, 127707
	Σ 2935	A B	22 43.1 −08 19	6.8, 7.9	306	2.4″	147	A5V	215114, 146271
	Σ 2939		22 45.3 −09 39	7.4, 9.3	62	10″	84	A7III	215449, 146296
	Hu 291		22 47.3 −16 09	7.6, 9.8	327	2.7″	34.0	G5V	215696, 165289
★	Σ 2944	A B	22 47.8 −04 14	7.3, 7.7	302	1.9″	32.0	G2V G4	215812, 146315
			High CPM, solar type binary, YO/− color. AB: P = 1,160 y, orbit r = 130 AU; slowly closing. (397 measures; 2012)							
	β 178		22 55.2 −04 59	6.0, 7.8	322	0.6″	83	G4III	216718, 146388
			Binary with a red giant primary. AB: P = 96 y, orbit r = 40 AU; closing along edgewise orbit. (2009)							
	Arn 80		23 02.5 −11 16	7.9, 9.9	270	79″	109	K0	217681, 165454
			CPM double with K giant, group n.p.; CPM double Σ 2970 (m.9, 8″) 3′ s.p. (2004)							
	β 384		23 02.6 −18 33	6.9, 9.2	59	1.1″	100	A2V	217684, 165456
★	Σ 2988		23 12.0 −11 56	7.9, 8.0	98	3.5″	210	G8III	218928, 165551
			Pretty matched double with G giant; easily found in dark field. AB: ps = 990 AU. (2012)							
	β 181	A B	23 13.8 −13 24	7.3, 9.3	306	1.5″	300	K4III	219151, 165571
91 psi 1	Σ II 12	A [B C]	23 15.9 −09 05	4.4, [10.5, 10.7]	311, [104]	50″, [0.6″]	45.9	K1III	219449, 146598
			High CPM, solar type 1+2 triple with K giant; brilliant in field. MSC system mass 5.5M$_\odot$. BC (β 1220): P = 84 y, orbit r = 22 AU; closing along edgewise orbit. AB: ps = 2,300 AU. (2012)							
★ 94	Σ 2998	(AaAb) B	23 19.1 −13 28	5.3, 7.0	349	13″	21.1	G5IV	219834, 165625
			Local, solar type (2)+1 triple, visual double. MSC 2.3M$_\odot$. Aa,Ab: P = 6.3 y, orbit r = 4 AU, e = 0.19. (241 measures; 2012)							
96	h 5394		23 19.4 −05 07	5.6, 10.4	20	11″	34.2	F3V	219877, 146639
	h 3184		23 20.9 −18 33	7.3, 8.4	283	5.6″	166	G8IV	220065, 165830
105 ome 2	β 279	(AaAb) B	23 42.7 −14 33	4.5, 9.9	89	5.5″	45.5	B9.5V	222661, 165842
★ 107 i 2	H II 24		23 46.0 −18 41	5.7, 6.5	136	6.6″	73	A9IV F2V	223024, 165867
			A type double; dark field. From the 1782 William Herschel catalog. AB: ps = 650 AU. (2012)							
	Tok 354		23 48.2 −16 42	7.6, 10.1	61	109″	57	F7V	223271, 165890
	β 729		23 55.4 −17 50	7.7, 11.7	346	11″	56	F7V	224135, 165951

Aquila   Aql   *Charts 18, 19*

Label	Catalog ID	Components	Coordinates (J2000)	Mag.	θ	ρ	Dist.	Spectral type(s)	HD, SAO No.
5	Σ 2379	(AaAb) B	18 46.5 −00 58	5.9, 7.0	121	13″	92	A2V	173654, 142606
			A type (2)+1 spectroscopic triple; rich field. MSC has 7.3 M$_\odot$. Aa,Ab: P = 34 y, e = 0.33, periastron 2023. AB: ps = 1,600 AU. (2012)							
	β 265		18 50.2 +11 31	7.4, 9.2	228	1.4″	520	A1V GIII	174485, 104570
	Σ 2404		18 50.8 +10 59	6.9, 7.8	182	3.5″	390	K5III+K3	174569, 104170
FF	Ho 91	(AaAb) B	18 58.2 +17 22	5.4, 10.1	144	6.8″	470	F8II	176155, 104296
	Σ 2432	A B	19 01.8 +12 32	6.8, 10.4	87	15″	230	B8IV	176873, 104379
	Σ 2446	A B C	19 05.8 +06 33	7.0, 8.9, 11.2	153, 346	9.4″, 36″	66	F5	177749, 124257
	Σ 2449		19 06.4 +07 09	7.2, 7.7	290	7.9″	106	F2V	177904, 124265
	Σ 2447	A B	19 06.6 −01 21	6.8, 9.6	343	14″	135	B5V	177880, 143029
	Σ 2471		19 10.9 +08 07	7.5, 10.6	122	8.6″	94	A9V	179123, 124355
	OΣΣ 178		19 15.3 +15 05	5.7, 7.6	267	89″	210	G8II	180262, 104655

Label	Catalog ID	Components	Coordinates (J2000)	Mag.	θ	ρ	Dist.	Spectral type(s)	HD, SAO No.
	OΣ 368	A B	19 16.0 +16 10	7.5, 8.5	219	1.1″	240	A9IV	180451, 104666
	Σ 2489	(AaAb) B	19 16.4 +14 33	5.7, 9.3	347	8.3″	126	B9.5V	180555, 104668
23	Σ 2492	A B C	19 18.5 +01 05	5.3, 8.3, 13.5	2, 70	3.2″, 10″	114	G9III	180972, 124487
24	Σ I 40	A B	19 18.8 +00 20	6.5, 6.8	316	7.0′	127	K0III	181053, 124492
	Σ 2501	A B	19 22.1 −04 44	7.8, 9.7	22	20″	98	F5	181806, 143315
	Σ 2519		19 28.2 −09 32	8.3, 8.5	125	12″	91	F0	183107, 143431
	Σ 2533		19 30.1 −00 27	7.4, 10.0	211	22″	72	A3V	183518, 143469
V822	D 20	(AaAb) B	19 31.3 −02 07	7.2, 9.6	65	1.3″	290*	B5V	183794, 143494
	Σ 2543		19 36.2 +06 00	6.8, 10.5	151	11″	200	G8III	184853, 124835
	Σ 2545	A B	19 38.7 −10 09	6.8, 8.5	326	3.8″	86	A9III	185298, 162843
			A type double; forms an optical "double double" with CPM pair Σ 2547 (m.8, 21″) 13′ s.f. AB: ps = 440 AU. (2012)						
	Σ 2562	A B	19 42.8 +08 23	7.0, 8.7	251	28″	66	F8V G0V	186226, 124969
	Σ I 45	A (BaBb)	19 43.1 −08 18	7.1, 7.6	147	98″	72	F4V F5V	186158, 143722
	Σ 2570	(AB) C	19 44.9 +10 47	7.6, 9.8	280	4.5″	950*	B3IV	186587, 105207
★ 52 pi	Σ 2583	A B	19 48.7 +11 49	6.3, 6.8	103	1.3″	158	A3V F9III	187259, 105282
			A type double with F giant; subtle O/B color. AB: ps = 280 AU. (317 measures; 2013)						
	β 148	A B	19 52.0 −10 21	7.7, 8.4	228	0.7″	87	F2	187774, 163059
			Solar type binary. AB: P = 783 y, orbit r = 120 AU; slowly closing. (2009)						
	Σ 2590	A B	19 52.3 +10 21	6.5, 10.3	310	14″	330	B7V	187961, 105355
	Σ 2596		19 54.0 +15 18	7.3, 8.7	298	2.0″	83	F8V	188328, 105396
★ 57	Σ 2594	(AaAb) (BaBb)	19 54.6 −08 14	5.7, 6.4	171	36″	148	B7Vn B8V	188293, 143898
			High mass (2)+(2) spectroscopic quadruple, visual double; splendid in rich field, BW/BW color enhanced by magnification. AB: ps = 7,190 AU. (2013)						
	Σ 2597	A B	19 55.3 −06 44	6.9, 8.0	101	0.6″	82	F2V	188405, 143911
			Solar type binary. AB: P = 425 y, orbit r = 90 AU; widening. (2010)						
	β 1475	A B	19 57.9 −09 04	7.8, 9.5	112	10″	250	K2	188938, 143951
	β 266		19 57.9 +11 25	7.4, 11.5	167	16″	198	A3	189093, 105474
	AC 12		19 58.4 −02 14	7.5, 8.3	299	1.5″	101	F5	189073, —
	Σ 2613	(AaAb) B	20 01.4 +10 45	7.5, 8.0	354	3.6″	81	F5V F5V	189783, 105560
	H I 93	A B	20 01.7 −00 12	7.7, 8.4	297	1.8″	190	A0	189759, 144002
	Σ 2621		20 04.6 +09 14	8.4, 8.6	224	5.5″	230	B8V A1V	190434, 125416
	β 56		20 05.1 −04 18	8.0, 9.1	187	1.3″	77	F5	190437, —
	OΣΣ 198	A B	20 06.6 +07 35	7.1, 7.6	186	64″	121	A2V A3V	190849, 125456
	β 428		20 06.7 +12 56	7.6, 9.1	356	0.8″	134	F4III	190887, 105700
			Solar type double with F giant; faint rich field. h 1476 4′ s.p. (2008)						
	Σ 2628		20 07.8 +09 24	6.6, 8.7	340	2.8″	39.4	F3V	191104, 125478
	Σ 2635	A B	20 10.1 +08 27	6.7, 10.2	80	7.8″	63	F7V	191533, 125517
★	Σ 2644		20 12.6 +00 52	6.9, 7.1	205	2.5″	181	B9	191984, 125566
			Nearly matched, high mass double. No orbit or linear solution. (211 measures; 2012)						
	Σ 2643		20 12.8 −03 00	7.1, 9.4	79	3.1″	133	A0V	192007, 144173
★	S 740		20 14.2 +06 35	7.8, 8.1	191	44″	61	G4IV	192344, 125597
			Solar type matched double; YO/W color. Near 100% probability the pair is physical. AB: ps = 3,620 AU. (2012)						
	Σ 2654		20 15.2 −03 30	7.0, 8.1	232	15″	41.4	F2V F4V	192461, 144212
	Σ 2656		20 15.6 +07 49	7.3, 11.3	234	9.7″	123	A7III	192622, 125620
	Rst 4659	A B	20 16.8 −03 30	7.2, 9.2	7	1.7″	65	F5	192791, 144237
	Σ 2661		20 19.9 −02 15	7.9, 9.2	341	24″	200	A0	193330, 144303
★	S 749	A B	20 27.5 −02 06	6.8, 7.5	188	60″	45.8	F7V+G2V	194765, 144450
			Solar type double; charming field. A characteristically wide system from the catalog of James South; near 100% probability the CPM pair is physical. AB: ps = 3,710 AU. (2012)						

91

Label	Catalog ID	Components	Coordinates (J2000)	Mag.	θ	ρ	Dist.	Spectral type(s)	HD, SAO No.
	Ara		Ara					*Chart 29*	
★ R	h 4866	(AaAb) B	16 39.7 −57 00	7.2, 7.8	120	3.3″	124	B9V	149730, 244037
		High mass (2)+1 triple with Algol type variable; rich field, K type giant 4′ n. AB: ps = 550 AU. Neglected. (1999)							
	HDS 2364		16 39.8 −52 00	7.7, 9.0	163	22″	31.5	G4V+K2V	149813, 244042
	Slr 12	A B C	16 39.9 −47 47	8.0, 8.1, 10.8	159, 40	1.4″, 36″	70	F5V	149901, 227020
	B 1818		16 40.8 −60 27	6.3, 9.1	36	1.6″	31.3	F2III	149837, 253651
	Slr 21		16 41.1 −47 45	7.4, 9.4	321	1.7″	250	B5III	150083, 227044
	Δ 206	(Aa1Aa2Ab) B [(CaCb) G]	16 41.3 −48 46	5.7, 8.4, [6.3, 9.6]	9, 263, [221]	1.7″, 10″, [4.3″]	1820*	O3V+O6V	150136, 227049
		Distant, high mass, spectroscopic multiple system within NGC 6193; listed as a single system in WDS but divergent CPM suggests the (2+1)+1 quadruple (Mlo 8) is unrelated to the (2)+1 triple (Sna 1). System mass of A triple ~135 M⊙. Aa1, Aa2: *P* = 2.7 d. Aa,Ab: *P* = 8.2 y. (2012)							
	h 4896		16 56.2 −46 51	7.8, 8.0	23	3.9″	330	B7V	152541, 227453
	λ 316		17 00.4 −48 39	6.3, 7.7	172	1.0″	107	G5IV	153221, 227542
	Hld 131		17 01.1 −56 33	6.4, 9.9	131	2.3″	200	A	153201, 244356
	h 4901		17 01.1 −58 51	8.3, 8.5	130	2.8″	950	B3V	153123, 244353
	CorO 206	A B	17 02.9 −50 10	7.3, 8.1	231	8.2″	150*	A0V	153579, 244386
	Δ 213		17 10.3 −46 44	7.0, 8.3	168	8.2″	460	B1I B1.5V	154873, 227682
	h 4920	(AaAb) B	17 13.0 −58 36	7.0, 9.2	323	3.0″	60	F4IV	155099, 244518
★ 41	BrsO 13	A B	17 19.1 −46 38	5.6, 8.9	257	10″	8.80	G8V M0V	156274, 227816
		Local, very high CPM, solar type binary with faint M type companion; Y/O color, faint rich field. AB: ps = 120 AU. (2013)							
	h 4931		17 20.6 −59 26	7.8, 7.8	256	0.9″	510	A0V	156335, 244638
	CorO 213	A B	17 22.8 −58 28	6.9, 9.3	284	9.3″	62	A5V	156751, 244675
	Shy 297		17 26.4 −48 37	7.1, 8.9	314	2.1′	47.3	F6V K1V	157555, 227954
★	h 4949	A B C	17 26.9 −45 51	5.6, 6.5, 7.1	252, 312	2.1″, 103″	195	B7V B9.5V A0V	157661, 227972
		High mass 2+1 triple; exemplary specimen in a rich field. MSC system mass 12.2 M⊙. AC: ps = 27,100 AU. (2007)							
	I 106		17 37.3 −49 15	7.3, 8.3	35	1.0″	310	B8III	159439, 228194
	R 303		17 45.1 −54 08	7.9, 9.0	101	2.7″	390	A1V A1V	160818, 245592
	CapO 17		17 45.6 −50 47	8.1, 8.1	182	1.1″	97	F7V	160996, 245004
	h 4982		17 50.5 −48 17	7.0, 9.3	58	41″	270*	K3III	161880, 228465
★ V539	h 4978	(AaAb) B	17 50.5 −53 37	5.7, 9.2	268	12″	300	B2V B3V	161783, 245065
		High mass (2)+1 triple with Algol type eclipsing variable; sparse field. MSC has 27.2 M⊙. AB: ps = 4,860 AU. Neglected. (2000)							
	Rmk 22		17 57.2 −55 23	7.0, 7.9	95	2.4″	210	K0III A2	163028, 245131
	h 5015		18 08.5 −45 46	6.2, 9.6	259	3.5″	270	B8II	165493, 228734
	Aries		Ari					*Chart 7*	
1	Σ 174		01 50.1 +22 16	6.3, 7.2	165	2.9″	180	G3III	11154, 74966
	Σ 178		01 52.0 +10 49	8.2, 8.2	204	3.0″	137	F1V	11386, 92669
★ 5 gam	Σ 180	A B	01 53.5 +19 18	4.5, 4.6	2	7.5″	50	A1 B9V	11503, 92681
		Mesarthim. Brilliant, high mass matched pair with alp2 CVn type variable. First noticed as double by Robert Hooke (1664). AB: ps = 505 AU; pair CD (β 512, m.14, 3.6′ following) is unrelated. (374 measures; 2014)							
★ 9 lam	H V 12	(AaAb) B	01 57.9 +23 36	4.8, 6.7	48	37″	39.5	F0IV F7V	11973, 75051
		Solar type (2)+1 spectroscopic triple, visual double with bet Lyr type eclipsing variable; striking in a featureless field, stars f. unrelated. AB: ps = 1,970 AU. (2013)							
	Σ 194		01 59.3 +24 50	7.6, 9.5	278	1.3″	193	A3	12101, 75071

Appendix A: The target list

Label	Catalog ID	Components	Coordinates (J2000)	Mag.	θ	ρ	Dist.	Spectral type(s)	HD, SAO No.
10	Σ 208	A B	02 03.7 +25 56	5.8, 7.9	343	1.4″	48.7	F8IV	12558, 75114
		Solar type binary; dark field, m.10 G giant 9′ f. System mass 2.6 M☉. AB: P = 325 y, orbit r = 70 AU, e = 0.59, apastron 2094. (224 measures; 2012)								
14	H VI 69	A B C	02 09.4 +25 56	5.0, 8.0, 8.0	34, 279	93″, 105″	89	F2III	13174, 75171
	Σ 240		02 17.3 +23 52	8.3, 8.6	53	4.8″	112	F0	14066, 75263
VW	OΣΣ 27	A B C	02 26.8 +10 34	6.7, 8.3, 11.8	31, 155	74″, 60″	128	A3	15165, 92952
	Hjl 1019	A [Ba Bb]	02 29.2 +23 28	8.0, [10.2, 10.5]	323, [49]	3.6′, [0.4″]	95	A5m	15385, 75398
30	Σ I 5	(AaAb) (BC)	02 37.0 +24 39	6.5, 7.0	275	38″	41.8	F5V F7V	16246, 75471
	Σ 287		02 39.0 +14 52	7.4, 9.6	73	6.8″	270	G5	16480, –
33	Σ 289	(AbAb) B	02 40.7 +27 04	5.3, 9.6	0	28″	71	A3V	16628, 75510
	Σ 291	A B C	02 41.1 +18 48	7.7, 7.5, 9.5	117, 242	3.3″, 65″	740	B9.5V	16694, 93052
	β 306	A B	02 43.9 +25 38	6.4, 10.4	19	3.0″	104	A3V	16955, 75539
	Σ 300		02 44.6 +29 28	7.9, 8.1	314	3.1″	92	F0IV	17007, 75544
	Σ 305	A B	02 47.5 +19 22	7.5, 8.3	307	3.6″	33.6	G0V	17332, 93105
		Solar type binary; sparse field. Low quality orbit AB: P = 720 y, orbit r = 100 AU, slowly decreasing θ. (320 measures; 2012)								
★ 42 pi	Σ 311	(AaAb) B C	02 49.3 +17 28	5.3, 8.0, 10.7	119, 112	3.2″, 24″	240	B6V	17543, 93127
		High mass (2)+1+1 spectroscopic and occultation quadruple; dark field. MSC system mass 8.7 M☉. AC: ps = 7,770 AU. (2012)								
	OΣ 46	A B C	02 50.0 +30 32	6.8, 10.8, 12.7	75, 169	4.5″, 20″	62	F0	17572, 55920
	Σ 326	A B C	02 55.6 +26 52	7.7, 10.0, 13.9	220, 266	4.8″, 44″	23.5	G5	18143, 75644
		Local, high CPM solar type 2+1 triple; Y/O color, sparse or dark field, the s.f. of 2 stars. AB: ps = 150 AU: there is both an orbit and linear solution, and CPM divergence is small. (2012)								
	β 525		02 58.9 +21 37	7.5, 7.5	272	0.6″	152	A3	18484, 75671
		A type binary, sparse field. AB: P = 242 y, orbit r = 65 AU, slowly increasing θ. (2010)								
★ eps	Σ 333	A B C	02 59.2 +21 20	5.2, 5.6, 12.7	209, 192	1.4″, 2.4′	102	A2V A2V	18519, 75673
		Matched A type 2+1 triple. AB: P = 1,216 y; orbit r = 220 AU, closing. (447 obs.) AC: ps = 14,700 AU. (2014)								
	OΣ 49	A B	03 00.5 +18 00	6.8, 9.9	50	2.3″	122	A0	18654, 93229
★ 52	Σ 346	A B C	03 05.4 +25 15	6.2, 6.2, 10.8	254, 358	0.4″, 5.1″	165	B7V	19134, 75723
		High mass, visually tiny 2+1 triple. System mass 8.0 M☉. AB: P = 227 y, orbit r = 80 AU, e = 0.73, widening to apastron 2052. (2009)								
	Σ 376		03 20.3 +19 44	8.3, 8.4	250	7.1″	123	A2V	20682, 93392
	Σ 375	A B	03 20.4 +23 41	7.6, 9.9	316	2.7″	105	A7IV	20655, 75873
	Σ 381		03 23.3 +20 58	7.6, 8.8	107	1.1″	30*	G5	20947, 75906
UX	Tok 13	(Aa1Aa2Ab) B	03 26.6 +28 43	6.6, 10.0	129	96″	52	G5IV	21242, 75927
		Solar type (2+1)+1 quadruple, visual binary with RS CVn type variable (see Introduction). MSC gives 3.0 M☉. Aa,Ab: P = 12.4 y, e = 0.77. AB: ps = 6,700 AU. (2001)								
★	Σ 394	(AaAb) B	03 28.0 +20 28	7.1, 8.2	163	6.8″	140*	A3 G5	21437, 75940
		A type (2)+1 triple, visual double; dark field. MSC system mass 6.6 M☉. AB: ps = 1,280 AU. (2011)								

Auriga — Aur — *Charts 3, 7, 8*

Label	Catalog ID	Components	Coordinates (J2000)	Mag.	θ	ρ	Dist.	Spectral type(s)	HD, SAO No.
★ 4 ome	Σ 616	A B	04 59.3 +37 53	5.0, 8.2	4	4.7″	52	A1V	31647, 57548
		A type double; the s.p. of two stars in very dark field. AB: ps = 330 AU. (2013)								
5	OΣ 92	(AaAb) B	05 00.3 +39 24	6.0, 9.5	284	4.1″	60	F5V	31761, 57559
9	H VI 35	(AaAb) B C	05 06.7 +51 36	5.0, 12.2, 10.0	82, 61	5.2″, 91″	26.3	F0V	32537, 25019
	OΣΣ 61	A B	05 09.7 +29 48	6.7, 8.5	245	68″	53	F8V	33185, 76989
★	Σ 644	(AaAb) B	05 10.3 +37 18	7.0, 6.8	221	1.7″	490	B2II K3	33203, 57704
		Distant, high mass (2)+1 spectroscopic triple; B supergiant with K companion; mixed, rich field. AB: ps = 1,120 AU. (2012)								
	Σ 648	A B D	05 11.0 +32 02	8.1, 8.9, 13.4	59, 73	4.4″, 46″	42.4	G5	33334, 57719
16	OΣ 103	(AaAb) B	05 18.2 +33 22	4.8, 10.6	55	4.1″	71	K3.5III	34334, 57853
		Solar type (2)+1 spectroscopic triple, visual binary with K giant. MSC gives 7.0 M☉. Aa,Ab: P = 1.2 y, e = 0.10; AB: ps = 390 AU. (2002)								

Label	Catalog ID	Components	Coordinates (J2000)	Mag.	θ	ρ	Dist.	Spectral type(s)	HD, SAO No.
	Σ 684		05 22.2 +45 05	7.7, 9.3	141	1.5″	670*	B8III	34788, 40280
	Es 576	A [C D]	05 24.4 +42 37	8.1, [8.9, 9.8]	237, [89]	43″, [0.8″]	170*	A2	35101, 40311
	Σ 698	A B	05 25.2 +34 51	6.7, 8.3	350	31″	125	K2III	35295, 57999
	Σ 699	A B	05 25.6 +38 03	7.9, 8.6	345	8.9″	310	A1V	35313, 58006
24 phi	β pm 79	A C	05 27.6 +34 29	5.2, 10.9	72	62″	139	K3III	35620, 58051
	Σ 719	A B C	05 30.1 +29 33	7.5, 8.8, 9.4	335, 354	1.3″, 14″	170	G5	36044, 77210
	OΣΣ 63		05 30.8 +39 50	6.5, 7.7	277	76″	175	G9III	36041, 58129
	Σ 711	A B	05 31.5 +54 39	7.8, 9.7	226	7.9″	39.3	G1 K2	35961, 25232
		High CPM, solar type double; Y/– color. The pair Ca,Cb (m.12) 3′ s.p. and m.12 double 3′ p. are unrelated. AB: ps = 420 AU. (2007)							
	Σ 718	A B	05 32.3 +49 24	7.5, 7.5	73	7.7″	83	F5	36146, 40400
	Σ 736		05 37.1 +41 50	7.5, 8.6	0	2.6″	44.2	F8	36929, 40485
★ 26	Σ 753	(AB) C	05 38.6 +30 30	5.5, 8.4	269	12″	174	G8III A1V	37269, 58280
		High mass (2)+1 triple with G giant; sparse field. MSC gives 7.4 M⊙. AB: P = 53 y, e = 0.65. AC: P = ~46,000 y, ps = 2,820 AU. (2010)							
	OΣ 112		05 39.9 +37 57	7.9, 8.2	47	0.9″	280*	B9	37384, 58301
		High mass double; O/– color. Spectroscopic parallax suggests ps = 340 AU. Tiny SEI 367 (m.11, 12″) 5′ f. (2012)							
	Eng 22	A B	05 41.3 +53 29	6.3, 9.8	69	100″	12.3	K1V M0.5V	37394, 25319
		Local, high CPM, low mass double; YO/– primary, wide m.11 pair 9′ f. in dark field. AB: ps = 1,660 AU, near 100% probability pair is physical. (2011)							
	Σ 768		05 43.3 +41 07	7.5, 10.3	221	19″	230	B8	37841, 40560
	β 560		05 47.4 +29 39	7.8, 8.2	125	1.7″	80	F8	38491, 77555
	OΣ 117		05 48.2 +30 32	7.1, 10.0	32	11″	280	K5	38583, 58451
	Σ 796	A B	05 49.9 +31 47	7.2, 8.2	62	3.7″	168	A3	38819, 58484
	Σ 799		05 52.2 +38 34	7.3, 8.3	162	0.8″	210	B8	39114, 58520
	β 1053		05 53.5 +37 20	6.9, 8.8	359	1.9″	69	F5	39315, 58535
★ 37 the	OΣ 545	A B	05 59.7 +37 13	2.6, 7.2	306	4.1″	51	A0	40312, 58636
		Mahasim. A type double with alp2 CVn type variable; faint rich field. AB: ps = 280 AU. (2013)							
	Lep 22	A [B C]	06 04.5 +44 16	6.7, [9.1, 12.0]	290, [38]	3.2′, [3.9″]	33.1	F8+K5	40979, 40830
	OΣ 128	A [B C]	06 04.5 +51 34	6.4, [9.6, 10.6]	14, [309]	40″, [0.6″]	135	A7III	40873, 25548
	OΣ 131		06 07.4 +36 16	7.0, 9.5	277	1.5″	300	B9II	41523, 58762
	OΣ 132		06 08.2 +37 59	7.2, 9.6	332	1.8″	185	A2V	41637, 58776
	Webb 5		06 09.7 +43 08	7.1, 9.2	216	44″	260	A0	41847, 40898
★ 41	Σ 845		06 11.6 +48 43	6.2, 6.9	359	7.5″	102	A1V A6V	42127, 40925
		Splendid A type matched double; W/W color, rich field. System mass ~4.9 M⊙. AB: ps = 1,030 AU. (2012)							
	Σ 862		06 12.1 +29 30	7.6, 10.8	339	6.6″	560	G2Ib	42454, 78095
	Σ 872	A B	06 15.7 +36 09	6.9, 7.4	216	11″	54	F4IV	43017, 58906
	Σ 888	(AB) C	06 20.0 +28 26	7.5, 9.6	263	2.9″	133	A6V	43885, 78233
		A type (2)+1 triple, visual binary. MSC system mass 6.1 M⊙. AB: P = 104 y, e = 0.05, apastron 2061. AC: ps = 520 AU. (2008)							
	Σ 918	A B	06 34.0 +52 28	7.3, 8.2	334	4.8″	91	A3	46048, 25811
	OΣ 147	A B [C D]	06 34.3 +38 05	6.8, 8.7, [10.6, 11.0]	75, 120, [109]	43″, 46″, [0.5″]	290	K0	46296, 59230
	Σ 928	A B	06 34.7 +38 32	7.9, 8.6	131	3.5″	189	F5	46359, 59239
	Σ 929		06 35.4 +37 43	7.4, 8.4	24	5.9″	184	G5	46482, 59259
	Σ 941	A B	06 38.7 +41 35	7.3, 8.2	82	1.9″	340	B9	47046, 41232
		Distant, high mass double; CPM pair Σ 933 (m.8, 26″) 30′ s. AB: ps = 870 AU. (2012)							
54	OΣ 152	(AaAb) B	06 39.6 +28 16	6.2, 7.9	36	0.9″	260	B7III	47395, 78593
	Es 586		07 30.3 +41 36	7.8, 11.2	22	14″	24*	K0	59083, 41819

Label	Catalog ID	Components	Coordinates (J2000)	Mag.	θ	ρ	Dist.	Spectral type(s)	HD, SAO No.
	Boötes		Boo						*Charts 4, 10*
1	Σ 1772	A B D	13 40.7 +19 57	5.8, 9.6, 7.4	133, 1	4.5″, 3.5′	101	A1V	119055, 82942
	β 115	A B	13 45.3 +09 03	7.5, 10.4	257	1.6″	55	G5	119825, 120096
	Σ 1785		13 49.1 +26 59	7.4, 8.2	183	3.0″	13.4	K4V K6V	120476, 83011
		Local, high CPM binary; O/O color, system mass ~1.4 M$_\odot$. AB: P = 156 y, orbit r = 33 AU, e = 0.45, periastron 2073. (862 measures; 2013)							
	S 656		13 50.4 +21 17	6.9, 7.4	208	87″	106	G0	120651, 83022
	Σ 1793		13 59.1 +25 49	7.5, 8.4	243	4.7″	154	A5V	122080, 83108
	Bgh 50		14 04.8 +25 49	7.0, 8.9	32	99″	44.0	F5 K0	123033, 83152
	Σ 1812	(AB) C D	14 12.4 +28 43	7.9, 9.5, 12.1	108, 154	14″, 72″	130*	F2V	124346, 83219
★ 17 kap 1,2	Σ 1821	A (BaBb)	14 13.5 +51 47	4.5, 6.6	234	14″	50	A8IV F1V	124675, 29046
		Asellus Tertius. Radiant A type 1+(2) binary with del Sct type variable (usually an A or F type, imperceptibly and rapidly pulsating star); sparse field. AB: P = 6,100 y, orbit r = 565 AU; at cusp of edgewise orbit. Long period spectroscopic binary Ba,Bb: P = 4.9 y. (2014)							
	OΣ 279		14 13.8 +12 00	6.8, 9.1	255	2.2″	171	K2III	124517, 100922
	Σ 1816		14 13.9 +29 06	7.4, 7.8	96	0.4″	113	F0 A2	124587, 83235
		Mixed A/solar type binary; brighter of two stars in dark field. AB: P = 1,340 y (?), orbit r = 185 AU, slowly closing. (222 measures; 2011)							
15	Kui 66		14 14.8 +10 06	5.4, 8.4	108	1.0″	81	K1III	124679, 100934
	Σ 1829		14 15.5 +50 26	8.1, 8.6	151	5.6″	161	F5	125020, 29061
★ 21 iot	Σ I 26	(AaAb) B	14 16.2 +51 22	4.8, 7.4	32	39″	29.1	A7IV K0V	125161, 29071
		Asellus Secundus. A type (2)+1 spectroscopic triple with del Sct type variable; kap 1,2 Boo 30′ n.p. Aa,Ab: P = 0.6 d. AB: ps = 1,530 AU. (2014)							
	Σ 1825		14 16.5 +20 07	6.5, 8.4	154	4.3″	32.6	F6V	125040, 83259
	OΣ 281		14 20.3 +08 35	7.7, 9.7	166	1.5″	51	G5	125608, 120406
	Σ 1834		14 20.3 +48 30	8.1, 8.3	109	1.8″	75	F9V	125796, 45000
		Matched solar type binary, sparse field. System mass 2.7 M$_\odot$. AB: P = 376 y, orbit r = 75 AU, e = 0.89, apastron 2091. (240 measures; 2012)							
★	Σ 1835	A (BC)	14 23.4 +08 27	5.0, 6.8	195	6.1″	66	A0V F2V	126129, 120426
		A type 1+(2) triple, visual binary; BC (β 1111): system mass 2.6 M$_\odot$, P = 40 y, e = 0.25, apastron 2018; AC: ps = 540 AU. (2013)							
	Σ 1838		14 24.1 +11 15	7.5, 7.7	334	9.4″	37.1	F8V G1V	126246, 101009
	Σ 1843	A B	14 24.6 +47 50	7.7, 9.2	186	20″	92	F4V	126531, 45045
23 the	OΣ 580		14 25.2 +51 51	4.1, 11.5	182	70″	14.5	F7V	126660, 29137
		Asellus Primus. Local, high CPM, unequal (q ~0.2) solar type double with del Sct type variable; dark field. AB: ps = 1,370 AU. (2003)							
	Σ 1850	A (BaBb)	14 28.6 +28 17	7.1, 7.6	260	26″	350	A1V A1V	127067, 83374
	Σ 1854	A B	14 29.8 +31 47	6.1, 10.6	256	26″	110	A0V	127304, 64178
	Σ 1858	A B	14 33.6 +35 35	8.1, 9.0	38	3.0″	36.8	G5	128041, 64213
★	Σ 1863		14 38.0 +51 35	7.7, 7.8	62	0.7″	82	F4V	128941, 29224
		Matched solar type binary; galaxy NGC 5707 5′ p. AB: P = 538 y, orbit r = 90 AU, widening. (202 measures; 2013)							
★ 29 pi 1	Σ 1864	(AaAb) (BaBb)	14 40.7 +16 25	4.9, 5.8	111	5.5″	94	B9V A6V	129174, 101138
		High mass (2)+(2) spectroscopic quadruple, visual double; W/W color, pretty brightness contrast, discovered by C. Mayer (1777) and first measured by W. Herschel (1779). AB: ps = 700 AU. (372 measures; 2013)							
★ 30 zet	Σ 1865	A B	14 41.1 +13 44	4.5, 4.6	297	0.5″	54	A0V A0V	129246, 101145
		Matched A type binary; splendid W/W color, subarcsecond resolution test. System mass 5.4 M$_\odot$. AB: P = 124 y, orbit r = 32 AU, e = 0.998 (highest known eccentricity), closing to less than 0.4″ by 2017, reappears s.f in 2029. (743 measures; 2012)							

Label	Catalog ID	Components	Coordinates (J2000)	Mag.	θ	ρ	Dist.	Spectral type(s)	HD, SAO No.
	Σ 1871		14 41.6 +51 24	8.0, 8.1	310	1.8″	99	F3V	129600, 29246
	Σ 1870		14 42.9 +08 05	7.5, 10.0	231	4.8″	198	F2	129538, 120618
	Σ 1873		14 44.8 +07 42	8.0, 8.4	93	6.4″	230	G5III	129868, 120635
★ 36 eps	Σ 1877	A (BaBb)	14 45.0 +27 04	2.6, 4.8	343	2.9″	62	K0II	129989, 83500
		Izar. F.W. von Struve's Pulcherrima ("most beautiful"), discovered by W. Herschel (1779) and admired for its O/B color. A "giant type" 1+(2) triple with K supergiant primary and spectroscopic binary component; despite AB: ps = 240 AU and a noticeable arc in the measures, there is no orbit. (455 measures; 2012)							
	OΣ 285	A B C	14 45.5 +42 23	7.8, 8.7, 12.4	86, 288	0.5″, 55″	84	F6V	130188, 45208
		Solar type 2+1 triple; MSC system mass 3.1 M⊙. AB: P = 88 y, orbit r = 27 AU, apastron 2016 (273 measures); AC: ps = 6,230 AU. (2011)							
	Σ 1879	A B C	14 46.3 +09 39	7.8, 8.5, 12.1	82, 233	1.8″, 38″	43.2	G2V	130145, 120651
		Solar type 2+1 triple; sparse field. AB: system mass 2.1 M⊙, P = 226 y, orbit r = 43 AU, e = 0.70, periastron 2119. (308 measures; 2013)							
	Σ 1884		14 48.4 +24 22	6.6, 7.5	54	2.0″	85	F8IV	130603, 83535
	A 1110	A B C	14 49.7 +07 59	7.7, 7.9, 12.0	244, 201	0.7″, 20″	132	F5 F8III	130726, 120683
	Σ 1886	A B	14 51.0 +09 43	7.6, 9.7	225	7.4″	29.4	K0	131023, 120697
★ 37 xi	Σ 1888	A B	14 51.4 +19 06	4.8, 7.0	306	5.7″	6.71	G8V K5V	131156, 101250
		Local, solar type binary with BY Dra type variable; Y/O color. System mass 1.6 M⊙. AB: P = 152 y, orbit r = 33 AU, periastron 2061. (1414 measures; 2013)							
	OΣ 288		14 53.4 +15 42	6.9, 7.6	160	1.0″	47.6	F9V	131473, 101273
		Solar type binary; dark field. AB: P = 313 y, orbit r = 65 AU, closing. (399 measures; 2013)							
	OΣ 289		14 56.0 +32 18	6.2, 10.2	110	4.6″	96	A2V	132029, 64408
	Sh 191		14 59.6 +53 52	6.9, 7.6	342	40″	118	F1V F1V	132909, 29372
BX	OΣ 291		15 00.6 +47 17	6.3, 9.6	156	35″	170	B9	133029, 45326
★ 44 i	Σ 1909	A (BaBb)	15 03.8 +47 39	5.2, 6.1	63	1.2″	12.5	F7V K4V	133640, 45357
		Local, high CPM, solar type 1+(2) spectroscopic triple with W UMa type eclipsing variable (see Introduction). AB: P = 225 y, orbit r = 46 AU, e = 0.51 (805 measures). Ba,Bb: P = 0.27 d. (2013)							
	Σ 1910		15 07.5 +09 14	7.4, 7.5	211	3.9″	31.6	G2V G3V	134066, —
	Es 2648	A C B	15 12.7 +48 35	7.3, 10.8, 11.3	310, 341	0.7″, 27″	154	K0	135364, 45436
49 del	Σ I 27	(AaAb) B	15 15.5 +33 19	3.6, 7.9	79	105″	37.3	G8IIICN	135722, 64589
★ 51 mu 1,2	Σ I 28	(AaAb) [Ba Bb]	15 24.5 +37 23	4.3, [7.1, 7.6]	171, [5]	109″, [2.3″]	34.7	F2IVa G0V	137391, 64686
		Alkalurops. Splendid, bright (2)+2 quadruple; visual triple. MSC system mass 5.2 M⊙. Aa,Ab: P = 0.82 y, e = 0.27; Ba,Bb: P = 257 y, orbit r = 50 AU, e = 0.58, periastron 2120. (2013)							
	OΣ 296	A B	15 26.4 +44 00	7.8, 9.1	274	2.1″	88	G5	137805, 45541
	Ku 108	A [B C]	15 27.7 +42 53	7.6, [9.7, 11.0]	319, [100]	41″, [0.4″]	32.1	G5	138004, 45551
★	OΣ 298	A B C	15 36.0 +39 48	7.2, 8.4, 7.8	183, 328	1.2″, 2.0′	22.3	K1V K3V	139341, 64800
		Local, high CPM, solar type 2+1 triple. AB: system mass 1.8 M⊙, P = 56 y, orbit r = 18 AU, e = 0.59, apastron 2021. (532 measures; 2013)							
	OΣ 301		15 46.2 +42 28	7.5, 10.4	29	3.7″	220	K0	141204, 45718

Caelum — Cae — Chart 21

Label	Catalog ID	Components	Coordinates (J2000)	Mag.	θ	ρ	Dist.	Spectral type(s)	HD, SAO No.
	I 271		04 21.8 −42 47	7.8, 10.4	143	2.5″	250	K1III	27843, 216764
	h 3650		04 26.6 −40 32	7.0, 8.2	184	3.0″	127	A1V	28358, 216803
★	Skf 811	A (BC)	04 49.2 −42 23	7.3, 7.3	134	2.5′	166	A3V+A2F2	30848, 217012
		High mass, wide 1+2 triple or visual double; sparse field, unrelated m.9 star 1′ n. AB: ps = 22,900 AU. (2010)							
	Bvd 48	A B C	04 51.9 −34 14	6.8, 8.7, 8.9	160, 309	52″, 100″	59	F6IV+G5	31142, 195364
★ gam 1	Jc 9		05 04.4 −35 29	4.7, 8.2	305	3.2″	56	K3III	32831, 195532
		Double with K type giant; pretty color contrast with giant F1III gam 2 Cae 15′ s. AB: ps = 240 AU. (2001)							

Label	Catalog ID	Components	Coordinates (J2000)	Mag.	θ	ρ	Dist.	Spectral type(s)	HD, SAO No.
	Camelopardalis	Cam						*Charts 1, 3*	
	OΣ 52	A B	03 17.5 +65 40	7.1, 7.4	58	0.5″	169	A2V	20104, 12686
		A type binary, subarcsecond resolution test. Low quality orbit, AB: *P* = 350 y, orbit *r* = 80 AU, decreasing *θ*. (2010)							
	Σ 373	A B C	03 22.1 +62 44	7.7, 10.0, 7.8	118, 112	20″, 116″	123	F8	20588, 12721
	Σ 374		03 24.2 +67 27	7.8, 9.0	297	11″	178	F8	20711, 12738
	Σ 384	A B C	03 28.5 +59 54	8.1, 8.9, 10.6	272, 342	1.9″, 117″	760	F8	21224, —
		Distant, solar type 2+1 triple; O/B color. Rich field with HLM 2 (m.9, 5″) and CS Cam (Σ 385, m.4, 2″) 5′ n.f. (2007)							
	Σ 385		03 29.1 +59 56	4.2, 7.8	162	2.3″	600	B9Ia	21291, 24054
	Σ 389	(AaAb) B	03 30.2 +59 22	6.4, 7.9	71	2.7″	95	A2V	21427, 24062
	Σ 396	A B	03 33.5 +58 46	6.4, 7.7	245	20″	155	A4III	21769, 24093
	Σ 400	A B	03 35.0 +60 02	6.8, 8.0	267	1.6″	53	F3V	21903, 24111
		Solar type "diamond ring" binary; Y/– color. System mass 2.8 M$_\odot$. AB: *P* = 288 y, orbit *r* = 65 AU, *e* = 0.67, apastron 2061. (2012)							
	OΣΣ 36	A (BC)	03 40.0 +63 52	6.9, 8.3	71	46″	42.5	F5V G8V	22399, 12854
	Σ 419	A [B C]	03 42.8 +69 51	7.8, [7.8, 9.4]	72, [112]	3.0″, [0.4″]	310	A5IV	22553, 12873
	Σ 421		03 46.2 +71 37	7.1, 10.7	237	12″	183	G9III	22912, 4984
	S 436		03 49.3 +57 07	6.5, 7.2	76	58″	149	A0Vn	23594, 24244
gam	h 2200	A C	03 50.4 +71 20	4.6, 9.1	86	107″	110	A2IVn	23401, 5006
	h 1139		03 51.7 +70 30	7.5, 9.6	177	48″	148	A3	23602, 5015
	OΣ 67		03 57.1 +61 07	5.3, 8.1	49	1.7″	450	K3I	24480, 12968
	Σ 511		04 17.9 +58 47	7.4, 8.7	77	0.5″	126	A2V	26839, 24521
		A type binary; pretty field. System mass 3.0 M$_\odot$. AB: *P* = 343 y, orbit *r* = 70 AU, *e* = 0.31, slowly widening. (2010)							
	Skf 1204		04 20.0 +78 05	7.1, 10.0	100	80″	250	K0+F	26415, 5153
	Arg 100	(AB) C	04 23.0 +59 37	6.2, 9.3	58	33″	142	A4V	27402, 24577
	Σ 531		04 26.8 +55 39	7.7, 8.8	324	1.0″	159	F4IV	27856, 24614
★ 1	Σ 550	A B	04 32.0 +53 55	5.8, 6.8	308	11″	1430*	BOIII+BOIV	28446, 24672
		DL Cam. Distant, high mass, spectroscopic double with bet Cep type variable; notable color, rich field. AB: ps = 15,700 AU. (2012)							
	Opi 5	A C	04 38.1 +71 28	7.7, 8.5	279	112″	185	A2	28760, 5263
★ 2	Σ 566	(AB) C D	04 40.0 +53 28	5.6, 7.5, 13.2	174, 224	0.8″, 21″	44.5	A8V	29316, 24744
		A type (2)+2 quadruple, visual triple; unrelated to optical pair D 4 (m.9, 6″) 3.8′ f. AB: *P* = 27 y, *e* = 0.86, periastron 2015; AB,C: *P* = 480 y, orbit *r* = 65 AU. (2009)							
	Hu 612		04 47.8 +53 18	7.1, 8.5	359	0.7″	127	F2	30136, 24830
		Solar type binary; sparse field. AB: *P* = 310 y, orbit *r* = 64 AU, widening to apastron. (2007)							
	Σ 584		04 50.1 +66 32	7.6, 9.4	122	12″	250	K0	30164, 13278
7	Σ 610	(AaAb) B C	04 57.3 +53 45	4.5, 7.9, 11.3	202, 242	0.6″, 26″	114	A1V	31278, 24929
	OΣ 88		04 57.3 +61 45	7.2, 8.3	307	0.8″	240	G0	31151, 13317
	Σ 618	A B	05 03.6 +63 05	7.7, 8.0	211	33″	32.2	G0	31865, 13348
		High CPM solar type double; AB: ps = 1,430 AU. CPM double Σ 617 (DE, m.9, 13″) 5′ south is unrelated. (2003)							
	β 749		05 07.5 +55 32	7.5, 9.2	241	1.3″	142	F8	32606, 25029
★	A 841	A [B C]	05 10.0 +75 41	7.3, [10.1, 10.9]	343, [228]	48″, [0.6″]	149	A2III	32230, 5442
		1+2 triple with A type giant; sparse field, large aperture needed to resolve BC. AB: ps = 9,650 AU. (2004)							
	Σ 633		05 10.7 +63 36	6.8, 10.6	342	12″	86	F0	32893, 13394
	Σ 638		05 14.3 +69 49	7.5, 9.1	222	5.1″	103	K1IV	33164, 13412
	Σ 677	A B C	05 24.7 +63 23	7.9, 8.5, 13.0	118, 227	1.1″, 6.8″	47.2	G0	34839, 13482
		Solar type, probable 2+1 triple; YO/– color. AB: *P* = 362 y, orbit *r* = 70 AU, decreasing *θ*. (2012)							
	Σ 676		05 24.8 +64 44	8.1, 8.9	268	1.3″	200	F8	34804, 13481
19	Hu 1107		05 37.3 +64 09	6.2, 9.8	56	1.5″	111	A0V	36570, 13550
	Σ 3115		05 49.1 +62 48	6.6, 7.5	338	0.8″	123	A4V	38284, 13618
		A type binary; small star group 6′ s. and s.p. AB: *P* = 977 y, orbit *r* = 165 AU; decreasing *θ*. (2008)							

Label	Catalog ID	Components	Coordinates (J2000)	Mag.	θ	ρ	Dist.	Spectral type(s)	HD, SAO No.
	Σ 780	A B C	05 51.0 +65 45	7.0, 8.2, 10.2	104, 148	3.8″, 12″	980	F8	38475, 13627
	Σ 922	A B	06 38.3 +64 44	7.7, 10.8	137	11″	78	F4V	46463, 13901
	Mlr 318		06 42.5 +66 12	7.3, 9.3	308	1.7″	70	F8	47215, 13929
	Σ 973	A B	07 04.1 +75 14	7.2, 8.2	32	13″	35.7	G0	51067, 6050
	Hzg 4		07 19.1 +66 44	7.8, 11.8	35	8.5″	53	F8	55745, 14145
	Σ 1051	A B C	07 26.6 +73 05	7.6, 9.1, 7.8	297, 84	1.1″, 32″	146	F2IV F0IV	57044, 6187
	Es 1895		07 31.6 +62 30	7.0, 10.7	289	9.9″	159	A0	58917, 14221
	Σ 1122		07 45.9 +65 09	7.8, 7.8	186	15″	70	F2	61907, 14311
	Σ 1127	A B C	07 47.0 +64 03	7.0, 8.5, 9.7	341, 176	5.3″, 13″	182	A2	62195, 14326
	OΣ 188		08 22.2 +74 49	6.5, 10.5	195	10″	143	G8III G0V	69054, 6511
	Σ 1625	A B	12 16.2 +80 08	7.2, 7.8	219	15″	500	F1V F3V	106799, 2009
★	Σ 1694	A (BaBb)	12 49.2 +83 25	5.3, 5.7	324	21″	179	A1IIIh A0V +A2V	112028, 2102
			Radiant, nearly matched 1+(2) spectroscopic triple, visual double with A type giant; dark field. AB: ps = 5,070 AU. (2011)						
	OΣ 258		12 54.2 +82 31	7.3, 10.6	71	10″	167	K0	112651, 2112

Cancer   Cnc   *Charts 8, 15*

Label	Catalog ID	Components	Coordinates (J2000)	Mag.	θ	ρ	Dist.	Spectral type(s)	HD, SAO No.
	Skf 1808		07 55.8 +10 27	7.0, 8.9	273	76″	300	B9+F2	64745, 97378
	Σ 1171		08 01.0 +23 35	6.5, 10.0	326	2.1″	88	K1III	65757, 79864
	OΣ 186		08 03.3 +26 16	7.7, 7.9	74	1.0″	188	A4V	66176, 79893
17	Σ 1177		08 05.6 +27 32	6.7, 7.4	350	3.5″	290	B9V	66684, 79928
	Σ 1187	(AaAb) B	08 09.5 +32 13	7.2, 8.0	21	3.0″	65	F2	67501, 60604
			Solar type (2)+1 triple; visual double. MSC system mass 2.9 M☉. AB: ps = 260 AU. (293 measures; 2012)						
	Wrh 27	(AaAb) B	08 09.7 +25 34	7.6, 10.7	97	29″	650		67613, 79987
★ 16 zet 1,2	Σ 1196	A B (CDaDb)	08 12.2 +17 39	5.3, 6.3, 5.9	28, 64	1.1″, 6.6″	25.1	F8V	68255, 97645
			Tegmine. Local, high mass 2+(1+2) quintuple, visual triple. One of the finest visual triple stars in the sky, compact and nearly matched. AC first noted by Flamsteed (1680), AB by W. Herschel (1781); D detected astrometrically by O. Struve (1874); Da,Db reported as an M dwarf binary by Hutchings *et al.* (2000). MSC system mass 11.7 M☉. Orbits retrograde, viewed nearly face on, but inclinations differ by 28°. AB: *P* = 60 y, orbit *r* = 22 AU, *e* = 0.32, apastron 2018; AC: *P* = 1,115 y, orbit *r* = 190 AU. CD: *P* = 17.3 y, orbit *r* = 4.6 AU, *e* = 0.08, periastron 2018. (1155 measures; 2013). See Figure 1.						
	Σ 1202		08 13.6 +10 51	7.4, 9.6	306	2.7″	77	F7V	68615, 97662
	Lep 30		08 15.6 +11 26	7.7, 9.8	238	32″	38.6	G5	69056, 97681
			High CPM, solar type double; star n.f. unrelated. AB: ps = 1,670 AU. (2009)						
	Ho 524	Aa Ab B	08 16.0 +18 42	7.5, 7.5, 10.5	34, 343	0.4″, 4.4″	185	K0	69072, 97684
	OΣ 191		08 24.8 +20 09	7.4, 8.6	192	38″	188	A5	70826, 80164
24 ups	Σ 1224	A (BC)	08 26.7 +24 32	6.9, 7.5	47	6.1″	80	F0V F7V	71153, 80184
			Solar type 1+(2) triple, visual binary. AB: ps = 660 AU; BC: *P* = 22 y, orbit *r* = 12 AU, *e* = 0.08, periastron 2020. (253 measures; 2013)						
	Σ 1245	(AaAb) (BaBb)	08 35.9 +06 37	6.0, 7.2	25	9.8″	25.1	F8V G5V	72945, 116929
			Local, solar type (2)+(2) quadruple, visual double; sparse field. MSC system mass 3.2 M☉. AB: ps = 330 AU. (2013)						
	OΣΣ 94		08 36.2 +13 47	7.4, 8.1	133	43″	300	A0	72965, 97952
	S 570	A B C	08 39.1 +19 41	7.5, 9.6, 9.4	84, 345	58″, 3.0′	100*	A8V	73449, 97999
	S 571	(AB) (CaCb) (DaDb) E	08 39.9 +19 33	7.3, 7.5, 6.7, 11.8	157, 242, 3	45″, 93″, 35″	180*	Am K0III	73618, 98013
			Striking, A type (2)+(2)+(2)+1 CPM, occultation and spectroscopic septuple system, visual quadruple; in NGC 2632 (Praesepe). Σ 1254 9′ n.f. (2012)						

Label	Catalog ID	Components	Coordinates (J2000)	Mag.	θ	ρ	Dist.	Spectral type(s)	HD, SAO No.
39,40	Eng 37	A (BaBb) D (CaCb)	08 40.1 +20 00	6.5, 6.6, 8.8, 9.0	152, 111, 309	2.5′, 2.3′, 2.2′	187	K0III	73665, 80333
		Fragile, possibly unstable 1+(2)+1+(2) CPM sextuple, visual quadruple with K giant, in NGC 2632 (Praesepe). AC: ps = 37,600 AU. (2013)							
★	Σ 1254	(AaAb) (BaBb) C D E	08 40.4 +19 40	6.5, 10.4, 7.6, 9.2, 12.5	54, 343, 44, 154	21″, 63″, 83″, 16.2″	220	K0III G9III	73710, 98021
		Another fragile CPM septuple with K giant Algol type variable, at north edge of NGC 2632 (Praesepe). AD: ps = 24,600 AU. (2012)							
41 eps	S 574	(AaAb) B	08 40.5 +19 33	6.3, 7.5	250	2.2′	180	A5m	73731, 98024
	Cbl 32		08 46.2 +27 36	7.4, 10.7	174	41″	99	K1IV	74669, 80409
★ 48 iot	Σ 1268		08 46.7 +28 46	4.1, 6.0	305	31″	102	G7III A3V	74739, 80416
		Wide CPM double with late G type giant; YO/B color, discovered by C. Mayer (1779). AB: ps = 4,270 AU. (2013)							
	Σ 1276	A B	08 47.2 +11 10	8.3, 8.6	354	12″	980	A0	74926, 98120
	Σ 1283		08 49.9 +14 50	7.7, 8.5	123	17″	440	F0	75355, 98148
	OΣ 195		08 54.0 +08 25	7.7, 8.3	139	9.8″	56*	F8	76037, 117244
	Σ 1291	A B C	08 54.2 +30 35	6.1, 6.4, 9.2	311, 202	1.5″, 55″	141	G7III	75959, 61125
		Probable solar type 2+1 triple with G giant; AC: ps = 10,460 AU. (274 measures; 2013)							
64 sig 3	Sh 100	A B C	08 59.5 +32 25	5.3, 9.0, 10.1	294, 217	90″, 2.6′	91	G8III	76813, 61177
66	Σ 1298	A B	09 01.4 +32 15	6.0, 8.6	139	4.1″	143	A2V	77104, 61202
	Σ 1311	A B	09 07.5 +22 59	6.9, 7.1	198	7.9″	58	F4V+F5V	78175, 80643
	OΣΣ 97		09 08.5 +27 33	8.3, 8.3	238	52″	42.6	G0V	—, 80655
	Σ 1322		09 12.7 +16 32	8.3, 8.7	53	1.7″	300*	A4IV	79127, 98430
	Σ 1332		09 17.3 +23 39	7.9, 8.1	29	5.8″	71	F6V F7V	79872, 80738
IP	Σ 3121	A B	09 17.9 +28 34	7.9, 8.0	219	0.4″	17.3	K0	79969, 80745
		Local, high CPM, low mass binary; AB: system mass 1.5 M$_\odot$, P = 34 y, orbit r = 12 AU, e = 0.32; less than 0.2″ at periastron (2015), but widening to 0.5″ by 2018. (403 measures; 2010)							

Canes Venatici   CVn   *Chart 9*

Label	Catalog ID	Components	Coordinates (J2000)	Mag.	θ	ρ	Dist.	Spectral type(s)	HD, SAO No.
	Σ 1606	(AaAb) B	12 10.8 +39 53	7.4, 7.9	151	0.5″	120	A8III	105824, 44064
		A type (2)+1 triple with rare A giant, visual binary; galaxy NGC 4145 9′ p. MSC system mass 4.5 M$_\odot$. Aa,Ab: P = 75y, e = 0.37, periastron 2030; AB: P = 1,431 y, orbit r = 240 AU, e = 0.78, widening. (258 measures; 2012)							
	Σ 1609	A B	12 11.7 +50 50	8.0, 9.6	204	11″	185	F2	105980, 28290
	Σ 1624	A B	12 16.7 +39 36	7.3, 10.2	153	5.9″	109	A2.5V	106784, 62930
	Σ 1632	(AaAb) B	12 20.2 +37 54	6.8, 10.0	193	11″	165	K0III F9V	107341, 62953
★ 12 alp 1,2	Σ 1692	(AaAb) (BaBb)	12 56.0 +38 19	2.9, 5.5	228	19″	35.2	F2V A0pec	112413, 63257
		Cor Caroli. Fine solar/A type (2)+(2) spectroscopic quadruple, visual double; resolved in all apertures with a subtle color contrast that varies with observer. Parallax equal within errors and CPM unite the pair. B is prototype of alpha2 CVn variables: mostly A type stars with silicon and rare earth absorption lines, strong magnetic fields and surface light fluctuations. AB: ps = 900 AU. (2014)							
	Σ 1723		13 08.2 +38 44	8.7, 10.1	11	6.4″	73	G2IV	114146, 63362
	OΣ 261		13 12.0 +32 05	7.4, 7.6	339	2.6″	73	F6V	114723, 63396
		Solar type matched binary; snug in small apertures, dark field. AB: P = 861 y, orbit r = 130 AU. (249 measures; 2013)							
★ 25	Σ 1768	A B	13 37.5 +36 18	5.0, 7.0	96	1.7″	61	A7IV	118623, 63648
		A type binary; YW/– color. System mass 3.8 M$_\odot$. AB: P = 228 y, orbit r = 60 AU, e = 0.80, periastron 2092. (486 measures; 2013)							
	Σ 1769	(AaAb) B C	13 38.0 +39 11	7.9, 10.4, 9.3	45, 259	1.6″, 56″	46.0	G5	118742, 63656
		Solar type (2)+1+1 quadruple; linear solution for AB but no CPM discrepancy. Aa,Ab: P = 11.6 d. AC: ps = 3,480 AU. (2013)							
	S 654	A B **C**	13 47.0 +38 33	5.6, 8.9, 8.9	239, 296	72″, 34′	97	K0III F8V	120164, 63739

Label	Catalog ID	Components	Coordinates (J2000)	Mag.	θ	ρ	Dist.	Spectral type(s)	HD, SAO No.
	Canis Major		CMa						*Chart 22*
	B 104		06 12.3 −25 15	8.0, 8.8	184	1.1″	113	F0	42899, 171379
	Skf 58	A (BaBb) C	06 13.8 −23 52	6.5, 12.5, 13.0	177, 171	25″, 2.7′	16.7	G6.5V	43162, 171428
			Local, solar type 1+(2)+1 quadruple with BY Dra type variable; pretty field. A(B): ps = 560 AU. AC: ps = 3,650 AU. (2010)						
	β 18		06 16.7 −12 03	7.1, 8.4	285	1.8″	162	F2V	43628, 151303
	B 110		06 17.6 −24 27	7.6, 11.5	48	6.8″	64	F3V	43879, 171521
FR	Σ 3116	A B	06 21.4 −11 46	5.6, 9.7	23	3.9″	340	B1Ve B9V	44458, 151401
	β 568		06 23.8 −19 47	6.9, 8.2	153	0.8″	410	B8III	44953, 151453
	β 753		06 28.7 −32 22	5.9, 7.6	43	1.2″	260	B4Vne	45871, 196861
	h 3863		06 29.4 −22 35	7.0, 8.7	126	3.0″	185	A1V	45941, 171831
	h 3871		06 34.1 −29 38	7.1, 8.2	354	7.7″	133	A1V	46813, 171964
6 nu 1	Sh 73		06 36.4 −18 40	5.8, 7.4	264	18″	108	G5III	47138, 151694
★	H II 60	(AaAb) B	06 36.7 −22 37	6.4, 9.3	335	8.7″	320	B5V	47247, 172021
			Distant, high mass (2)+1 triple with Algol type eclipsing variable; Y/B color with pretty brightness contrast. AB: ps = 3,760 AU. Neglected. (1999)						
	β 19		06 42.0 −16 00	7.1, 9.0	169	3.9″	530	B8III	48287, 151807
	β 195	A B	06 42.5 −23 14	7.1, 9.6	215	5.9″	640	B3V	48425, 172196
	S 534		06 42.8 −22 27	6.3, 8.3	144	18″	54	F2V	48501, 172204
★ 9 alp	AGC 1	A B	06 45.1 −16 43	-1.5, 8.5	83	9.6″	2.64	A1Vm D	48915, 151881
			Sirius. Local, very high CPM, A type spectroscopic binary with white dwarf in close orbit; infamously difficult brightness contrast, good seeing and observing in twilight are helpful. From variations in proper motion, Bessel (1844) identified Sirius and Procyon as "genuine binary systems, each consisting of a visible and an invisible star." Sirius B ("the Pup") was detected by Alvan G. Clark (1862) as it neared apastron; at 0.98 M$_\odot$, it is the most massive white dwarf known. AB: P = 50y, orbit r = 20 AU, e = 0.59, apastron 2019. (642 measures; 2013). See Figure 1.						
HP	h 3891		06 45.5 −30 57	5.7, 8.2	220	5.4″	580	B2III	49131, 197177
	β 20		06 48.8 −16 13	7.8, 9.9	28	3.3″	240	K1III	49649, 151957
	S 538		06 49.6 −24 09	7.2, 8.2	4	27″	250	A2	49868, 172383
★	β 324	A B C	06 49.7 −24 05	6.6, 7.9, 8.3	211, 282	1.8″, 30″	250	A1V	49891, 172389
			A type 2+1 triple; comoving "double double" with S 538 (above) 5′ s. in rich field. AC: ps = 7,500 AU. (2008)						
HZ	H V 108	A (BC)	06 50.4 −31 42	5.8, 7.7	66	43″	199	B6Vne	50123, 197263
	β 325		06 51.8 −26 35	7.9, 9.1	38	1.7″	1180*	B2V	50379, 172461
18 mu	Σ 997	A C	06 56.1 −14 03	5.3, 10.3	289	87″	380	G5III A2	51250, 152123
	S 541		06 56.6 −22 39	7.5, 8.5	45	23″	290	K1III	51457, 172606
★ 21 eps	CapO 7		06 58.6 −28 58	1.5, 7.5	162	7.9″	124	B2II	52089, 172676
			Adhara. High mass double with B giant; visually resembles Sirius, but component is probably A type star. (2008)						
	I 183		07 00.8 −25 39	7.4, 9.9	139	3.9″	520	B2IV	52596, 172750
GU	Hu 112	(AaAb) B	07 01.8 −11 18	7.0, 7.7	198	0.6″	440	B2Vne	52721, 152255
	h 3914		07 02.1 −23 30	7.5, 11.0	314	11″	280*	G8III	52925, 172808
	Σ 1016		07 04.6 −11 31	7.4, 9.5	149	5.4″	1960*	B0V	53456, 152324
FN	Σ 1026	A B C	07 06.7 −11 18	5.7, 6.9, 9.0	111, 350	0.6″, 18″	930	B0.5IV	53974, 152394
	β 575	A B C	07 14.8 −15 29	8.3, 7.9, 9.9	297, 1	0.5″, 16″	93	F5V	56012, 152609
	BrsO 2		07 17.0 −30 54	6.3, 7.8	183	38″	132	A9II	56731, 197789
	h 3949		07 18.6 −30 48	7.7, 7.9	77	3.0″	610	B2V	57120, 197827
★	Lal 53		07 19.3 −22 03	7.6, 7.7	166	3.8″	119	A4V	57190, 173466
			Pretty matched pair adrift in Milky Way streams, s.p. NGC 2367. HDS 1026 (m.9, 17″) 3.6′ n.f. AB: ps = 450 AU. (2000)						
	Δ 129		07 24.7 −31 49	5.4, 9.7	311	2.1″	200	G8II	58535, 197964
			Solar type double, open cluster Collinder 140 in wide field; probably unrelated to CD (B 1540, m.8, 0.9″), 98″ n.p. Neglected. (1991)						
	β 199	A B	07 25.1 −21 10	7.2, 8.1	24	1.7″	—	B1Ib	58510, 173684
	Stone 17		07 26.1 −18 22	7.6, 9.6	76	5.0″	220	A0V	58698, —
	β 332	A B	07 27.9 −11 33	6.2, 7.4	173	0.7″	490	G8Ib	59067, 152909

Label	Catalog ID	Components	Coordinates (J2000)	Mag.	θ	ρ	Dist.	Spectral type(s)	HD, SAO No.
	Canis Minor	CMi						*Chart 15*	
	Σ 1074	A B	07 20.5 +00 24	7.4, 7.8	172	0.7″	230	B9.5V	57275, 115294
	Σ 1103		07 30.6 +05 15	7.1, 8.6	244	4.2″	188	B9	59538, 115532
	OΣ 176	A B	07 38.5 +00 30	7.2, 9.2	221	1.6″	490	B9	61275, 115733
	Σ 1126	(AaAb) B	07 40.1 +05 14	6.6, 7.0	174	1.0″	300*	A0III	61563, 115773
		A type (2)+1 spectroscopic triple, 15′ following Procyon (alp CMi). AB frequently measured, but no orbit (281 measures; 2012)							
	Σ 1134	(AaAb) B	07 43.5 +03 29	7.1, 10.4	146	9.4″	51	F8	62323, 115851
	Σ 1137		07 46.6 +04 08	8.0, 9.1	133	2.9″	440	F5	62968, 115910
	OΣΣ 88	A B	07 49.0 +00 40	7.5, 8.9	5	58″	70	F2	63436, 115967
	Σ 1149		07 49.5 +03 13	7.8, 9.2	41	22″	130	G0	63536, 115981
	OΣ 182		07 52.7 +03 23	7.8, 7.9	12	1.0″	210	A2	64165, 116064
	OΣ 185		07 57.3 +01 08	7.1, 7.3	17	0.4″	62	F7V	65123, 116165
		Matched solar type binary; a pretty field of stars. AB: system mass 2.9 M⊙, *P* = 58 y, orbit *r* = 21 AU, *e* = 0.67, widening until 2020. (2013)							
	Σ 1175		08 02.4 +04 09	7.9, 9.1	285	1.4″	40.6	G5	66177, 116263
	Σ 1182		08 05.4 +05 50	7.5, 8.8	74	4.7″	330	B9	66801, 116324
	Capricornus	Cap						*Chart 25*	
6 alp 2	h 608	A [B C]	20 18.1 −12 33	3.7, [11.2, 11.5]	196, [245]	6.6″, [1.2″]	32.4	G9III	192947, 163427
★ 9 bet 1,2	Σ I 52	(AaAb) (BaBb) C	20 21.0 −14 47	3.2, 6.1, 8.8	267, 133	3.4′, 4.4′	100	F8V A0	193495, 163481
		Dabih. High mass (2)+(2)+1 occultation and interferometric quintuple, wide visual triple. MSC 18.1 M⊙. Aa,Ab (Bla 7): *P* = 3.8 y, *e* = 0.43. Ba,Bb (Bar 12): *P* = 8.7 d, is possibly eclipsing. AB: ps = 27,500 AU. (2012)							
10 pi	β 60	(AaAb) B	20 27.3 −18 13	5.1, 8.5	146	3.2″	167	B8II	194636, 163592
11 rho	Sh 323	A B	20 28.9 −17 49	5.0, 6.9	191	1.6″	30.3	F3V	194943, 163614
12 omi	Sh 324		20 29.9 −18 35	5.9, 6.7	238	22″	66	A3Vn A7V	195094, 163626
	h 2975		20 33.5 −22 14	7.5, 11.6	22	9.9″	47.0	F8	195680, 189425
	Σ 2699	A B	20 36.9 −12 44	8.0, 9.2	197	9.4″	91	F2V	196310, 163730
	β 674	A B	20 44.8 −20 54	8.0, 9.6	99	1.7″	250	K1III	197523, 189638
	β 153		20 47.3 −26 25	7.4, 9.0	254	1.8″	123	A1m F0	197889, 189697
	S 763	A B	20 48.4 −18 12	7.2, 7.8	293	16″	104	G8III	198063, 163895
	h 5226		20 50.1 −27 22	7.3, 8.8	67	19″	139	K0III	198278, 189747
	h 3003		20 53.0 −23 47	6.6, 8.6	194	1.5″	108	K0III	198732, 189801
	h 616		21 55.0 −11 58	8.1, 10.5	275	31″	137	A2m	208208, 164738
	Carina	Car						*Charts 27, 28*	
	I 157		06 46.7 −54 42	6.6, 9.3	349	1.9″	127	G6III	49705, 234704
	Δ 39		07 03.3 −59 11	5.8, 6.8	86	1.4″	151	B9IV	53921, 234890
	h 3941		07 09.4 −60 23	7.3, 8.3	273	0.5″	220	G5III	55527, 249754
	h 3952		07 16.2 −54 03	7.4, 10.6	278	16″	163	K1III	56960, 235059
	Rmk 6		07 20.4 −52 19	6.0, 6.5	26	9.2″	31.7	F2IV	57852, 235110
	Hu 1583		07 41.3 −60 00	8.4, 8.7	241	1.1″	290	B9IV	62597, 235419
	CorO 58	A B	07 46.2 −59 49	8.2, 8.3	45	23″	38.9	G8V G8V	63581, 235503
		Solar type matched binary, bound or comoving with m.8 C (Shy 194, HD 62850) 47′ n.p., all three are pre main sequence (very young) stars. AC: ps = 81,000 AU. Neglected. (1999)							
	h 4014		07 48.6 −63 41	7.9, 9.1	156	11″	280	B9V	64186, 249972
	CorO 60		07 49.6 −55 05	7.5, 9.1	55	3.9″	280	A1V	64162, 235543

Label	Catalog ID	Components	Coordinates (J2000)	Mag.	θ	ρ	Dist.	Spectral type(s)	HD, SAO No.
	LDS 198	(AaAb) B	07 57.8 −60 18	5.6, 9.9	78	60″	16.2	G0V	65907, 250035
		Local, high CPM, solar type (2)+1 spectroscopic triple, visual double. AB: ps = 970 AU. (2000)							
	h 4031		07 58.4 −60 51	7.1, 7.7	357	5.5″	290	B6 + B8	66066, 250047
	MlbO 2		08 05.0 −60 23	7.7, 8.6	353	1.6″	330	A0	67515, 250102
★	Rmk 8		08 15.3 −62 55	5.3, 7.6	69	4.1″	78	A2V	69863, 250164
		Beautiful A type double; rich field, optical pair h 4077 (m.9, 19″) 5′ n.f. AB: ps = 430 AU. (2010)							
	Hu 1443	A (BC)	08 39.1 −55 57	7.8, 8.9	19	0.9″	186	F3V	74045, 236148
	h 4128		08 39.2 −60 19	6.8, 7.5	203	1.1″	137	A0V	74148, 250291
	h 4130		08 40.7 −57 33	6.5, 8.3	240	4.0″	67	A3V	74341, 236179
★	Rmk 9	A B	08 45.1 −58 43	6.9, 6.9	292	4.2″	260	B7III	75086, 236241
		Sparkling high mass double with B type giant; with possible small stellar group in rich field. AB: ps = 1,470 AU. (2010)							
	Slr 9		08 47.2 −63 49	8.0, 8.9	358	1.3″	220	A0V	75541, 250328
	h 4178		09 04.8 −57 51	6.5, 9.6	161	3.4″	310	A8III	78293, 236578
	h 4190		09 11.5 −57 58	6.6, 10.1	23	8.2″	300	B3IV	79421, 236707
	h 4206	(AB) C	09 17.4 −74 54	5.3, 9.6	345	7.0″	108	A1V	80951, 256599
	h 4213		09 25.5 −61 57	5.8, 9.6	330	8.8″	77	A4V	81830, 250575
	Syo 1		09 32.8 −57 06	7.1, 11.1	19	11″	2800	B6II	82919, 237090
	R 123		09 33.3 −57 58	7.5, 7.6	34	1.9″	670	B8III	82988, 237097
	Rss 208	A B C	09 36.1 −64 57	8.0, 11.7, 6.6	204, 68	12″, 7.9′	81	F5V	83359, 250629
		Solar type 2+1 triple; part of a comoving group of six stars (Shy 543, 546 & 548) within 3.5 pc. AC: ps = 51,500 AU. (2000)							
	h 4241		09 42.2 −66 55	6.3, 10.3	305	34″	160	A0V	84416, 250672
★ ups	Rmk 11		09 47.1 −65 04	3.0, 6.0	126	5.0″	440	A8Ib	85123, 250695
		Distant double with A type supergiant; CPM double h 4252 (m.8, 12″) 5′ s.f. in faint rich field. AB: ps = 2,970 AU. (2010)							
	Rmk 12		09 55.1 −69 11	6.9, 8.9	213	9.2″	180	B9V	86388, 250750
	Hrg 47		10 03.6 −61 53	6.3, 7.9	352	1.2″	500	B7IVne	87543, 250795
★	h 4295	A B C	10 09.5 −68 41	6.6, 6.5, 11.5	104, 40	0.7″, 26″	210	A0IV	88473, 250844
		Matched A type 2+1 triple; pretty field. AC: ps = 7,370 AU. (2000)							
	Slr 17	A B	10 13.0 −65 10	7.0, 9.9	343	3.4″	310	B5V	88894, 250872
	CapO 48	A B	10 15.1 −67 17	7.7, 9.1	341	1.9″	128	K1III	89205, 250894
	h 4306		10 19.1 −64 41	6.3, 6.5	313	2.6″	80	A1V	89715, 250917
	R 141		10 20.1 −67 10	7.5, 8.3	44	1.8″	380	B6V	89891, 250923
	Gli 293		10 20.7 −71 09	8.1, 8.6	208	28″	270	A0V	90037, 256702
	TDS 7279	A B C	10 27.1 −62 00	9.2, 11.5, 7.6	52, 316	0.5″, 64″	71	G5	307693, 250959
	R 151		10 31.1 −68 54	7.6, 9.7	190	2.9″	196	A0V	91423, 250999
	Δ 93	A B	10 34.9 −64 08	7.5, 8.4	39	24″	220	A0V	91906, 251022
	Δ 94		10 38.8 −59 11	4.9, 7.5	20	15″	720	K5III	92398, 238295
	R 152		10 38.9 −64 30	7.2, 8.9	20	2.1″	153	B9Vn	92467, 251043
	NZO 19		10 39.4 −67 10	7.7, 9.8	345	5.4″	240	G8III	92571, —
★	I 860	A B C	10 42.0 −63 30	8.0, 8.1, 11.5	73, 73	0.5″, 10″	123	A3IV	92896, 251075
		Exquisite, miniature, matched A type 2+1 triple, unresolved in small apertures. AB: ps = 83 AU. AC: ps = 1,660 AU. (2000)							
	Δ 97	A B C E	10 43.2 −61 10	6.6, 7.9, 8.1, 10.0	175, 17, 83	13″, 83″, 81″	570	B3III	93010, 251085
		Distant comoving or unstable bound group of massive B giant and A type stars. AC: ps = 47,300 AU. Neglected. (1998)							
	Δ 99	A [Ba Bb]	10 44.3 −70 52	6.3, [6.4, 9.0]	75, [48]	63″, [0.6″]	103	A5IV	93344, 256750
★ eta	h 4366	(AaAb...) B...G	10 45.1 −59 41	6.5, 11.1, 10.0	317, 67	1.6″, 38″	2300*	pec (~O3I)	93308, 238429
		Distant, high mass and unstable luminous blue variable (LBV) star associated with the eta Carinae star forming region and young star cluster of high mass stars (Trumpler 16), surrounded by ejecta from a nova eruption in 1837–1853 – "an old nova in a nebula with nuclei" (WDS notes) – with 12 or more possible components (not listed here) within a 5″ radius. A singular object. (2000)							
	R 161		10 49.4 −59 19	6.1, 7.4	292	1.0″	230	B9.5IV	93943, 238514

Label	Catalog ID	Components	Coordinates (J2000)	Mag.	θ	ρ	Dist.	Spectral type(s)	HD, SAO No.
	I 418		10 53.0 −63 05	8.0, 8.2	203	2.2″	350	A8IV	94466, 251162
	h 4383		10 53.7 −70 43	6.4, 7.1	288	1.5″	230	B6V B6V	94650, 256770
	h 4393		10 57.3 −69 02	6.6, 8.7	132	8.7″	270	B7V	95122, 251195
	R 164	A B	10 59.2 −61 19	6.5, 10.0	76	3.9″	118	B8IV	95324, 251205
	Gli 159		11 05.1 −59 43	7.7, 9.2	275	18″	980	B1Ib	96261, 238749
	BrsO 19	A [B C]	11 11.7 −60 27	7.8, [9.0, 9.4]	13, [27]	53″, [0.4″]	550*	B1	97400, 251307
	I 231		11 12.1 −71 13	7.1, 9.6	2	2.6″	240	B7Vn	97535, 256809
	R 163	A (BC)	11 17.5 −59 06	7.2, 7.6	58	1.6″	240	B9.5III	98278, 238939

Cassiopeia — Cas — Chart 2

Label	Catalog ID	Components	Coordinates (J2000)	Mag.	θ	ρ	Dist.	Spectral type(s)	HD, SAO No.
	OΣ 485	(AaAb) C	23 02.7 +55 14	6.5, 10.5	80	56″	260	B9III	217833, 35092
	OΣ 490	A B	23 10.2 +57 27	7.2, 9.2	297	1.2″	240	G3III	218803, 35193
	OΣ 495		23 24.1 +57 32	7.5, 7.8	123	0.4″	300	B2V	220562, 35386

High mass binary; sparse field. AB: P = 300 y, orbit r = 90; widening. Large aperture challenge. (2005)

Label	Catalog ID	Components	Coordinates (J2000)	Mag.	θ	ρ	Dist.	Spectral type(s)	HD, SAO No.
★ AR	Sh 355	(AB) (Ialb) F [C D]	23 30.0 +58 33	4.9, 9.9, 10.6, [7.2, 9.1]	207, 338, 269, [215]	3.9′, 67″, 76″, [1.4″]	190	B3IV	221253, 35478

High mass comoving (2)+(2)+1+2 group of three binaries with Algol type variable as primary star, and Σ 3022 (m.8, 21″) 10′ s.f. AC: ps = 14,400 AU; MSC system mass >21.5 M⊙, probable cluster remnant. (In WDS, compare with β 442 in Cygnus.) A unique and remarkable system. (2002)

Label	Catalog ID	Components	Coordinates (J2000)	Mag.	θ	ρ	Dist.	Spectral type(s)	HD, SAO No.
	OΣ 498	A B	23 31.3 +52 25	7.6, 10.9	244	17″	76	F6V	221377, 35501
	OΣ 499	A (BC)	23 33.2 +57 24	7.6, 9.5	75	9.9″	107	G5	221625, 35873
	Σ 3037	A B E	23 46.1 +60 28	7.4, 9.2, 9.7	211, 63	2.6″, 110″	360	K0	223070, 20832
V650	OΣ 507	A B	23 48.7 +64 53	6.8, 7.8	319	0.7″	169	A0	223358, 20866

A type binary with alp2 CVn type variable; faint field. AB: P = 566 y, orbit r = 170 AU. (2011)

Label	Catalog ID	Components	Coordinates (J2000)	Mag.	θ	ρ	Dist.	Spectral type(s)	HD, SAO No.
	Gui 40	A B	23 50.4 +51 37	6.5, 12.1	164	21″	40.6	F3V	223552, 35823
8 sig	Σ 3049	A B	23 59.0 +55 45	5.0, 7.2	326	3.1″	1340	B1V	224572, 35947
	Σ 3057		00 04.9 +58 32	6.7, 9.3	297	3.8″	840	B3V	225257, 21062
V640	Σ 3062	(AaAb) B	00 06.3 +58 26	6.4, 7.3	352	1.5″	21.5	G3V	123, 21085

Local solar type (2)+1 spectroscopic triple with eclipsing binary (P = 1.1 d); small group s.f. AB: P = 107 y, orbit r = 31 AU, e = 0.45, periastron 2049. (595 measures; 2013)

Label	Catalog ID	Components	Coordinates (J2000)	Mag.	θ	ρ	Dist.	Spectral type(s)	HD, SAO No.
	Ary 8	A B (CaCb)	00 10.6 +58 45	8.1, 8.6, 7.5	100, 43	39″, 105″	590*	B3IV	594, 21166
	Σ 7		00 11.6 +55 58	8.0, 8.5	211	1.3″	600	B8V	709, 21191
	OΣ 1		00 11.8 +66 08	7.5, 9.5	212	1.6″	260	A0	724, 11030
	Σ 10	A B	00 14.8 +62 50	8.0, 8.6	176	18″	170*	A2V	1026, 11062
	OΣ 9	A B C	00 26.2 +56 47	6.9, 9.7, 9.9	51, 3	2.1″, 22″	180	G3III	2170, 21395
14 lam	OΣ 12		00 31.8 +54 31	5.3, 5.6	210	0.2″	116	B8V B9V	2772, 21489

Matched high mass binary. AB: P = 246 y, orbit r = 31 AU, e = 0.69; will appear "rodlike" in large aperture as it approaches periastron in 2025. (270 measures; 2010)

Label	Catalog ID	Components	Coordinates (J2000)	Mag.	θ	ρ	Dist.	Spectral type(s)	HD, SAO No.
	Mri 31		00 34.6 +62 35	8.1, 9.0	9	117″	63		3068, 11270
	β 108	A B	00 34.6 +62 54	7.8, 10.6	6	4.5″	54	F5	3067, 11269
	Σ 48	A B	00 42.7 +71 22	7.8, 8.1	334	5.4″	280	A	3891, 4191
21 YZ	H N 122	(AaAb) B	00 45.7 +74 59	5.7, 10.6	160	36″	93	A2IV	4161, 4216
	Σ 59	A B	00 48.0 +51 27	7.2, 8.1	148	2.2″	210	B9.5IV	4536, 21716

High mass double; shares rich field with optical pair H 5 82 (m.7, 56″), 20′ s. (2012)

Label	Catalog ID	Components	Coordinates (J2000)	Mag.	θ	ρ	Dist.	Spectral type(s)	HD, SAO No.
★ 24 eta	Σ 60	A B	00 49.1 +57 49	3.5, 7.4	324	13″	5.95	G1V+M	4614, 21732

Achird. Local, very high CPM, solar type binary with RS CVn type variable; YW/O color, rich field. AB: system mass 1.6 M⊙, P = 480 y, orbit r = 70 AU, e = 0.50. (1,056 measures; 2013)

Label	Catalog ID	Components	Coordinates (J2000)	Mag.	θ	ρ	Dist.	Spectral type(s)	HD, SAO No.
	Σ 65		00 52.8 +68 52	8.0, 8.0	220	3.3″	260	A2	4947, 11440
	Σ 70	A B	00 53.8 +52 41	6.3, 9.5	248	8.0″	88	A0	5128, 21814

A type double; CD (m.11, 1.5″) 1′ s. appears unrelated. (2011)

Label	Catalog ID	Components	Coordinates (J2000)	Mag.	θ	ρ	Dist.	Spectral type(s)	HD, SAO No.
	A 2901		01 01.5 +69 22	7.1, 7.8	60	0.4″	270	B9	5839, 11526
		High mass binary. AB: *P* = 1,517 y, orbit *r* = 270 AU, increasing θ. (2006)							
	β 396		01 03.6 +61 04	6.1, 8.6	67	1.3″	610	F0II	6130, 11551
	OΣ 23	A B	01 10.1 +51 45	8.1, 8.6	191	14″	79	F8	6872, 22050
	β 235	Aa Ab	01 10.6 +51 01	7.5, 7.8	138	0.8″	105	F5V	6918, 22060
		Solar type binary; doubles p. (β 235 BC) and f. (OΣ 24) in rich field. Aa,Ab: *P* = 278 y, orbit *r* = 80 AU, closing. (2011)							
V761	β 258	A B	01 13.2 +61 42	6.5, 8.8	262	1.5″	200	B9V	7157, 11637
	Σ 115	A B	01 23.4 +58 09	7.1, 7.3	160	0.4″	59	F5V	8272, 22230
		Solar type binary; bright field. AB: system mass 2.4 M⊙, *P* = 222 y, orbit *r* = 49 AU, *e* = 0.93, widening to apastron 2096. (2012)							
	Σ 114		01 24.1 +72 51	7.2, 9.7	358	3.7″	570	A0	8226, 4393
	h 2045		01 29.0 +74 12	7.3, 12.6	83	27″	56	G5	8730, 4407
	h 1088	A B	01 42.3 +58 38	6.3, 9.8	169	20″	600	B7III Am	10293, 22520
V773	Arn 55	(AaAb) B D	01 44.3 +57 32	6.3, 8.7, 9.9	332, 45	0.6″, 2.7′	85	A3V	10543, 22566
		A type (2)+1+1 quadruple with eclipsing binary, visual triple; splendid field. MSC gives 6.1 M⊙. (AB): *P* = 193 y; AD: ps = 18,600 AU. (2010)							
	Σ 170		01 55.5 +76 13	7.5, 8.2	243	3.1″	158	A5	11316, 4512
	Σ 182	A B	01 56.4 +61 17	8.3, 8.4	124	3.6″	550*	B7V	11669, 12046
48	β 513	A B C	02 02.0 +70 54	4.7, 6.7, 13.2	308, 52	0.6″, 23″	35.3	A3V	12111, 4554
		A type 2+1 spectroscopic triple; MSC system mass 4.6 M⊙. AB: *P* = 61 y, orbit *r* = 22 AU, *e* = 0.36, periastron 2025. (215 measures; 2011)							
V779	Σ 185	(AaAb) B	02 02.2 +75 30	6.8, 8.6	9	1.1″	192	A0	12013, 4550
	Σ 191		02 03.2 +73 51	6.2, 9.1	195	5.1″	119	A5III	12173, 4559
		Double with rare A giant; Y/B color, delicate brightness contrast, rich field. Σ 184 (m.9, 17″) 11′ n.p. (2006)							
	OΣΣ 26	A B	02 19.7 +60 02	7.0, 7.3	202	63″	240	A2V	14172, 23194
		HIP 10855/56. Prominent CPM arcminute pair; in rich field n. of Double Cluster (χ/h Per). AB: ps = 15,100 AU. (2012)							
V559	Σ 257	(AaAb) B	02 25.7 +61 33	7.5, 8.2	71	0.4″	210	B8IV	14817, 12277
		High mass (2)+1 triple with Algol type eclipsing binary. MSC has 9.5 M⊙. Aa,Ab: *P* = 1.6 d. AB: *P* = 836 y, orbit *r* = 170 AU, widening. (2007)							
★ iot	Σ 262	(AaAb) B (CaCb)	02 29.1 +67 24	4.6, 6.9, 9.1	230, 115	2.9″, 7.1″	40.7	A5p F5V	15089, 12298
		Pretty, compact (2)+1+(2) quintuple, visual triple with alp2 CVn type variable; subtle color contrast among components. Connection to C components uncertain. MSC gives a system mass of 4.9 M⊙. Aa,Ab: *P* = 47 y. AB: *P* = 620 y, orbit *r* = 120 AU. (255 measures; 2013)							
	OΣΣ 28		02 39.0 +62 35	6.7, 7.6	148	68″	125	B9	16218, 12369
	Σ 302		02 50.0 +64 38	7.5, 10.3	170	5.1″	1510	B8II	17327, 12455
	Σ 306	A B	02 51.1 +60 25	7.4, 9.1	93	2.2″	880	O5Ve O7V	17505, 12470
		Distant, high mass spectroscopic binary; O/B color, rich field with cluster IC 1848. AB: ps = 2,600 AU. (2008)							
	Σ 312	A B C	02 56.2 +72 53	8.2, 8.9, 10.1	45, 131	1.8″, 43″	66	G0	17785, 4776
	Σ 317		02 58.1 +69 12	7.9, 9.8	84	4.1″	80	F2	18056, 12530
	OΣΣ 31		03 00.9 +59 40	7.3, 8.0	230	74″	240	B9	18473, 23761
	Σ 349		03 10.8 +63 47	7.9, 8.6	323	5.9″	84	F5V	19440, 12635

Centaurus   Cen   *Charts 23, 28*

Label	Catalog ID	Components	Coordinates (J2000)	Mag.	θ	ρ	Dist.	Spectral type(s)	HD, SAO No.
V815	h 4409		11 07.3 −42 38	5.2, 7.7	252	1.3″	84	A	96616, 222581
★	R 165		11 13.1 −47 03	7.5, 7.5	253	2.8″	89	G0IV	97547, 222645
		Solar type matched double; s.f. of 2 m.8 stars in mixed field, from the H.C. Russell catalog. AB: ps = 335 AU. (2000)							
	h 4423		11 16.5 −45 53	7.0, 7.3	278	2.6″	52	F3V	98096, 222687
		Solar type matched double; YW/YW color, ethereal faint field, m.7 star 8′ n.p. AB: ps = 180 AU. (2000)							
	h 4426		11 21.6 −43 33	7.2, 10.3	173	13″	430	K0III	98779, 222738

Label	Catalog ID	Components	Coordinates (J2000)	Mag.	θ	ρ	Dist.	Spectral type(s)	HD, SAO No.
	Skf 2009		11 22.3 −58 23	6.7, 9.6	328	54″	152	G8III+F8	98897, 239003
★	BrsO 5		11 24.7 −61 39	7.7, 8.8	242	4.4″	13.7	K5V+M0V	99279, 251393
		Local, high CPM, rare low mass binary; rich field. AB: P = 399 y, orbit r = 80 AU, e = 0.67, widening. Neglected. (1996)							
	Skf 1217		11 25.0 −35 55	8.1, 9.0	346	51″	380	K0II+K0	99241, 202424
	h 4438		11 27.5 −39 53	7.3, 10.7	197	23″	137	A0V	99627, 202462
★	I 78		11 33.6 −40 35	6.1, 6.2	99	0.7″	118	A2IV	100493, 222863
		A type matched double; pretty field, K giant HR 4447 14′ n.p. AB: ps = 110 AU. Neglected. (1998)							
	Rss 267		11 36.5 −61 40	7.1, 9.0	315	73″	3630*	B1Ib	100943, 251480
	Gli 165		11 36.6 −60 54	8.1, 9.1	4	1.9″	260	B8V	100942, 251482
V871	I 422	(AaAbAc) B C	11 38.3 −63 22	7.1, 7.4, 9.9	116, 6	0.4″, 1.7″	2400*	O7N	101205, 251511
	h 4460		11 39.2 −57 44	7.2, 8.2	176	8.6″	510	A0V	101312, 239261
	Howe 70		11 39.5 −37 26	8.2, 8.4	107	3.4″	104	F5V	101327, 202675
	CapO 11	(AaAb) B	11 40.6 −62 34	6.9, 7.4	219	2.6″	—	O9.5	101545, 251533
	Hd 212		11 41.2 −61 08	7.5, 7.9	322	1.0″	200	K1III	101629, 251540
	Gli 169		11 51.8 −64 36	7.5, 9.0	226	4.3″	800	B4V	103066, 251616
★	Hld 114		11 55.0 −56 06	7.4, 7.8	172	3.3″	30.2	G3IV	103493, 239487
		Solar type double; splendid rich field. AB: ps = 135 AU. (2000)							
	BrsO 20		11 55.0 −62 35	7.7, 8.9	266	19″	640*	M3III	103515, 251641
	I 80		11 55.4 −41 54	8.0, 8.2	94	1.3″	116	A5V	103567, 223115
	I 895		12 01.7 −41 50	8.1, 8.2	318	0.8″	134	A3III	104472, 223175
89	λ 143		12 03.6 −39 01	7.1, 7.7	28	0.5″	46.0	F7V	104747, 203084
		Solar type high CPM binary. System mass 2.4 M$_\odot$. AB: P = 111 y, orbit r = 31 AU, e = 0.58, periastron 2024. (2013)							
	h 4491		12 03.8 −44 07	8.1, 8.8	42	23″	55	G2III	104760, 223195
	h 4492		12 03.8 −54 43	7.5, 11.2	273	16″	81	F3V	104764, 239624
	h 4500		12 06.6 −37 52	6.7, 9.1	31	50″	151	K1III	105173, 203137
	I 423		12 11.0 −45 25	6.8, 10.6	166	2.7″	400	K0III	105852, 223266
	R 192	A B	12 11.2 −52 13	7.9, 9.7	100	3.2″	124	A6V	105874, 239730
★ D	Rmk 14	(AaAb) B	12 14.0 −45 43	5.8, 7.0	243	2.8″	175	K3III	106321, 223297
		Solar type (2)+1 spectroscopic triple, visual double with K giant; O/Y color, faint field. AB: ps = 660 AU. Neglected. (1995)							
	Slr 10		12 15.0 −36 13	7.8, 9.8	244	2.0″	150	A3IV	106488, 203275
	h 4518		12 24.7 −41 23	6.5, 8.4	208	10″	190	K3III	107998, 223417
	Rss 16		12 30.2 −53 36	8.1, 8.6	152	50″	370*	A0IV	108771, 240002
gam	h 4539	(AB) C D	12 41.5 −48 58	(2.4, 2.9), 14.4, 3.9	114, 304	58″, 45′	39.9	A1IV	110304, 223603
		Muhlifain. A type (2)+1+1 quadruple, visual triple. AB: system mass 6.3 M$_\odot$, P = 83 y, orbit r = 34 AU, e = 0.81, apastron 2056; B reappears n.f. in 2021. AD (tau Cen): ps = 108,500 AU. (2013)							
	h 4546		12 44.9 −52 45	7.9, 9.5	222	15″	88	F0V	110770, 240206
	I 83		12 56.7 −47 41	7.4, 7.7	234	0.8″	81	F5V	112361, 223774
		Solar type binary, sparse field, from the R.T.A. Innes 1927 catalog. AB: P = 191 y, orbit r = 42 AU, near apastron. (2010)							
★	CorO 143	A B	12 58.2 −54 11	7.4, 8.7	112	17″	560	A0V	112532, 240424
		Distant A type double; field of faint doubles and tiny stellar groups. AB: ps = 9,520 AU. Neglected. (1999)							
	CapO 13		13 00.3 −48 36	7.2, 9.2	68	5.1″	150	G8IV	112851, 223827
	Δ 164		13 01.8 −30 50	7.6, 11.4	291	6.1″	320	K4III	113129, 204050
	I 917		13 06.6 −46 02	8.1, 8.4	281	1.3″	123	F3V	113766, 223904
	R 213		13 07.4 −59 52	6.6, 7.0	21	0.7″	170	G2	113823, 240566
	Skf 1128		13 09.3 −41 01	7.7, 10.8	312	55″	170	K0III	114164, 223935
	CorO 149		13 09.3 −51 41	7.7, 10.3	257	13″	110	A0V	114139, 240601
V831	I 424	(AaAbB) C	13 12.3 −59 55	4.6, 8.4	7	1.9″	116	B8V	114529, 240645
		High mass (2+1)+1 quadruple with bet Lyr type variable; faint field, CorO 152 8′ n.f. MSC system mass 12.1 M$_\odot$. (AB): P = 27 y, closing. (2013)							
	LDS 436		13 12.5 −34 45	7.8, 9.9	221	31″	77	F5V	114692, 204235

Label	Catalog ID	Components	Coordinates (J2000)	Mag.	θ	ρ	Dist.	Spectral type(s)	HD, SAO No.
	Mug 3		13 12.7 −31 52	5.6, 10.1	331	8.5″	36.1	G0V M4	114729, 204237
	CorO 152	(Aa1Aa2Ab) B	13 12.9 −59 49	6.3, 9.4	146	25″	42.2	G0V	114630, 240653
			Solar type (2+1)+1 spectroscopic quadruple, visual double; stars n.f. unrelated. MSC 3.3 M$_\odot$. Aa1,Aa2: P = 4.2 d. Aa,Ab: P = 32 y, orbit r = 13 AU, widening. AB: ps = 1,420 AU. (2000)						
	MlbO 3	(AaAb) (BaBb)	13 14.7 −63 35	7.0, 9.1	39	1.7″	470	O9II	114886, 252220
			Distant, high mass (2)+(2) spectroscopic quadruple, visual binary with O supergiant (C, E, F below limit magnitude); rich field. MSC system mass 40.1 M$_\odot$. AB: ps = 1,080 AU. (2014)						
	h 4576		13 16.1 −57 04	7.1, 10.0	127	5.6″	140	F2IV	115113, 240693
	I 233	A B	13 16.8 −41 17	7.3, 9.7	109	3.5″	189	K0II	115279, 224025
	h 4578		13 17.6 −37 01	7.6, 10.8	148	9.3″	169	A3IV	115431, 204325
	Rst 632		13 19.6 −63 45	7.7, 10.6	302	4.9″	64	F5V	115602, 252261
	h 4579		13 21.4 −64 03	7.9, 8.6	98	5.1″	56	G0	115863, 252277
	Skf 900		13 22.0 −43 03	7.2, 10.0	108	46″	151	A3IV F8	116081, 224089
			A type CPM double; ps = 9,400 AU. Tiny, m.11 trapezium 6′ f. (2000)						
★ J	Δ 133	(AB) C	13 22.6 −60 59	4.5, 6.2	345	61″	109	B2.5Vn	116087, 252284
			High mass (2)+1 triple, visual double; faint rich field. B is V790 Cen, a Cepheid variable. (2010)						
	Slr 18	A B C	13 22.9 −47 57	6.7, 7.2, 10.8	243, 62	0.7″, 37″	300	A4V	116197, 224095
	R 218		13 28.1 −43 46	7.5, 9.7	169	2.4″	260	K1III	116991, 224157
	λ 180	A B	13 31.4 −42 28	6.8, 9.2	234	3.6″	145	K0III	117483, 224214
	Skf 1958		13 31.6 −32 08	7.5, 10.5	165	113″	162	F2V	117548, 204552
	Hrg 86		13 32.5 −62 21	8.1, 8.6	239	1.5″	310	B9IV	117540, 252356
	λ 184		13 37.8 −35 04	7.5, 9.6	302	2.8″	34.0	G3V	118465, 204673
★	R 223		13 38.1 −58 25	6.6, 9.9	13	2.6″	158	K1III	118384, 241013
			Double with K type giant; rich field, s. of 3 m.7 stars, small group n. AB: ps = 550 AU. Neglected. (1991)						
	h 4600		13 39.2 −49 00	7.8, 9.3	117	16″	200	K2III	118651, 224311
★ Q	Δ 141		13 41.7 −54 34	5.2, 6.5	163	5.4″	83	B8Vn	118991, 241076
			High mass double; faint rich field, double HDS 1926 (m.9, 11″) 3′ s.f. AB: ps = 610 AU. (2010)						
	h 4608		13 42.3 −33 59	7.4, 7.5	9	4.3″	76	F5	119191, 204749
	Howe 95	A B	13 43.8 −40 11	7.5, 7.9	185	1.0″	152	F2V	119415, 204777
	Δ 142		13 44.0 −59 14	6.5, 7.6	90	33″	260	B8V	119283, 241114
	Howe 94		13 48.9 −35 42	6.6, 10.2	359	11″	29.5	G3IV	120237, 204867
			High CPM, solar type double; mixed field. AB: ps = 440 AU. (2013)						
	CapO 61	A B	13 51.5 −48 18	7.4, 7.4	130	30″	75	G6III	120592, 224485
N	Rmk 18		13 52.1 −52 49	5.2, 7.5	288	18″	80	B8V	120642, 241239
★ 4 h	H N 51	(AaAb) (BaBb)	13 53.2 −31 56	4.7, 8.5	185	15″	195	B4IV	120955, 204944
			High mass, spectroscopic (2)+(2) quadruple; sparse bright field. MSC system mass 12.3 M$_\odot$. AB: ps = 3,950 AU. (2013)						
y	Howe 28	A B	13 53.5 −35 40	6.3, 6.4	313	1.0″	52	F4V	120987, 204955
			Solar type binary. AB: P = 292 y, orbit r = 70 AU; slowly increasing θ. A "Dawes" (matched m.6) resolution test. (2011)						
	BrsO 9		13 54.6 −50 42	8.1, 8.3	76	18″	450	F2V F4V	121092, 241280
	Howe 74		13 55.3 −32 06	7.2, 9.8	117	5.8″	175	G5	121287, 204988
	R 227	A B	13 56.3 −54 08	6.5, 7.5	10	1.9″	140	A1V	121336, 241309
	β 1197		14 03.0 −31 41	6.5, 7.8	221	2.4″	38.2	F8V	122510, 205123
	WFC 145		14 06.0 −41 11	4.3, 8.5	78	85″	156	B2V	122980, 224673
	h 4642	A B [D E]	14 07.1 −63 27	6.9, 12.6, [10.6, 12.4]	12, 333, [10]	9.7″, 26″, [0.6″]	410	K4III	122938, 252599
	Slr 19		14 07.7 −49 52	7.1, 7.4	323	1.2″	46.5	G3V	123227, 224702
	Pol 3		14 11.0 −46 55	8.2, 8.8	53	3.8″	152	F3IV	123794, 224745
	CorO 167	(AaAb) (BaBb) C	14 15.0 −61 42	6.9, 8.7, 13.1	157, 42	2.8″, 2.5″	—	O8nke	124314, 252665
	BrsO 10	(AB) H	14 16.5 −57 18	7.2, 10.0	115	30″	260	B9IV	124620, 241583

Label	Catalog ID	Components	Coordinates (J2000)	Mag.	θ	ρ	Dist.	Spectral type(s)	HD, SAO No.
★	Pol 9	(AB) [C D]	14 20.8 −42 25	7.2, [8.9, 9.0]	211, [202]	78″, [1.7″]	132	A1V	125494, 224851
		A type (2)+2 quadruple, visual triple; the two binaries are joined by parallax. AC: ps = 13,900 AU. (2013)							
★	Δ 159	(AaAb) B	14 22.6 −58 28	5.0, 7.6	157	9.1″	118	G8III F5V	125628, 241673
		Solar type (2)+1 spectroscopic triple, visual double with G type giant; brightest of three stars in very rich field. AB: ps = 1450 AU. (2010)							
	β 1112		14 33.2 −30 43	6.2, 9.9	13	2.7″	101	K0III	127624, 205681
★ alp	Rhd 1	(AaAb) B (CaCb)	14 39.7 −60 50	0.0, 1.3, 11.1	263, 225	5.1″, 185′	1.35	G2V K1V	128620, 252838
		Rigil Kent. Local, very high CPM, solar type (2)+1+(2) spectroscopic quintuple, visual triple; striking Y/O color and the closest double system to the Sun. "Beyond comparison the finest double star in the sky" (J. Herschel). Discovered by J. Richaud (1689), first system to be visually measured (by L. Feuille, 1709). AB: system mass 2.0 M$_\odot$, P = 80 y, orbit r = 24 AU, e = 0.52, periastron 2035 (446 measures); the C binary is faint Proxima Centauri (HIP 70890, m.11, 0.3″), an M5V flare star (V645 Cen) discovered by R.T.A. Innes (1905) at θ = 212°, ρ =150′, ps = 16,400 AU. (2013). See Figure 1.							
	Skf 1973		14 41.0 −36 08	5.6, 9.4	148	83″	190	A0+F0V	128974, 205823
	β 414		14 41.9 −30 56	6.8, 7.6	347	1.0″	350	A	129161, 205841
	h 4702	(AB) C	14 48.5 −35 50	6.9, 9.4	216	9.8″	187	K0III	130311, 205964
	I 226	A B	14 54.4 −34 09	7.1, 11.2	222	3.2″	98	A1V	131399, 206071
	I 84	A B	14 54.9 −36 26	7.3, 11.1	260	4.7″	160	A1V	131461, 206083
	I 227	A B	14 56.5 −34 38	8.1, 8.4	106	0.4″	61	F8V	131751, 206111
		Solar type binary. AB: P = 40 y, orbit r = 16 AU, widening. Large aperture challenge. (2010)							
	h 4718		14 57.6 −35 23	7.4, 8.7	63	1.8″	280	M1III	131921, 206128
	h 4722	A B	14 59.5 −30 43	7.1, 9.3	337	8.5″	102	A3m F0	132347, 206173

Cepheus   Cep   *Charts 1, 2*

Label	Catalog ID	Components	Coordinates (J2000)	Mag.	θ	ρ	Dist.	Spectral type(s)	HD, SAO No.
★ 1 kap	Σ 2675	A B	20 08.9 +77 43	4.4, 8.3	120	7.2″	97	B9III	192907, 9665
		High mass double with B type giant; brighter of two stars in sparse field. AB: ps = 940 AU. (2003)							
	β 152		20 42.3 +57 23	7.2, 8.8	83	1.1″	152	A4IV	197618, 32812
	Σ 2751		21 02.2 +56 40	6.2, 6.9	357	1.6″	280	B8III	200614, 33078
	Σ 2780	(AaAb) B C E	21 11.8 +59 59	6.1, 6.8, 8.9, 10.3	213, 211, 45	1.0″, 121″, 62″	1315*	B0II	202214, 33210
		Distant, high mass quintuple with B type supergiant; in small stellar group. Aa,Ab: P = 57 y, e = 0.79, apastron 2029. (2011)							
	H I 48		21 13.7 +64 24	7.2, 7.3	243	0.5″	42.8	G2IV G2IV	202582, 19257
		Matched solar type binary; good subarcsecond resolution test. System mass 2.9 M$_\odot$. AB: P = 82 y, orbit r = 30 AU, e = 0.81, apastron 2044. (2012)							
	Σ 2783		21 14.1 +58 18	7.7, 8.1	352	0.7″	128	A3V	202519, 33251
		A type binary; several stars, incl. two doubles, 5′ f. (part of Trumpler 37). AB: P = 1,760 y, orbit r = 215 AU. (2011)							
	Bvd 135		21 15.7 +68 21	7.9, 9.3	310	25″	75	F3V F9V	202986, 19281
	Σ 2801		21 18.5 +80 21	7.9, 8.6	270	2.1″	82	F6V	204129, 3547
	Σ 2790	(AaAb) B	21 19.3 +58 37	5.9, 9.3	45	4.6″	590	M1I+B2	203338, 33318
	Pop 1233	A B C	21 22.3 +57 34	8.2, 10.7, 8.7	16, 192	1.2″, 83″	96	G5	203802, 33367
8 bet	Σ 2806	(AaAb) B	21 28.7 +70 34	3.2, 8.6	248	14″	210	B1IV	205021, 10057
		Alfirk. High mass, B type (2)+1 triple, visual double. MSC has system mass 19.1 M$_\odot$. Aa,Ab: P = 83 y, e = 0.73, apastron 2039. (2013)							
★	Σ 2816	(AaAb) B C D	21 39.0 +57 29	5.7, 13.3, 7.5, 7.5	317, 120, 338	1.8″, 12″, 20″	610	O6	206267, 33626
		Distant, high mass (2)+1+1+1 quintuple, difficult visual quadruple; lovely field with Σ 2819 (below) n.f., both in IC 1396. (2001)							
	Σ 2819		21 40.4 +57 35	7.4, 8.6	58	13″	310	F5V	206482, 33652
		Solar type double; rich field. Σ 2816 14′ s.p. (in IC 1396). (2011)							
	Σ 2836	(AaAb) B	21 49.1 +66 48	6.5, 10.4	156	12″	98	F4V	207826, 19665

Label	Catalog ID	Components	Coordinates (J2000)	Mag.	θ	ρ	Dist.	Spectral type(s)	HD, SAO No.
	Σ 2843	A B	21 51.6 +65 45	7.0, 7.3	150	1.3″	88	A1m	208132, 19686
	Σ 2845	A B	21 52.3 +63 06	8.1, 8.2	173	2.0″	750	B2V	208185, 19694
	OΣΣ 226	A B	21 53.1 +68 06	7.5, 8.9	245	76″	420	G8II	208411, 19712
EM	S 800	(AaAb) B	21 53.8 +62 37	7.1, 7.9	145	62″	490	B0.5V B1III	208392, 19718
	OΣ 457		21 55.5 +65 19	6.0, 8.2	245	1.3″	310	B2.5Ve	208682, 19742
	OΣ 458	A B	21 56.5 +59 48	7.2, 8.4	348	1.0″	350	A0V	208744, 33894
★	Σ 2873	(AaAb) (BaBb)	21 58.2 +82 52	7.0, 7.5	66	14″	29.9	F6IV	209942, 3673

Solar type (2)+(2) quadruple, visual double with eclipsing binary. MSC gives 3.8 M$_\odot$. Aa,Ab: P = 23.0 y, e = 0.4, periastron 2016. Ba,Bb (V376): P = 1.2 d. AB: ps = 565 AU. (2012)

Label	Catalog ID	Components	Coordinates (J2000)	Mag.	θ	ρ	Dist.	Spectral type(s)	HD, SAO No.
★ 17 xi	Σ 2863	(AaAb) B	22 03.8 +64 38	4.5, 6.4	268	8.1″	29.6	A3m	209790, 19827

Al Kurah. A type (2)+1 astrometric and spectroscopic triple, visual binary. Aa,Ab: P = 2.2 y, e = 0.48. (272 measures; 2013)

Label	Catalog ID	Components	Coordinates (J2000)	Mag.	θ	ρ	Dist.	Spectral type(s)	HD, SAO No.
★	Sti 2618		22 05.7 +57 08	7.6, 13.0	78	9.7″	107	F5	209991, 34046

Solar type double; optical pair Bar 57 (m.10, 47″) 5′ n. in rich field. AB: ps = 1,400 AU. (2009)

Label	Catalog ID	Components	Coordinates (J2000)	Mag.	θ	ρ	Dist.	Spectral type(s)	HD, SAO No.
	Ary 45		22 08.3 +69 59	7.9, 8.1	207	67″	240	B9.5V	210550, 19893
	Σ 2872	A [B C]	22 08.6 +59 17	7.1, [8.0, 8.0]	316, [297]	22″, [0.8″]	190*	B9.5V	210433, 34101
	Σ 2883		22 10.6 +70 08	5.6, 8.6	252	15″	33.6	F2V	210884, 19922
	β 436	A B	22 10.9 +57 56	7.4, 12.0	327	19″	105	A5	210760, 34141
	Σ 2879	A B	22 11.0 +63 24	8.0, 8.3	234	0.8″	940	B5	210808, 19921
	Σ 2893		22 12.9 +73 18	6.2, 7.9	348	29″	210	K0II+A3V	211300, 10284
	Σ 2896		22 18.5 +63 13	7.8, 8.6	241	22″	2400*	B0.5V	211880, 19993
	OΣ 470		22 21.0 +66 58	7.4, 9.8	352	4.3″	105	A7V	212278, 20023
	Σ 2903		22 21.8 +66 42	7.1, 7.8	95	4.1″	270	A7V G0III	212391, 20034
★ DO	Kr 60	A B	22 28.0 +57 42	9.9, 11.4	329	1.5″	4.00	M4 M5	239960, 34788

Local, high CPM, low mass binary; the preceding corner of faint stellar triangle. AB: P = 45 y, orbit r = 10 AU, e = 0.41, apastron 2037. One of the nearest binary stars, with faint absolute magnitudes (M1=11.9, M2=13.4) and tiny masses (A: 0.22 M$_\odot$, flare star B: 0.18 M$_\odot$), but with a close orbit that can be resolved in small apertures. All other "components" in WDS are field stars. Extremely rare type of system. (286 measures; 2013)

Label	Catalog ID	Components	Coordinates (J2000)	Mag.	θ	ρ	Dist.	Spectral type(s)	HD, SAO No.
★ 27 del	Σ I 58	A (CaCb)	22 29.2 +58 25	4.2, 6.1	191	41″	270	F5I B7V	213306, 34508

High mass 1+(2) spectroscopic triple, visual double; the eponymous Cepheid (solar type, supergiant and slowly pulsating) variable star (P = 5.4 d). Optical pair H IV 31 (m.8, 25″) 6′ p. Estimated system mass 9.2 M$_\odot$. AC: ps = 14,900 AU. (2012)

Label	Catalog ID	Components	Coordinates (J2000)	Mag.	θ	ρ	Dist.	Spectral type(s)	HD, SAO No.
	Σ 2923	A B	22 33.3 +70 22	6.3, 9.2	47	9.6″	126	A0V	214019, 10418
	OΣΣ 236	(AaAbB) D	22 36.1 +72 53	7.6, 8.4	137	42″	84	F5	214511, 10429

Solar type (2+1)+1 quadruple. MSC gives 4.4 M$_\odot$. AB (β 1092): P = 51 y, e = 0.66, apastron 2027. AD: ps = 4,760 AU. (2008)

Label	Catalog ID	Components	Coordinates (J2000)	Mag.	θ	ρ	Dist.	Spectral type(s)	HD, SAO No.
	OΣ 481		22 43.8 +78 31	7.5, 9.5	276	2.3″	390	A0II	215730, 10613
	OΣ 480		22 46.1 +58 04	7.7, 8.6	117	31″	210	F8	215714, 34785

Solar type CPM double; in rich field, near NGC 7380. AB: ps = 8,780 AU. (2010)

Label	Catalog ID	Components	Coordinates (J2000)	Mag.	θ	ρ	Dist.	Spectral type(s)	HD, SAO No.
	OΣ 482	A B	22 47.5 +83 09	4.9, 9.6	38	3.5″	96	K3III	216446, 3794
	Σ 2948		22 49.6 +66 33	7.3, 8.6	4	2.7″	260	B6Vn	216227, 20267
	Σ 2950	A B C	22 51.4 +61 42	6.0, 7.1, 11.1	276, 355	1.2″, 39″	73	G8III	216380, 20281
	OΣΣ 238	(AaAb) B	22 52.7 +67 59	7.0, 7.6	280	68″	59	F2	216606, 20295
V453	Σ 2953	(AaAb) B	22 52.8 +60 55	7.6, 9.5	137	8.3″	250	A0V G0III	216572, 20292
	Σ 2963		22 54.3 +76 20	8.0, 8.5	2	1.9″	123	A3	216886, 10541
	OΣ 484	(AB) C	22 56.2 +72 50	7.6, 10.4	255	31″	179	A2	217085, 10560

High mass, A type (2)+1 triple; easily found in sparse field. AB: P = 143 y, orbit r = 65 AU implies 13.4 M$_\odot$. AC: ps = 7,490 AU. (2007)

Label	Catalog ID	Components	Coordinates (J2000)	Mag.	θ	ρ	Dist.	Spectral type(s)	HD, SAO No.
	Σ 2971		22 56.7 +78 30	8.0, 8.9	4	5.4″	119	G5	217294, 10562
	Σ 2977	A B	23 06.5 +61 26	6.8, 10.3	354	1.9″	113	F5V	218375, 20424
	Σ 2984	A B	23 07.4 +70 40	7.6, 9.8	294	4.4″	290	G8III	218535, 10623
33 pi	OΣ 489	(AaAb) B C	23 07.9 +75 23	4.6, 6.8, 12.2	355, 244	1.1″, 58″	76	G2III	218658, 10629

High mass solar type (2)+1+1 astrometric and spectroscopic quadruple; faint field. MSC has 8.8 M$_\odot$. Aa,Ab: P = 1.5 y. AB: P = 163 y, orbit r = 60 AU, apastron 2016. (2010)

Label	Catalog ID	Components	Coordinates (J2000)	Mag.	θ	ρ	Dist.	Spectral type(s)	HD, SAO No.
	OΣ 492		23 08.9 +82 35	7.7, 11.5	229	8.9″	190	K2	219014, 3851
★ 34 omi	Σ 3001	A B C	23 18.6 +68 07	5.0, 7.3, 12.9	221, 356	4.4″, 44″	62	K0III	219916, 20554
		2+1 triple with K giant; pretty Y/B color. MSC gives 3.8 M$_\odot$. AB: P = 1,505 y, orbit r = 195 AU, widening. (226 measures; 2012)							
	Σ 3017	A B	23 27.7 +74 07	7.6, 8.5	21	1.3″	106	F1V	221071, 10742
	Σ 3051		00 02.8 +80 17	7.7, 9.5	24	17″	149	F2V	225020, 2
	Σ 2		00 09.3 +79 43	6.7, 6.9	17	0.9″	105	A4IV	431, 4048
		A type matched binary; faint field. System mass 3.6 M$_\odot$. AB: P = 540 y, orbit r = 105 AU, e = 0.72, widening. (217 measures; 2012)							
	Σ 13		00 16.2 +76 57	7.0, 7.1	51	0.9″	192	B8V	1141, 4071
		High mass, matched binary; unidentified m.11 double 7′ n.f. Low quality orbit AB: P = 971 y, orbit r = 190 AU. (207 measures; 2006)							
	Σ 26	A B C	00 21.4 +67 00	7.5, 8.8, 9.9	156, 114	0.7″, 13″	166	B8.5V	1658, 11128
		Solar type 2+1 triple; sparse field. MSC system mass 8.8 M$_\odot$. AB: P = 335 y, orbit r = 65 AU, widening. (2009)							
U	Knott 1	(AaAb) B	01 02.3 +81 53	6.9, 11.8	62	14″	250	B7V G8III	5679, 168
	OΣ 28	A B	01 19.1 +80 52	7.6, 8.8	296	0.9″	156	F3V	7471, 218
	OΣ 34		01 49.9 +80 53	7.6, 8.1	285	0.5″	199	A0V	10648, 291
		A type binary; dark field with three faint binaries. Preliminary orbit AB: system mass 9.5 M$_\odot$, P = 196 y, orbit r = 140 AU (?), closing. (2007)							
	OΣ 37		02 10.5 +81 29	7.0, 9.2	206	1.2″	168	A3	12543, 345
	S 405	(AaAb) B	02 12.8 +79 41	6.5, 7.2	277	55″	141	A5III	12927, 4594
	Σ 320		03 06.1 +79 25	5.7, 9.2	231	4.7″	210	M2.5III	18438, 4810
	Σ 460		04 10.0 +80 42	5.6, 6.3	143	0.8″	116	G8III A4V	25007, 650
		High mass binary with G giant and A type companion. AB: system mass 6.3 M$_\odot$, P = 372 y, orbit r = 100 AU, increasing in $θ$. (2012)							

Cetus Cet *Charts 12, 13, 20*

Label	Catalog ID	Components	Coordinates (J2000)	Mag.	θ	ρ	Dist.	Spectral type(s)	HD, SAO No.
	β 393		00 18.3 −21 08	6.9, 8.4	29	0.7″	182	B9.5IV	1431, 166174
	h 1957	A (BC)	00 21.9 −23 00	7.7, 9.2	26	6.1″	59	G0	1766, 166213
	Shy 384	A (BaBb)	00 29.3 −05 55	7.8, 9.5	51	14.0′	57	G0 G5	2567, 128781
	h 1981	A [B C]	00 31.0 −10 05	6.9, [9.7, 9.1]	88, [318]	79″, [0.4″]	149	A5IV	2760, 147317
	Σ 39	(AB) (CaCb)	00 34.5 −04 33	7.1, 8.7	44	17″	77	G0IV	3125, 128831
		Solar type (2)+(2) quadruple, visual binary; dark field. MSC gives 4.3 M$_\odot$. AB: P = 2,980 y, orbit r = 210 AU. (2011)							
★	β 395		00 37.3 −24 46	6.6, 6.2	101	0.6″	15.4	G8V	3443, 166418
		Local, very high CPM, solar type spectroscopic binary; isolated in dark field. AB: P = 25 y, orbit r = 10 AU (among the smallest orbits resolved in amateur instruments), e = 0.22, periastron in 2023, closing. (2011)							
	Mlf 1		00 45.7 −16 25	6.5, 10.4	196	2.9″	89	F2V	4338, 147436
	Stone 3	A (BD)	00 52.2 −22 37	7.6, 8.4	243	2.0″	53	F8V	5059, 166640
	WNO 1		00 53.2 −24 47	6.6, 9.2	7	5.0″	74	F6IV	5156, 166651
	h 2004		00 57.6 −19 00	7.0, 7.2	238	3.5″	142	A2Vn	5617, 147537
	S 390		00 58.2 −15 41	7.8, 7.9	216	6.4″	59	dF6 dF7	5659, 147543
	Σ 81		01 00.1 −02 01	7.3, 10.1	67	18″	90	F5	5861, 129072
	Σ 91		01 07.2 −01 44	7.4, 8.6	314	4.3″	48.7	F9V	6651, 129128
37	Σ I 3	(AaAb) B	01 14.4 −07 55	5.2, 7.9	330	51″	23.4	F6V G9V	7439, 129193
		Local, high CPM solar type 2+1 triple, visual double; YW/O color. MSC system mass 2.4 M$_\odot$. AB: ps = 1,610 AU. (2012)							
★ 42	Σ 113	A (BC)	01 19.8 −00 31	6.5, 7.0	22	1.7″	101	A2V G4III	8036, 129235
		A type 1+(2) triple with G giant. BC: system mass 2.6 M$_\odot$, P = 27 y, e = 0.22, periastron 2022. (339 obs.) A,BC: ps = 230 AU. (2013)							

Label	Catalog ID	Components	Coordinates (J2000)	Mag.	θ	ρ	Dist.	Spectral type(s)	HD, SAO No.
	h 2036		01 20.0 −15 49	7.4, 7.6	339	2.3″	46.2	G0IV	8071, 147741
	h 2043		01 22.5 −19 05	6.5, 8.7	72	5.0″	62	F6IV	8350, 147767
	Se 1		01 23.6 −24 21	6.8, 9.1	89	3.1″	81	A8III	8487, 167011
	β 1163		01 24.3 −06 55	6.6, 7.0	217	0.4″	46.6	F3V	8556, 129277
			Solar type spectroscopic binary. AB: P = 16 y, orbit r = 9 AU, e = 0.93, now disappearing to periastron 2021, it returns to view 2027. (2011)						
	Σ 120		01 25.0 −05 57	6.8, 10.0	278	7.4″	112	A0	8627, 129283
	β 399	A B	01 27.8 −10 54	6.4, 8.8	303	1.5″	118	K4III	8921, 147819
			A (typically difficult) Burnham double with K giant. AB: ps = 240 AU. Neglected. (1991)						
	h 3437		01 28.1 −17 16	7.4, 9.4	247	12″	57	F2III	8957, 147822
	Σ 150		01 43.4 −07 05	7.7, 8.2	196	36″	140*	A	10606, 129472
	β 6	A B	01 44.7 −06 46	6.7, 8.9	163	2.3″	189	G5III	10725, 129482
53 chi	Eng 8		01 49.6 −10 41	4.7, 6.8	250	3.2′	23.2	F3III G1V	11171, 148036
			Local, solar type double with F giant; dominant in dark field. B is EZ Cet, BY Dra type variable. AB: ps = 6,010 AU. (2012)						
	Σ 186		01 55.9 +01 51	6.8, 6.8	248	0.8″	39.9	F9V	11803, 110235
			Matched solar type binary; dark field. AB: system mass 2.4 M$_\odot$, P = 166 y, orbit r = 40 AU, e = 0.73, periastron 2058. (435 measures; 2012)						
	Gal 315		01 57.2 −10 15	6.5, 11.1	134	30″	32.9	G5IV+K7	11964, 148123
			High CPM, solar type double; dark field, unidentified m.9 double 9′ s.p. AB: ps = 1,330 AU. (2007)						
58	β 7	A B	01 58.0 −02 04	6.6, 10.4	18	2.8″	150	A0	12020, 129588
AA	H II 58	(AaAb) B	01 59.0 −22 55	7.3, 7.6	303	8.8″	111	A7V+G0	12180, 167451
61	H V 102	(AaAb) B	02 03.8 −00 20	6.0, 10.8	194	44″	111	G5II+G5V	12641, 129667
	LDS 3346		02 07.6 −00 37	6.9, 10.5	342	82″	36.9	G2V	13043, 129706
			High CPM solar type double; m.12 double 4′ s.f., faint stars n.p. unrelated. AB: ps = 3,030 AU. (2007)						
66	Σ 231	(AaAb) B C	02 12.8 −02 24	6.2, 7.7, 11.5	235, 54	17″, 2.5′	39.7	F8V G1V	13612, 129752
	β 8		02 21.4 +08 53	8.0, 9.2	225	1.5″	120	F0	14562, —
	β 517	A B	02 24.9 −03 54	6.9, 12.1	248	11″	139	K0	15005, 129887
AB	H III 80	(AaAb) B C	02 26.0 −15 20	5.9, 9.1, 10.5	295, 31	12″, 110″	77	A6V	15144, 148386
	Kui 8		02 28.0 +01 58	7.1, 7.6	39	0.5″	167	K0III	15328, 110542
			Binary with K giant; dark field. AB: P = 653 y, orbit r = 130 AU; slowly closing from cusp of edgewise orbit. (2008)						
	Σ 280		02 34.1 −05 38	8.0, 8.0	346	3.7″	178	K1III	15994, 129981
79	Mug 2	A B	02 35.3 −03 34	6.8, 12.0	187	6.2″	39.0	G5IV+M2V	16141, 129992
78 nu	Σ 281	(AaAb) B	02 35.9 +05 36	5.0, 9.1	80	8.4″	104	G3III	16161, 110635
	h 3511		02 36.0 −21 24	7.2, 8.7	99	15″	340	G2III	16263, 167903
	OΣΣ 30		02 39.0 +08 55	7.7, 9.6	214	69″	280	A0	16500, 110658
84	Σ 295		02 41.2 −00 42	5.8, 9.7	301	3.6″	22.5	F6V	16765, 130055
			Local, solar type double with del Sct type variable; dark field. AB: ps = 110 AU. (2012)						
★ 86 gam	Σ 299	A B C	02 43.3 +03 14	3.5, 6.2, 10.2	298, 306	2.1″, 14.1′	24.4	A1V	16970, 110707
			Kaffaljidhma. Local, A type 2+1 CPM triple; pretty brightness and separation contrast in dark field. AB: ps = 69 AU (229 measures); AC: ps = 27,800 AU. (2013)						
	Σ 323		02 52.7 +06 28	7.8, 7.9	278	2.7″	196	B9	17907, 110807
	Σ 330		02 57.2 −00 34	7.3, 9.1	192	8.6″	190*	G8III	18384, 130205
	Σ 334		02 59.4 +06 39	7.9, 8.2	309	1.2″	125	F0	18570, 110883
94	h 663	A B	03 12.8 −01 12	5.1, 11.0	207	2.3″	22.6	F8V	19994, 130355
			Local, solar type binary. AB: P = 1,420 y (?), ps = 70 AU, closing in poorly measured orbit. (2002)						
	Σ 367		03 14.0 +00 44	8.1, 8.2	131	1.2″	66	F8	20115, 111062
	β 1039	(AB) C	03 17.4 +07 39	7.4, 7.8	38	2.6′	46.2	F9IV	20430, 111104
95	AC 2	A B (C)	03 18.4 −00 56	5.6, 8.0	259	1.2″	67	G9IV	20559, 130408
			Solar type 2+(1) triple (C is m.16); Y/B color, dark field. MSC gives 1.8 M$_\odot$. AB: P = 282 y, orbit r = 65 AU, closing. (2011)						

Label	Catalog ID	Components	Coordinates (J2000)	Mag.	θ	ρ	Dist.	Spectral type(s)	HD, SAO No.
	Chamaeleon		Cha					*Chart 30*	
	h 4109		08 22.8 −76 26	7.2, 8.2	130	26″	185	A0V	71972, 256507
del 1	I 294		10 45.3 −80 28	6.2, 6.5	85	0.8″	107	K0III	93779, 258592
eps	h 4486	A B C	11 59.6 −78 13	5.3, 6.0, 6.6	211, 40	0.4″, 2.2′	111	B9Vn+A	104174, 256894
S	h 4590		13 33.2 −77 34	6.6, 9.2	133	22″	35.2	F6V	117360, 257060
	Circinus		Cir					*Chart 28*	
	Δ 145		13 54.6 −66 54	7.8, 8.9	48	24″	174	B9V	120891, 252509
	h 4632	A B	13 58.5 −65 48	6.4, 9.5	13	6.4″	157	K0III	121557, 252534
	Don 632		13 59.5 −66 43	7.9, 10.1	268	3.8″	169	K0III	121707, 254347
	NZO 52		14 40.8 −66 57	7.9, 8.5	59	2.2″	98	F6V	128573, 252840
★ alp	Δ 166	A B	14 42.5 −64 58	3.2, 8.5	226	15″	16.6	A	128898, 252853
	Local, A type double with alp2 CVn type variable; faint rich field, double B 823 (m.8, 1.4″) 8′ s.f. (2013)								
	Don 680		14 49.4 −67 14	7.5, 9.8	243	2.2″	24.9	K0V	130042, 252899
	Local, solar type double; mixed field. AB: P = 261 y, orbit r = 50 AU, widening. (2013)								
	h 4707		14 54.2 −66 25	7.5, 8.1	276	1.1″	38.5	G0V	130940, 252945
	High CPM, solar type binary. AB: P = 346 y, orbit r = 60 AU, widening. (2010)								
	Gli 213		15 01.3 −67 59	7.1, 9.3	334	5.1″	320	B4V	132127, 252988
	h 4735		15 12.8 −60 24	7.6, 10.5	31	7.4″	89	F5V	134450, 253056
	I 329	A B C	15 14.0 −61 21	6.7, 7.7, 9.6	339, 297	0.9″, 45″	260	B5V	134657, 253064
	CorO 180		15 14.7 −59 49	7.6, 9.2	288	12″	172	K2III	134820, 242335
★ gam	h 4757		15 23.4 −59 19	4.9, 5.7	1	0.8″	138	B5IV F8	136415, 242463
	High mass binary. AB: P = 258 y (?), orbit r = 355 AU, decreasing θ in poorly measured, low quality orbit. (2012)								
	CapO 16	A B	15 29.5 −58 21	7.0, 8.0	32	2.4″	86	A8V	137583, 242593
	Columba		Col					*Charts 21, 22*	
	h 3728		05 08.5 −41 13	6.8, 10.3	260	10″	54	G3V	33473, 217198
	Δ 22		05 31.2 −42 18	7.2, 7.8	169	7.3″	101	A8V+	36648, 217374
	h 3781		05 38.6 −41 18	8.0, 9.4	136	16″	88	F3V	37718, 217448
	h 3825		06 02.1 −27 26	7.2, 10.5	338	32″	57	F5V	41172, 171158
	h 3849		06 19.8 −39 29	6.7, 8.1	54	40″	250	K2III	44404, 196693
	h 3858	A [B C]	06 25.5 −35 04	6.4, [7.6, 8.2]	48, [309]	2.2′, [3.8″]	168	K3III	45383, 196805
	h 3860		06 25.8 −40 59	7.3, 8.8	228	8.6″	150	A5V	45501, 217921
	I 4		06 30.7 −40 27	7.3, 7.5	306	0.8″	820*	B3IV	46288, 217981
	β 754	A B	06 34.7 −34 01	7.2, 7.7, 12.2	49, 51	1.0″, 49″	200	A4V	46973, 196959
	β 755	A B C	06 35.4 −36 47	5.9, 6.9, 11.5	260, 302	1.5″, 21″	156	A0V	47144, 196978
	CapO 6		06 35.4 −38 49	8.2, 8.2	241	1.1″	182	A0V	47146, 196979
	UC 1454		06 35.6 −36 16	7.2, 12.0	181	34″	162	A1V	47168, 196982
	Shy 185	(AaAb) (BaBb)	06 35.9 −36 05	6.4, 7.3	129	4.8′	39.6	G0V G1V	47230, 196987
	Solar type (2)+(2) quadruple, visual binary. Aa,Ab (Fin 19): P = 29 y, e = 0.44, periastron 2022 (2013); Ba,Bb (Rst 4816): P = 14 y, e = 0.58, periastron 2018 (2013); AB: ps = 15,400 AU is physical. Neglected. (1999)								
	Rst 4819		06 37.2 −36 59	5.9, 7.5	4	0.5″	350	B7IV	47500, 197014
	Coma Berenices		Com					*Chart 9*	
2	Σ 1596		12 04.3 +21 28	6.2, 7.5	236	3.7″	101	F0IV	104827, 82123
	Cbl 141		12 05.4 +17 17	7.7, 11.0	312	44″	97	F5	104999, 99936
	Σ 1615	A (BaBb)	12 14.1 +32 47	7.0, 8.6	87	27″	127	G5	106365, 62904

Label	Catalog ID	Components	Coordinates (J2000)	Mag.	θ	ρ	Dist.	Spectral type(s)	HD, SAO No.
	β 27		12 20.1 +13 51	7.1, 10.5	108	3.6″	143	G9III	107288, 100048
	Σ 1633		12 20.7 +27 03	7.0, 7.1	246	9.0″	92	F3V F3V	107398, 82254
	Sh 143	(AaAb) C	12 22.5 +25 51	4.9, 8.9	168	59″	90	G0III+A3	107700, 82273
	Σ 1639	A B	12 24.4 +25 35	6.7, 7.8	324	1.8″	90	A7V F4V	108007, 82293
			A type binary; dark field. AB: P = 575 y, orbit r = 110 AU. (439 measures; 2013)						
AI	Σ I 21	A (BaBb)	12 28.9 +25 55	5.2, 6.6	250	2.4′	73	A0	108662, 82330
	Hjl 1069		12 30.3 +21 57	7.9, 10.1	312	115″	139	G7III	108863, 82342
★ 24	Σ 1657	A (BaBb)	12 35.1 +18 23	5.1, 6.3	270	20″	138	K0II A9V	109511, 100160
			Probable high mass 1+(2) spectroscopic triple with variable K supergiant; classic YO/B color contrast, dark field. AB: ps = 3,720 AU. (2012)						
	Hjl 1072		12 39.2 +16 29	7.6, 9.3	23	118″	102	F3V	110025, 100194
	Σ 1685	A B	12 51.9 +19 10	7.3, 7.8	201	16″	220	Am F8III	111844, 100307
★ 35	Σ 1687	(AaAb) B C	12 53.3 +21 15	5.2, 7.1, 9.8	194, 128	1.2″, 29″	87	G7III	112033, 82550
			High mass (2)+1+1 quadruple with G giant. MSC 6.9 $M_\odot$. AB: P = 359 y, orbit r = 105 AU, e = 0.15. (385 measures; 2012)						
	Met 9		12 54.7 +22 06	7.0, 9.3	51	1.7″	33.9	F8V M2V	112196, 82559
	Σ 1709		13 02.5 +23 30	7.9, 10.0	251	2.7″	160	F4V	113303, 82618
	OΣ 259		13 07.7 +24 01	8.2, 8.6	21	39″	35.4	G5V	114060, 82662
			High CPM, solar type double; W/W color in dark field. AB: ps = 1,860 AU, near 100% probability the pair is physical. (2012)						
★ 42 alp	Σ 1728	A B	13 10.0 +17 32	4.9, 5.5	12	0.6″	17.8	F5V F6V	114378, 100443
			Diadem. Local, high CPM, solar type, possible Algol type eclipsing binary in remarkably large but edge-on orbit; AB: P = 26 y, orbit r = 12 AU, last transit 2015, widening to 0.5″ in 2019 before next potential eclipse (duration ~1.5 d) in 2026. (657 measures; 2011)						
	Bgh 46	(AaAb) B	13 16.5 +19 47	6.5, 7.6	58	3.4′	85	A3 A2	115365, 82751
	β 800	A B	13 16.8 +17 01	6.7, 9.5	105	7.2″	11.1	K1V M1V	115404, 100491
			Local, high CPM, solar type binary, discovered by W. Herschel (1782). AB: P = 770 y, orbit r = 90 AU. (200 measures; 2012)						
	Σ 1737		13 21.8 +17 46	7.9, 10.3	220	15″	98	F0	116206, 100534
	OΣ 266		13 28.4 +15 43	8.0, 8.4	358	2.1″	51	F5	117190, 100583
			Solar type binary; wide double n.p. AB: P = 1,954 y, orbit r = 170 AU, increasing θ in poor quality orbit. (216 measures; 2013)						

Corona Australis   CrA   *Chart 25*

Label	Catalog ID	Components	Coordinates (J2000)	Mag.	θ	ρ	Dist.	Spectral type(s)	HD, SAO No.
★	h 5014		18 06.8 −43 26	5.7, 5.7	4	1.8″	41.8	A5V A5V	165189, 228708
			A type, matched binary, member of the beta Pictoris comoving group. AB: P = 450 y, orbit r = 85 AU, widening. (2010)						
	h 5023		18 10.8 −40 26	8.3, 8.6	276	8.8″	400	A2	166060, 228773
	I 1020		18 16.4 −40 28	8.2, 8.0	275	0.4″	520	B7V	167297, 228864
kap 1,2	Δ 222		18 33.4 −38 44	5.6, 6.2	358	22″	300	B9V A0III	170868, 210295
	I 250		18 41.2 −42 10	7.4, 8.6	115	1.1″	210	A2IV	172261, 229191
11	CorO 227	A B	18 43.8 −38 19	5.1, 10.0	214	29″	63	A2Vn	172777, 210501
	h 5074		18 59.2 −39 32	6.5, 11.8	246	16″	153	B9.5V	175855, 210786
	BrsO 14	(AaAb) (BaBb)	19 01.1 −37 04	6.3, 6.6	280	13″	230	B8V B9V	176270, 210816
TY	B 957	A B (Ca1Ca2Cb)	19 01.6 −36 53	7.3, 13.4, 9.6	137, 23	4.1″, 57″	128	B9IV	176386, 210828
★ gam	h 5084		19 06.4 −37 04	4.5, 6.4	0	1.4″	17.3	F8V F8V	177474, 210928
			Local, solar type binary. System mass 2.4 $M_\odot$. AB: P = 122 y, orbit r = 33 AU, e = 0.32, apastron 2062. (269 measures; 2013)						

Corona Borealis   CrB   *Chart 10*

Label	Catalog ID	Components	Coordinates (J2000)	Mag.	θ	ρ	Dist.	Spectral type(s)	HD, SAO No.
★	Σ 1932	A B	15 18.3 +26 50	7.3, 7.4	264	1.6″	36.0	F6V F6V	136176, 83756
			Matched solar type binary. AB: P = 203 y, orbit r = 43 AU, e = 0.65, apastron 2043. (547 measures; 2013)						
★ 2 eta	Σ 1937	A B (E)	15 23.2 +30 17	5.6, 6.0	198	0.7″	17.9	F8V G0V	137107, 64673
			Local, solar type 2+(1) spectroscopic triple. AB: P = 42 y, orbit r = 15 AU, e = 0.28 (1,094 measures). Component E (Kir 4, m.17, 3.2′), below the magnitude limit, is a brown dwarf; AE: ps = 4,680 AU. (2013) See Figure 1.						

Label	Catalog ID	Components	Coordinates (J2000)	Mag.	θ	ρ	Dist.	Spectral type(s)	HD, SAO No.
4 the	Cou 610		15 32.9 +31 22	4.3, 6.3	200	0.8″	115	B6Ve	138749, 64769
★	Σ 1964	(AaAb) B [C (DaDb)]	15 38.2 +36 15	8.1, 9.9, [8.1, 9.0]	79, 85, [21]	1.2″, 15″, [1.6″]	110	F4V F3V	139691, 64821

Solar type (2)+1+1+(2) sextuple; tiny visual quadruple. A,C & D discovered by F.W Struve (1829), B by W. Hussey (1905). MSC system mass 6.1 M⊙. Aa,Ab: *P* = 3.9 d. AB: *P* = 834 y, orbit *r* = 140 AU; Da,Db: *P* = 14.3 d. CD: *P* = 1,230 y, orbit *r* = 175 AU. A,CD: ps = 2,225 AU. (2012)

Label	Catalog ID	Components	Coordinates (J2000)	Mag.	θ	ρ	Dist.	Spectral type(s)	HD, SAO No.
★ 7 zet 1,2	Σ 1965	(AaAbAc) (BaBb)	15 39.4 +36 38	5.0, 5.9	306	6.4″	145	B7V B9V	139891, 64833

High mass, B type (2+1)+(2) quintuple, visual double. MSC system mass 17.5 M⊙. A is a (2+1) spectroscopic triple with periods of 1.7 and 251 d. CPM pair AB: ps = 1,250 AU. (324 measures; 2013)

Label	Catalog ID	Components	Coordinates (J2000)	Mag.	θ	ρ	Dist.	Spectral type(s)	HD, SAO No.
★ 8 gam	Σ 1967		15 42.7 +26 18	4.0, 5.6	111	0.6″	44.8	B9V A3V	140436, 83958

A type binary with del Sct type variable. System mass 4.3 M⊙. AB: *P* = 91 y, orbit *r* = 33 AU, *e* = 0.48, disappearing to periastron 2022; reappears in 2027. (487 measures; 2012)

Label	Catalog ID	Components	Coordinates (J2000)	Mag.	θ	ρ	Dist.	Spectral type(s)	HD, SAO No.
	Σ 1973		15 46.4 +36 27	7.6, 8.8	321	31″	80	F5	141186, 64893
	Ho 399		15 55.4 +29 32	7.7, 10.5	117	3.5″	155	A2V	142796, 84078
	WNO 47		16 04.9 +39 09	6.7, 12.9	280	70″	14.5	K0V	144579, 65065

Local, high CPM, low mass double. AB: ps = 13,70 AU. (2000)

Label	Catalog ID	Components	Coordinates (J2000)	Mag.	θ	ρ	Dist.	Spectral type(s)	HD, SAO No.
	OΣ 305	A B	16 11.7 +33 21	6.4, 10.2	268	5.0″	155	K2III	145802, 65129
	Σ 2022		16 12.8 +26 40	6.5, 10.0	153	2.2″	73	F2V	145976, 84247
	Σ 2029		16 13.8 +28 44	8.0, 9.6	186	6.1″	133	F4IV	146168, 84258
★ 17 sig	Σ 2032	(AaAb) B (EaEb)	16 14.7 +33 52	5.6, 6.5, 12.3	238, 241	7.2″, 11′	21.1	G0V G1V	146361, 65165

TZ CrB. Local, high CPM, solar type (2)+1+(2) quintuple, visual 2+1 triple with RS CVn type variable. MSC system mass 3.8 M⊙. AB: *P* = 726 y, orbit *r* = 110 AU. AE: ps = 18,100 AU. (1,071 measures; 2013)

Corvus			Crv					*Chart 23*	
	β 920		12 15.8 −23 21	6.9, 8.2	305	1.9″	70	F7V	106612, 180617
	β 921	A (BC)	12 17.9 −24 01	7.0, 10.7	221	3.4″	141	A0V	106955, 180662
7 del	Sh 145	A B	12 29.9 −16 31	3.0, 8.5	213	25″	26.6	B9.5V	108767, 157323
	β 28	A B	12 30.1 −13 24	6.5, 9.6	342	2.1″	24.7	F9V	108799, 157326

Local, solar type binary. System mass 1.8 M⊙. AB: *P* = 151 y, orbit *r* = 35 AU, widening. (2009)

	Σ 1659	A B	12 35.7 −12 02	7.9, 8.3	351	28″	145	G0	109556, 157384
★ 58	Σ 1669	(AaAb) (BaBb) C	12 41.3 −13 01	5.9, 5.9, 10.3	314, 229	5.2″, 60″	79	F5V F5V	110318, 157447

B = VV Crv. Solar type (2)+(2)+1 quintuple, visual matched 2+1 triple with Algol type (detached) eclipsing variable. MSC system mass 5.2 M⊙. Aa,Ab: *P* = 44 d; Ba,Bb: *P* = 1.5 d. AB: ps = 550 AU, AC: ps = 6,400 AU. (2010)

Crater			Crt					*Charts 16, 23*	
	Σ 1509		11 06.5 −13 25	7.4, 9.4	16	33″	210	K0III	96377, 156461
★ 15 gam	h 840		11 24.9 −17 41	4.1, 7.9	93	5.3″	25.2	A5V	99211, 156661

Local A type double; AB: ps = 180 AU. Despite proximity to Sun it has only 21 measures – neglected! (1968)

	H IV 112	A [B C]	11 29.2 −17 21	4.1, [8.5, 9.1]	331, [202]	28″, [0.9″]	78*	F5V	99878, 156707
	Jc 16	A B C	11 29.6 −24 28	5.8, 8.6, 8.9	82, 116	8.4″, 2.8′	137	A0V	99922, 179935
	Σ 3072		11 30.9 −06 43	7.7, 9.9	331	9.8″	110	F6V	100070, 138243

Crux			Cru					*Chart 28*	
	Hrg 74		12 09.4 −63 49	7.4, 9.5	163	2.3″	850	B M1e	105563, 251765
	BrsO 8		12 24.8 −58 07	7.8, 8.0	334	5.3″	173	G0V	107976, 239932

Label	Catalog ID	Components	Coordinates (J2000)	Mag.	θ	ρ	Dist.	Spectral type(s)	HD, SAO No.
★ alp 1	Δ 252	(AaAb) B (CaCb)	12 26.6 −63 06	1.3, 1.6, 4.8	112, 202	3.9″, 90″	99	B0IV B1V B4V	108248, 251904
		Acrux. Brilliant, wide, high mass (2)+1+(2) spectroscopic quintuple, visual triple with more possible components nearby; brilliant W/W/Y colors in rich field, discovered by Fontenay (1685). MSC system mass 49.5 M⊙. AB: ps = 520 AU; AC: ps = 12,000 AU. (2013)								
	CapO 12	(AaAb) (BC)	12 28.3 −61 46	7.3, 8.2	188	2.0″	48.4	G3V	108500, 251919
		Solar type (2)+(2) quadruple. MSC has 3.2 M⊙. BC: P = 28 y, orbit r = 11 AU. A,BC: P = 2,520 y, orbit r = 265 AU. (2013)								
	Hld 116	A B	12 38.1 −55 56	7.1, 8.9	182	1.9″	111	A1V	109808, 240104
	CorO 140		12 45.2 −62 13	7.9, 9.9	97	4.6″	3480	A3Ib	110786, 252014
★ mu 1,2	Δ 126	A B	12 54.6 −57 11	3.9, 5.0	24	37″	127	B2IV B5Ve	112092, 240366
		High mass double, joined by CPM and equivalent parallax; W/YW color, rich field. AB: ps = 6,340 AU. (2011)								

Cygnus Cyg *Charts 5, 11*

Label	Catalog ID	Components	Coordinates (J2000)	Mag.	θ	ρ	Dist.	Spectral type(s)	HD, SAO No.
	Σ 2479	A B C	19 08.3 +55 20	7.7, 9.5, 9.7	348, 28	0.5″, 6.5″	120	A5IV	179142, 31417
★	Σ 2486	A B	19 12.1 +49 51	6.5, 6.7	205	7.2″	24.5	G3V G3V	179957, 48193
		Local, high CPM, matched solar type binary; lovely rich field. AB: P = 3,100 y, orbit r = 310 AU. (284 measures; 2013)								
	OΣ 373		19 24.1 +46 26	7.6, 9.9	232	1.8″	280	B9.5II	182754, 48406
	Σ 2534		19 27.7 +36 32	8.2, 8.4	63	6.3″	900	B9III	183363, 68347
★6 bet 1,2	Σ I 43	(AaAb) Ac B	19 30.7 +27 58	3.4, 5.2, 4.7	101, 55	0.4″, 35″	133	K3II B9.5 B8	183912, 87301
		Albireo. High mass (2)+1+1 quadruple, visual triple, discovered by C. Mayer (1777). K supergiant with high mass companion and celebrated YO/B hues (Victorian observers described "topaz and sapphire"). A,Ac (MCA 55, P = 214 y), identified by McAlister (1976) with speckle interferometry, is widening and now can be visually resolved with skill, large aperture and good seeing. AB: ps = 6,280 AU. Hipparcos data imply a 12 pc separation but given the total system mass (11.5 M⊙) and parallaxes equal within errors, many consider this a physical system. (273 measures; 2012)								
	Arn 82	A B	19 36.4 +35 41	8.1, 8.4	34	44″	220*		185173, 68569
	OΣ 378	A B C	19 36.5 +41 01	7.7, 8.9, 7.8	285, 161	1.3″, 2.8′	350	A0	185266, 48640
★ 16	Σ I 46	(AaAb) B	19 41.9 +50 31	6.0, 6.2	133	40″	21.1	G1.5V	186408, 31898
		Local (2)+1 triple, long period visual double; near planetary nebula NGC 6826. AB: P = 13,500 y, orbit r = 740 AU. (580 measures; 2012)								
	OΣ 383	A B	19 42.9 +40 43	7.0, 8.3	15	0.8″	155	B9.5V	186465, 48756
	OΣ 384	A B	19 43.8 +38 19	7.6, 8.2	196	1.0″	680	B5V	186605, 68767
★ 18 del	Σ 2579	A B	19 45.0 +45 08	2.9, 6.3	218	2.5″	51	B9.5IV	186882, 48796
		Rukh. High mass binary; pretty brightness contrast, needs good seeing. AB: P = 918 y, orbit r = 175 AU. (462 measures; 2013)								
★ 17 chi	Σ 2580	A B [F G J]	19 46.4 +33 44	5.1, 9.3, [8.5, 8.6, 11.7]	70, 235, [157, 241]	26″, 13.0′, [3.0′, 21″]	21.2	F5V	187013, 68827
		Local, high CPM, solar type 2+[2+1] wide visual quintuple; color contrast, rich field. MSC system mass 3.3 M⊙. AB: ps = 740 AU; AF: ps = 22,300 AU. FG (Σ 2576): P = 232 y, orbit r = 44 AU, e = 0.77, apastron 2061. (2013)								
	OΣ 387		19 48.7 +35 19	7.1, 7.9	117	0.5″	64	F6V	187458, 68893
		Solar type binary; rich field. System mass 2.6 M⊙. AB: P = 165 y, orbit r = 41 AU, e = 0.06, decreasing θ to periastron 2035. (363 measures; 2012)								
	Lep 94		19 49.3 +41 35	7.5, 10.5	67	67″	53	F5	187637, 48869
	OΣ 390	A B	19 55.1 +30 12	6.6, 9.5	25	9.0″	260	B6V A5V	188651, 69079
★ 24 psi	Σ 2605	(AaAb) B	19 55.6 +52 26	5.0, 7.5	177	2.7″	86	A4Vn	189037, 32114
		Pretty A type (2)+1 triple, visual double. MSC system mass 4.9 M⊙. Aa,Ab (Yr 2): P = 54 y, orbit r = 12 AU. (2013)								
	Σ 2607	(AB) C	19 57.9 +42 16	6.6, 9.1	289	3.0″	330	A3V	189377, 49031
		Distant, A type (2)+1 triple. MSC 7.1 M⊙. AB: P = 270 y, orbit r = 85 AU; AC: ps = 1,340 AU. Neglected. (1996)								
	Σ 2606	A B	19 58.5 +33 17	7.7, 8.4	147	0.7″	78	F5IV	189378, 69186
	Σ 2609		19 58.6 +38 06	6.7, 7.6	22	2.0″	440	B5IV	189432, 69193

Label	Catalog ID	Components	Coordinates (J2000)	Mag.	θ	ρ	Dist.	Spectral type(s)	HD, SAO No.
	Σ 2611		19 58.9 +47 22	8.5, 8.5	208	5.4″	290	K0	189636, 49052
	OΣ 394	A B	20 00.2 +36 25	7.1, 10.3	295	11″	152	K1III	189751, 69238
	Σ 2624	(AaAb) (BaBb)	20 03.5 +36 02	7.1, 7.7	175	2.0″	860	O9.5IIIe	190429, 69324
	Sh 316	A [B D]	20 05.7 +35 36	7.8, [8.8, 13.1]	323, [85]	70, [17″]	490	O7IIIe	190864, 69391
	Es 25	A B D F	20 06.0 +35 46	7.9, 12.0, 8.7, 6.8	119, 236, 329	8.8″, 20″, 96″	740	B5n	227634, 69405
	A 382		20 08.0 +42 23	7.2, 9.5	96	1.7″	200	K0	191394, 49217
	OΣ 400	A B	20 10.2 +43 57	7.6, 9.8	332	0.7″	51	G3V	191854, 49262
			Solar type visual and spectroscopic binary; sparkling field. AB: P = 86 y, orbit r = 23 AU, e = 0.49, periastron 2056. (236 measures; 2012)						
	AC 17	A B	20 12.5 +51 28	6.2, 10.6	83	4.3″	106	K2.5III	192439, 32354
	Σ 2658	A B D	20 13.7 +53 08	7.2, 9.4, 12.9	105, 285	6.8″, 31″	44.3	F5V	192679, 32380
	OΣ 403	A B C	20 14.4 +42 06	7.3, 7.6, 9.8	171, 32	0.9″, 12″	580	B9IV	192659, 49345
	Σ 2663	A B	20 16.8 +39 42	8.2, 8.7	324	5.4″	1590	A0II	193063, 69749
	Ho 588	A [B C]	20 16.9 +31 30	6.9, [8.9, 12.6]	297, [29]	51″, [8.9″]	189	A0III	193010, 69743
	Σ 2666	(AaAbAc) B C D	20 18.1 +40 44	5.8, 8.2, 11.1, 10.4	245, 207, 181	2.8″, 34″, 50″	600	O9V B1.5V	193322, 49438
			High mass, distant (3)+1+1+1 system; possible cluster remnant, rich field, m.10 double 2′ n.f. MSC system mass 41.9 M⊙. Aa,Ab: P = 35.2 y, e = 0.49, periastron 2030. AD: ps = 40,500 AU. (2011)						
	Σ 2671	(AaAb) (BaBb)	20 18.4 +55 24	6.0, 7.5	338	3.7″	91	A2V	193592, 32455
	Σ 2668	(AB) C	20 20.3 +39 24	6.3, 8.5	281	3.4″	181	A2V A5V	193702, 69856
			A type (2)+1 triple, visual double. MSC gives 6.8 M⊙. AB: P = 83 y, closing. 5′ south are Es 2050, Sei 1090, Sei1091. (2008)						
	Ho 128	A B	20 22.9 +42 59	6.4, 8.8	358	1.4″	103	G8III	194220, 49550
	S 755	A B	20 30.9 +49 13	6.6, 9.7	278	60″	147	A2	195710, 49731
	OΣ 408		20 34.0 +34 41	6.8, 9.4	193	1.6″	330	B7V	196120, 70206
	Stu 13		20 34.9 +41 43	7.6, 10.1	196	27″	340	B9	196305, 49828
	Σ 2705	A B	20 37.7 +33 22	7.5, 8.5	262	3.1″	600	K0II	196673, 70300
	Ary 48		20 37.8 +32 24	8.2, 8.8	41	53″	70*	F8	—, 70298
	Σ 2717	A B	20 37.8 +60 45	7.3, 9.5	259	2.0″	260	G3III	196988, 18958
	Σ 2707	A (CaCb)	20 37.9 +47 57	7.9, 8.6	195	55″	172	A0	196808, 49879
V2130	OΣ 410	A B C	20 39.6 +40 35	6.7, 6.8, 8.7	4, 69	0.9″, 68″	280	B8III	197018, 49899
			Distant, high mass, nicely displayed 2+1 triple. AB: P = 1,408 y, orbit r = 210 AU; AC: ps = 25,700 AU. (2013)						
49	Σ 2716	(AaAb) B	20 41.0 +32 18	5.8, 8.1	44	2.6″	240	G2III	197177, 70362
★ 52	Σ 2726		20 45.7 +30 43	4.2, 8.7	70	6.0″	62	G9III	197912, 70467
			Solar type double with G giant; YO/− color, rich field with the Veil Nebula (NGC 6960). AB: ps = 500 AU. (2006)						
T	β 677	A B C	20 47.2 +34 22	4.9, 10.0, 11.2	120, 212	8.1″, 16″	136	K3III	198134, 70499
	OΣ 414	A B C	20 47.2 +42 25	7.4, 8.9, 9.9	94, 16	9.9″, 106″	350	B7V	198195, 50055
54 lam	OΣ 413	(AaAb) B	20 47.4 +36 29	4.7, 6.3	5	1.0″	240	B5Ve	198183, 70505
			High mass (2)+1 spectroscopic triple. Aa,Ab: P = 11.6 y, e = 0.52, periastron 2017; AB: P = 391 y, orbit r = 190 AU. (417 measures; 2013)						
	Σ 2731	A B	20 49.0 +39 47	7.7, 9.6	85	4.2″	500	B9IV	198436, 70541
	β 67		20 50.6 +30 55	6.9, 9.9	306	1.6″	120	A8III	198626, 70564
			Double with A type giant, at the center of the Veil Nebula NGC 6960. Neglected. (1999)						
	β 155	A B C	20 51.1 +51 25	7.4, 8.1, 12.6	39, 40	0.7″, 15″	125	A9IV	198834, 32933
	OΣ 419		20 54.7 +37 04	7.2, 10.0	25	1.7″	194	A0	199234, 70659
	OΣ 418		20 54.8 +32 42	8.2, 8.3	284	1.0″	61	G0	199220, 70660
			Solar type binary, matched m.8 arcsecond resolution test. AB: P = 787 y, orbit r = 110 AU. (2013)						
	OΣ 423		20 55.3 +42 31	7.1, 9.6	76	2.8″	420	B9	199355, 50226
			Distant, high mass twin of ?Σ 422 (m8, 2.6″) 3.6° n. (2011)						
	Σ 2741	A B	20 58.5 +50 28	5.9, 6.8	25	2.0″	330	B5Vn	199955, 33034

Label	Catalog ID	Components	Coordinates (J2000)	Mag.	θ	ρ	Dist.	Spectral type(s)	HD, SAO No.
60	OΣ 426	(AaAb) B	21 01.2 +46 09	5.4, 9.5	161	2.9″	470	B1Ve	200310, 50359
			V1931 Cyg. High mass double; Be star is variable, probably a spectroscopic binary, large brightness contrast. AB: ps = 1,840 AU. (2009)						
	Σ 2746		21 01.8 +39 16	7.9, 8.7	322	1.2″	250	F0	200370, 70808
	β 445	A B	21 03.5 +29 06	7.0, 11.1	109	4.8″	197	G8III	200578, 89415
	OΣΣ 214	A B	21 03.9 +41 38	6.4, 8.6	185	57″	96	F3IV	200723, 50409
	Σ 2757		21 04.6 +52 24	7.8, 9.2	262	1.9″	230	B9.5V	200943, 33110
★ 61	Σ 2758	A B	21 06.9 +38 45	5.2, 6.1	152	31″	3.49	K5V K7V	201091, 70919
			Local, high CPM, low mass binary; YO/YO pair in a sparkling low power field. Noted as binary by J. Bradley (1753); first measured parallax (by Bessel, 1838); known as "Piazzi's Flying Star" it was famous for its high proper motion (5.3″/yr). The A (northern) component is a 0.74 M$_\odot$ BY Dra type variable, with massive star spots that dim the brightness as they rotate into view; B is a 0.46 M$_\odot$ flare star that can abruptly brighten as flares erupt. AB: P = 678 y, orbit r = 145 AU, widening to a 34″ apastron in 2106; all other "components" in WDS are field stars. (1,667 measures; 2013)						
V389	Σ 2762	A B	21 08.6 +30 12	5.7, 8.1	304	3.3″	116	B9V	201433, 70968
	β 159		21 10.5 +47 42	6.6, 9.1	312	1.2″	420	B6IV	201836, 50536
	Sei 1445		21 12.5 +38 34	7.2, 10.5	23	28″	200*	B9	202088, 71065
	OΣ 432		21 14.3 +41 09	7.8, 8.1	115	1.3″	240	F8V	202403, 50604
★ 65 tau	AGC 13	A B (FaFb) (I)	21 14.8 +38 03	3.8, 6.6, 12.0	213, 184	0.9″, 90″	20.3	F3V+F7V	202444, 71121
			Local, high CPM 2+(2)+(1) quintuple, visual triple (I is m.16) with del Sct type variable. MSC system mass 3.0 M$_\odot$. AB P = 50 y, orbit r = 18 AU, e = 0.24, decreasing θ; Fa,Fb: P = 2.2 y, e = 0.43. (346 measures; 2013)						
	Skf 346		21 17.8 +52 29	7.7, 11.1	305	56″	107	F2	203046, 33295
	OΣ 434	A B	21 19.0 +39 45	6.7, 9.9	122	24″	144	B9V	203112, 71195
	Σ 2789	A B	21 20.0 +52 59	7.7, 7.9	114	6.9″	720	F8V	203380, 33334
			Distant, solar type matched double; faint field. Colorful, ambiguous S 786 (m.7, 48″) 6′ n.p. (2012)						
	OΣ 437	A B	21 20.8 +32 27	7.2, 7.4	20	2.5″	66	G4V	203358, 71230
			Solar type binary; several faint doubles visible in a pretty field. AB: P = 1,421 y, orbit r = 170 AU. (285 measures; 2013)						
	β 369	A B	21 26.5 +52 45	7.6, 11.8	31	16″	199	B9.5V	204401, 33431
	Arn 78		21 31.7 +48 29	7.6, 8.8	99	50″	260	A0m A1IV	205117, 50985
	Ho 603	A B	21 32.1 +34 12	7.5, 9.8	251	81″	97	F0	205075, 71434
	β 167		21 36.2 +30 03	6.4, 10.0	89	1.8″	111	G8III	205688, 89834
	h 1676		21 39.6 +47 12	7.8, 11.0	136	30″	580	K0	206268, 51158
	Σ 2820	A B	21 42.6 +42 26	7.5, 10.6	233	16″	154	A0	206673, 51214
★ mu	Σ 2822	(AaAb) B	21 44.1 +28 45	4.8, 6.2	318	1.7″	22.2	F6V G2V	206826, 89940
			Local, high CPM, solar type (2)+1 spectroscopic triple; a tiny "diamond ring." AB: P = 789 y, orbit r = 120 AU. (735 measures; 2013)						
	Σ 2832	A B	21 49.2 +50 31	7.8, 8.3	213	13″	152	B9IV	207661, 33781
V1942	β 694	A B	22 02.9 +44 39	5.7, 7.8	7	1.0″	151	A0IV	209515, 51595

Delphinus Del *Chart 19*

Label	Catalog ID	Components	Coordinates (J2000)	Mag.	θ	ρ	Dist.	Spectral type(s)	HD, SAO No.
	Σ 2665	A (BC)	20 19.4 +14 22	6.9, 9.6	12	3.3″	168	A0 G	193350, 105957
	Σ 2664		20 19.6 +13 00	8.1, 8.3	322	28″	410	K0	193391, 105967
	Ho 131	A B	20 28.3 +18 46	7.0, 10.6	331	3.5″	38.5	G1V	195019, 106138
	S 752	A B C	20 30.2 +19 25	6.8, 11.1, 7.3	126, 288	2.6″, 107″	390	B7IV	195358, 106177
1	β 63	A B	20 30.3 +10 54	6.2, 8.0	350	0.9″	230	A1eShell	195325, 106172
	Σ 2703	A B	20 36.8 +14 44	8.4, 8.4	290	25″	440	A5	196411, 106302
	Σ 2701		20 37.0 +12 03	8.3, 8.6	221	2.1″	118	G5	196423, 106305
7 kap	OΣ 533	(AaAb) C	20 39.1 +10 05	5.2, 8.6	100	3.5′	30.1	G1IV K2IV	196755, 126059
			High CPM, solar type (2)+1 triple; several faint stars nearby; component B is optical. Aa,Ab: P = 45 y, orbit r = 16 AU; now unresolved, m.9 Ab may widen to 0.5″ in 2022. AC: ps = 8,530 AU. (2001)						

Label	Catalog ID	Components	Coordinates (J2000)	Mag.	θ	ρ	Dist.	Spectral type(s)	HD, SAO No.
	OΣ 409	A B	20 40.3 +03 26	7.1, 10.2	84	17″	250	K0	196929, 126088
	Σ 2715		20 41.8 +12 31	7.8, 10.2	3	12″	92	F8	197179, 106390
	Σ 2718	A B	20 42.6 +12 44	8.3, 8.4	87	8.4″	145	F5	197312, 106409
	Σ 2723	(AaAb) B	20 44.9 +12 19	7.0, 8.3	146	1.2″	210	A3IV	197684, 106443
	Σ 2725	(AaAb) B	20 46.2 +15 54	7.5, 8.2	12	6.1″	36.5	K0	197913, 106466
			Solar type (2)+1 triple, visual double. MSC gives 2.6M☉. Low quality orbit AB: *P* = 2,850 y, orbit *r* = 270 AU. (260 measures; 2013)						
★ 12 gam 1,2	Σ 2727	A B	20 46.7 +16 07	4.4, 5.0	266	9.0″	38.7	F7V K1IV	197963, 106477
			Solar type binary; in rich field with double Σ 2725 (above) 14′ s.p., subtle Y/V color contrast. Low quality orbit AB: *P* = 3,250 y, orbit *r* = 395 AU. (515 measures; 2013)						
13	β 65		20 47.8 +06 00	5.6, 8.2	199	1.5″	131	A0V	198069, 126222
	Σ 2730		20 51.1 +06 23	8.4, 8.6	333	3.3″	45.5	K1III	198569, 126289
	Σ 2733		20 52.7 +07 20	8.4, 8.6	145	41″	370	A	198812, 126322
★	Σ 2735		20 55.7 +04 32	6.5, 7.5	281	2.0″	142	G6III	199223, 126373
			Probable A type double with G giant; sparse bright field. AB: ps = 385 AU. (2011)						
	OΣΣ 213		20 59.8 +16 49	6.7, 9.2	37	71″	48.1	F4III	199941, 106738
	h 1608	A B	21 04.9 +12 27	7.7, 11.4	257	20″	90	F2V	200745, 106819

Dorado Dor *Chart 27*

Label	Catalog ID	Components	Coordinates (J2000)	Mag.	θ	ρ	Dist.	Spectral type(s)	HD, SAO No.
	Shy 462	Aa Ab B	04 18.7 −52 52	6.2, 8.8, 7.8	106, 236	0.8″, 4.5′	72	F7IV F5V	27604, 233476
	Rmk 4		04 24.2 −57 04	6.9, 7.2	248	5.4″	27.7	G4V	28255, 233506
	Slr 6		04 25.5 −53 07	7.1, 9.1	105	0.9″	136	F0V	28349, 233512
	h 3658		04 28.9 −49 36	8.1, 8.5	122	5.7″	158	A	28689, —
★	h 3683		04 40.3 −58 57	7.3, 7.5	89	3.7″	30.6	G5V	30003, 233622
			Solar type binary; sparse field. AB: *P* = 326 y, orbit *r* = 75 AU, widening. (2013)						
	CapO 4		04 47.4 −61 29	7.4, 10.3	45	2.6″	69	F3V	30865, 249108
	LDS 135		05 02.3 −56 05	7.1, 10.8	144	79″	22.5	G5V	32778, 233796
			Local, high CPM solar type double, n.p. of three similar stars. AB: ps = 2,400 AU. (2010)						
zet	Shy 22		05 05.5 −57 28	4.8, 9.1	155	5.4′	11.7	F7V K7V	33262, 233822
			Local, solar type double; sparse field. AB: ps = 5,110 AU. (2010)						
	I 276		05 27.0 −68 37	6.7, 7.0	161	1.4″	78	F0IV	36584, 249281
			Solar type double; faint rich field, centered in front of Large Magellanic Cloud. AB: ps = 145 AU. (2008)						
	Δ 26	A B	06 12.2 −65 32	6.9, 8.1	120	21″	70	F6V F7IV	43618, 249477

Draco Dra *Charts 4, 5*

Label	Catalog ID	Components	Coordinates (J2000)	Mag.	θ	ρ	Dist.	Spectral type(s)	HD, SAO No.
★	Σ 1362		09 37.9 +73 05	7.0, 7.2	125	4.9″	71	F1V F2V	82685, 6915
			Matched solar type double; a classic Struve binary, beautifully displayed in dark field. AB: ps = 470 AU. (2007)						
	Σ 1437		10 34.2 +73 50	7.6, 10.4	291	24″	186	A3	91114, 7161
	OΣ 539	A C	11 15.2 +73 29	7.8, 11.3	326	6.5″	14.7	K4V+M2	97584, 7320
			Local, low mass double; dark field, B (m.8, 36″) is optical. AC: ps = 130 AU. (2011)						
	β 794	A B	11 53.7 +73 45	7.2, 8.3	50	0.5″	58	F8V	103246, 7445
			Solar type binary; system mass 2.4 M☉. AB: *P* = 77 y, orbit *r* = 24 AU, *e* = 0.50, apastron 2025. (2009)						
	Σ 1590		12 01.6 +70 51	7.4, 10.1	235	5.3″	157	K1III	104435, 7485
	Σ I 25	A B	13 13.5 +67 17	6.6, 7.1	296	3.0′	122	K2III	115136, 16018
	OΣΣ 123	A B	13 27.1 +64 44	6.7, 7.0	148	70″	72	F0	117200, 16078
	Σ 1860		14 33.9 +55 14	8.0, 9.0	112	1.0″	152	A5	128230, 29198
	Σ 1872	A B	14 41.0 +57 57	7.5, 8.3	49	7.6″	55	K0	129580, 29244

Label	Catalog ID	Components	Coordinates (J2000)	Mag.	θ	ρ	Dist.	Spectral type(s)	HD, SAO No.
DL	Σ 1878		14 42.1 +61 16	6.3, 9.2	314	4.1″	40.7	F4V	129798, 16466
	Σ 1882	A B C	14 44.1 +61 06	6.9, 9.2, 10.5	359, 223	12″, 7.2″	82	F3V	130173, 16478
		Solar type triple (probable 1+2); in same rich field with Σ 1878 (above). AB: ps = 1,330 AU. (2010)							
BV, BW	Σ 1927	(AaAb) (BaBb)	15 11.8 +61 51	8.1, 8.8	353	16″	71	F8V G3V	135421, 16636
		Solar type (2)+(2) spectoscopic quadruple, visual double; both components are W UMa type eclipsing contact binaries. Aa,Ab: P = 0.29 d. Ba,Bb: P = 0.35 d. AB: ps = 1,530 AU. (2013)							
	Hu 149		15 24.6 +54 13	7.5, 7.6	270	0.7″	230	K0	137588, 29514
TW	OΣ 299	(AaAb) B	15 33.9 +63 54	7.5, 9.9	25	3.4″	139	A5	139319, 16767
	β 946		15 47.6 +55 23	5.9, 9.5	129	2.3″	74	A3m	141675, 29668
	Σ 1984	A B	15 51.2 +52 54	6.9, 8.9	278	6.4″	127	A1V	142282, 29691
	Σ 2054	(AaAb) B	16 23.8 +61 42	6.2, 7.1	350	1.0″	155	G8III	148374, 17073
		(2)+1 triple, visual double with G giant; in field with OΣ 312 (below) 12′ s.f. AB: ps = 210 AU. (2013)							
14 eta	OΣ 312	A B	16 24.0 +61 31	2.8, 8.2	139	4.8″	28.2	G8III	148387, 17074
★ 17,16	Σ 2078	A B (CaCb)	16 36.2 +52 55	5.4, 6.4, 5.5	104, 193	3.1″, 90″	126	B9V A1V B9.5V+D	150117, 30013
		High mass 2+(2) quadruple; visual wide triple; Cb is white dwarf. MSC gives 11.4 M$_\odot$. AB: ps = 525 AU. AC (Σ I 30): ps = 15,300 AU. (2013)							
	β 953	(AB) (DaDb)	16 36.7 +69 48	8.0, 8.0	46	2.4′	107	F2V	150631, 17167
		Solar type (2)+(2) quadruple, visual wide matched binary. AB: system mass 2.1 M$_\odot$, P = 21 y, e = 0.44, periastron 2120; AD: ps = 20,800 AU. (2008)							
20	Σ 2118	A B	16 56.4 +65 02	7.1, 7.3	65	0.9″	70	F2IV	153697, 17285
		Solar type binary. System mass 2.3 M$_\odot$. AB: P = 422 y, orbit r = 75 AU, e = 0.14, apastron 2050. (286 measures; 2013)							
★ 21 mu	Σ 2130	A B	17 05.3 +54 28	5.7, 5.7	5	2.4″	27.4	F7V	154905, 30239
		Arrakis. Matched solar type binary; sparse field. System mass 2.8 M$_\odot$. AB: P = 812 y, orbit r = 125 AU. (812 measures; 2013)							
	Σ 2146	A B	17 13.1 +54 08	6.9, 8.8	224	2.6″	109	A9III	156162, 30299
	Σ 2155	A B	17 16.1 +60 43	6.9, 10.0	113	9.8″	182	F3III	156890, 17410
	Σ 2180		17 29.0 +50 52	7.8, 8.1	258	3.1″	111	A7IV	158868, 30413
	LDS 5227		17 29.7 +63 51	7.7, 8.4	288	3.2′	45.2	G0	159329, 17504
		Solar type, wide CPM double; sparse or dark field. Near 100% probability pair is physical. AB: ps = 11,700 AU. (2003)							
★ 24,25 nu 1,2	Σ I 35	(AaAb) (BaBb)	17 32.3 +55 10	4.9, 4.9	311	62″	30.5	A4m A6V	159560, 30450
		Kuma. Regal, matched A type (2)+(2) spectroscopic and CPM quadruple; fine binocular pair. AB: ps = 2,550 AU; near 100% probability pair is physical. (2012)							
★ 26	LDS 2736	A B C	17 35.0 +61 52	5.3, 8.5, 10.2	313, 161	0.9″, 12′	14.2	G0V	160269, 17546
		Local, high CPM, solar type 2+1 triple. AB: system mass 1.8 M$_\odot$, P = 76 y, orbit r = 22 AU, e = 0.18, periastron 2023. AB,C: ps = 14,100 AU; near 100% probability pair is physical. (2011)							
	Σ 2199		17 38.6 +55 46	8.0, 8.6	55	2.1″	121	F8V	160780, 30494
		Solar type binary; wide m.10 double 3′ n.f. Low quality orbit AB: P = 1,299 y, orbit r = 270 AU. (226 measures; 2013)							
	H I 41		17 39.7 +72 56	8.1, 8.5	335	1.0″	110*	F2	161692, 8876
	Σ 2218		17 40.3 +63 41	7.1, 8.4	309	1.4″	68	F8V	161285, 17599
★ 31 psi 1	Σ 2241	A B	17 41.9 +72 09	4.6, 5.6	16	30″	22.8	F5IV F8V	162003, 8890
		Dziban. Local, solar type, long period double; dominant in dark field, superb in any aperture. AB: ps = 925 AU. (2013)							
	Σ 2261	A B	17 58.1 +52 13	7.6, 10.0, 12.1	261, 45	9.6″, 59″	114	A2	164394, 30665
	Σ 2273	A B	17 59.2 +64 09	7.3, 7.6	283	21″	73	F4V F5V	164984, 17717
★ 40,41	Σ 2308	(AaAb) (BaBb)	18 00.2 +80 00	5.7, 6.0	231	19″	64	F7V F7V	166866, 8996
		Solar type (2)+(2) spectroscopic quadruple, visual matched binary; bright, wide, pretty. MSC system mass 5.2 M$_\odot$. Aa,Ab: system mass 5.6 M$_\odot$, P = 3.4 y, e = 0.975 (highest known spectroscopic eccentricity). Ba,Bb: P = 10.5 d. AB: ps = 1,640 AU; near 100% probability the pair is bound, probably a long period binary despite linear solution. (2013)							
	Σ 2284		18 01.4 +65 57	8.0, 9.5	191	3.6″	190	F7IV	165522, 17729
	Σ 2302	A B C	18 02.8 +75 47	7.0, 10.0, 9.7	249, 277	5.0″, 23″	141	A0V	166655, 8999
	Σ 2278	B C	18 02.9 +56 26	8.1, 8.5	38	34″	138	A9V	165501, 30716

Label	Catalog ID	Components	Coordinates (J2000)	Mag.	θ	ρ	Dist.	Spectral type(s)	HD, SAO No.
	Σ 2326		18 05.3 +81 29	8.0, 9.0	194	16″	190*	A8IV	168518, 2993
★ 43 phi	OΣ 353	(AaAb) B	18 20.8 +71 20	4.5, 5.9	266	0.5″	93	B8V A0	170000, 9084
		High mass (2)+1 spectroscopic triple with alp2 CVn type variable. AB: *P* = 308 y, orbit *r* = 90 AU, *e* = 0.75, widening. (2011)							
39	Σ 2323	A B (CaCb)	18 23.9 +58 48	5.1, 8.1, 8.0	345, 20	3.6″, 89″	56	A1V	170073, 30949
	UC 3571		18 24.1 +79 13	6.7, 12.3	73	46″	130	K0	171606, 9130
	OΣ 351	(AB) C	18 25.3 +48 46	7.9, 8.3	25	0.8″	145	G5	170109, 47482
		Solar type (2)+1 triple; poorly measured. AB: *P* = 300 y, orbit *r* = 60 AU. AC: *P* = 560 y, orbit *r* = 160 AU, closing. (2011)							
	OΣ 363		18 37.4 +77 41	7.5, 8.1	338	0.5″	110	F0IV	173831, 9199
		Solar type binary; resolution challenge, ~1° n.f m.6 star in sparse field. Low quality orbit AB: *P* = 642 y, orbit *r* = 105 AU. (2011)							
	Σ 2377	A B	18 38.4 +63 32	7.0, 9.8	339	17″	230	K2III	172923, 17963
	Σ 2368	A B	18 38.9 +52 21	7.6, 7.8	320	1.9″	260	A3	172712, 31086
	Σ 2403		18 44.3 +61 03	6.3, 8.4	278	1.1″	110	G8III	173949, 17995
	Σ 2452		18 53.6 +75 47	6.7, 7.4	218	5.1″	250	A1V	176795, 9286
	Σ 2433	A B	18 56.9 +56 45	7.2, 10.1	124	7.5″	141	F2V	176409, 31286
	Σ 2440	A B C	18 57.3 +62 24	6.6, 9.6, 10.9	122, 60	18″, 2.7′	95	G8III	176668, 18082
	Σ 2438		18 57.5 +58 14	7.0, 7.4	358	0.9″	125	A2IV	176560, 31292
		A type binary; system mass 3.6 M⊙. AB: *P* = 231 y, orbit *r* = 60 AU, *e* = 0.99, at cusp of highly eccentric orbit. (2013)							
	Σ 2450	A (BC)	19 02.1 +52 16	6.5, 9.5	299	5.2″	210	G8III	177483, 31337
	OΣ 369		19 07.1 +72 04	7.8, 7.9	10	0.7″	210	F7V	179729, 9323
	Σ 2509		19 16.9 +63 12	7.4, 8.2	328	1.8″	78	F6V	181566, 18257
		Solar type binary; unidentified m.11 double 5′ s.f. AB: *P* = 627 y, orbit *r* = 105 AU, widening. (2011)							
	Σ 2550	A B	19 27.0 +73 22	8.5, 8.4	251	1.9″	146	F2	184292, 9428
	Σ 2571	A B	19 29.5 +78 16	7.7, 8.3	19	11″	200	F0IV	185497, 9452
	Σ 2549	A B	19 31.2 +63 19	8.3, 12.0	329	2.1″	470	K0	184563, 19285
	Σ I 44		19 33.2 +60 10	6.5, 8.2	288	76″	310	K4III	184936, 18395
	Σ 2574		19 40.6 +62 40	7.7, 8.9	90	0.5″	129	F8V	186453, 18469
		Solar type binary; pretty field. AB: *P* = 685 y, orbit *r* = 170 AU. (2009)							
63 eps	Σ 2603		19 48.2 +70 16	4.0, 6.9	21	3.2″	45.4	G7III	188119, 9540
	Σ 2604		19 52.8 +64 11	6.9, 9.0	183	28″	240	G5	188772, 18575
	Σ 2640	A B	20 04.7 +63 53	6.3, 9.5	15	5.7″	79	A2III	191174, 18692
		Double with A type giant; CPM binary Σ 2642 (m.9, 1.8″) 12′ s.f. AC: ps = 16,600 AU. (2006)							
	Σ 2650		20 07.0 +66 18	7.0, 10.8	227	22″	185	A1V	191700, 18722
	Σ 2694		20 14.4 +80 32	6.9, 9.6	343	4.2″	130	B9.5IV	194375, 3364
75	β pm 211	A B C	20 28.2 +81 25	5.5, 11.3, 6.7	11, 282	110″, 3.3′	140	G9III	196787, 3408
74	OΣ 593	A B	20 29.5 +81 05	6.1, 8.7	36	3.6′	64	K0III F8V	196925, 3413

Equuleus Equ   *Chart 19*

Label	Catalog ID	Components	Coordinates (J2000)	Mag.	θ	ρ	Dist.	Spectral type(s)	HD, SAO No.
★ 1 eps	Σ 2737	(AaAb) B C	20 59.1 +04 18	6.0, 6.3, 7.1	284, 67	0.4″, 11″	54	F6IV	199766, 126428
		Solar type (2)+1+1 spectroscopic quadruple, visual triple; impressive in rich field. MSC system mass 3.5 M⊙. Aa,Ab: *P* = 2.0 d. AB: *P* = 101 y, orbit *r* = 35 AU, *e* = 0.71, periastron 2021. (479 obs.) AC: ps = 800 AU. (296 measures; 2012)							
2	Σ 2742		21 02.2 +07 11	7.4, 7.6	214	2.9″	81	F8	200256, 126482
		Pretty solar type matched double. No orbit or linear solution yet. AB: ps = 320 AU. (202 measures; 2011)							
	Σ 2765	A B	21 11.0 +09 33	8.5, 8.5	78	2.8″	152	A3IV	201686, 126601
	S 781	A B D	21 13.5 +07 13	7.4, 9.4, 7.2	347, 172	0.6″, 3.1′	96	A7V	202073, 126625
		A type 2+1 triple. MSC system mass 4.2 M⊙. AB: *P* = 113 y, orbit *r* = 34, slowly closing; AD: ps = 24,100 AU. (2008)							
	β 163	(AaAb) B	21 18.6 +11 34	7.3, 8.9	258	0.8″	47.7	G0V G6V	202908, 107015
		Solar type (2)+1 spectroscopic triple. System mass 2.3 M⊙. Aa,Ab: *P* = 4.0 d. AB: *P* = 79 y, orbit *r* = 25 AU, *e* = 0.87, apastron 2026. (213 measures; 2012)							

Label	Catalog ID	Components	Coordinates (J2000)	Mag.	θ	ρ	Dist.	Spectral type(s)	HD, SAO No.
	Σ 2786		21 19.7 +09 32	7.5, 8.2	189	2.7″	186	A3IV	203067, 126707
	Σ 2793	(AB) C	21 25.1 +09 23	7.4, 9.0	242	27″	136	A5IV	203943, 126783
	A type 2+1 triple, visual double. AB (β 164): *P* = 362 y, orbit *r* = 75 AU. (2008)									

Eridanus   Eri   *Charts 13, 21, 26*

Label	Catalog ID	Components	Coordinates (J2000)	Mag.	θ	ρ	Dist.	Spectral type(s)	HD, SAO No.
	Δ 4		01 38.8 −53 26	7.2, 8.5	104	11″	90	F5IV	10241, 232483
★ p	Δ 5		01 39.8 −56 12	5.8, 5.9	187	11″	6.76	K2V+K2V	10360, 232490
	Local, solar type matched double; dark field, at low end of solar mass. AB: ps = 100 AU. (2013)									
	I 455		02 08.2 −55 32	7.8, 11.3	199	4.9″	220	K1III	13307, 232645
	h 3527		02 43.3 −40 32	7.0, 7.2	41	2.3″	182	B9.5V	17098, 216019
	β 83	A B	02 46.0 −04 57	7.7, 9.6	14	1.0″	76	F2	17251, 130100
	Solar type double. AB: *P* = 716 y, orbit *r* = 180 AU, *e* = 0.15, apastron 2026. (2012)									
	β 10		02 50.4 −04 59	7.2, 10.4	100	2.8″	135	A0	17699, 130139
★ the	Pz 2	(AaAb) B	02 58.3 −40 18	3.2, 4.1	91	8.6″	49.9	A4III A1V	18622, 216113
	Acamar. Spectacular high mass (2)+1 spectroscopic binary with rare A giant; B may be optical! AB: ps = 580 AU. (2013)									
9 rho 2	β 11		03 02.7 −07 41	5.4, 8.9	63	1.4″	81	K0II	18953, 130254
	Σ 341		03 03.0 −02 05	7.6, 10.0	221	9.0″	53	F5	18975, 130256
	h 3548		03 03.8 −21 22	7.5, 11.2	125	12″	58	K0V	19096, 168268
	β 527	A B	03 06.1 −13 26	8.2, 8.9	98	1.3″	124	F5V	19313, —
	Cbl 120		03 10.7 −20 07	7.7, 11.1	313	59″	197	G9IV	19814, 168355
★	h 3556	A B C	03 12.4 −44 25	6.4, 7.4, 8.8	159, 189	0.6″, 3.9″	42.5	F7III A0V	20121, 216209
	A type 2+1 triple with F giant. MSC gives 5.8 M⊙. AB: *P* = 45 y, orbit *r* = 17 AU, *e* = 0.90, periastron 2022. AC: ps = 220 AU. (2013)									
	β 84		03 16.0 −05 55	6.4, 7.9	9	1.0″	200	B9V	20319, 130388
16 tau 4	Jc 1	A B	03 19.5 −21 45	3.9, 9.5	291	5.7″	93	M3III	20720, 168460
	β 12		03 24.4 −14 00	7.0, 9.1	281	2.4″	197	A2V	21160, 148943
	Σ 408		03 30.7 −04 16	8.2, 8.4	323	1.2″	113	A3	21789, 130529
	β 532	A B C	03 33.3 −10 04	8.6, 12.1, 7.4	272, 311	2.7″, 80″	185	F2	22096, 149032
	CII 2		03 33.6 −07 25	7.6, 8.0	139	66″	166	A9IV	22128, 130572
	Δ 15		03 39.8 −40 21	6.9, 7.7	330	7.7″	132	A3V	22986, 216431
	h 3589		03 44.1 −40 40	6.7, 9.3	349	5.0″	97	K1III	23508, 216459
★ f	Δ 16		03 48.6 −37 37	4.7, 5.3	216	8.4″	51	B9V A1V	24072, 194551
	High mass double with possible bet Lyr type variable; very bright in dark field. AB: ps = 580 AU. (2009)									
	β 401	A B	03 50.3 −01 31	6.5, 10.5	254	4.2″	43.7	F2	24098, 130762
	Solar type double, comoving with E (HD 22584, m.10) 3.2° s.p. AB: ps = 580 AU. AE (Shy 164): ps = 2.5 pc. (2009)									
★ 32	Σ 470	(AaAb) B	03 54.3 −02 57	4.8, 5.9	351	6.9″	96	G8III A2V	24555, 130806
	A type (2)+1 spectroscopic triple with G giant; sparse bright field. AB: ps = 895 AU. (2011)									
	h 3611		03 56.6 −39 55	8.0, 8.7	139	4.1″	129	A3	24988, 216564
	β 1004	A B	04 02.1 −34 29	7.3, 7.9	57	1.2″	52	G1V	25535, 194709
	I 152	A (BC)	04 04.9 −35 27	8.4, 8.7	76	1.1″	71	G2V	25926, 194747
	I 153		04 08.3 −32 51	8.1, 8.2	347	1.1″	127	A7V	26301, 194790
	Srt 2	(AaAb) B	04 11.6 −20 21	5.8, 7.7	334	62″	81	A1V F0	26591, 169206
	h 3628		04 12.5 −36 09	7.2, 8.0	50	50″	134	F3V	26758, 194831
39	Σ 516	A B	04 14.4 −10 15	5.0, 8.5	144	6.3″	74	K2III	26846, 149478
	h 3632		04 15.1 −30 04	7.8, 9.7	163	11″	140	A0V	27016, 194866
★ 40 omi 2	Σ 518	A [B C]	04 15.3 −07 39	4.5, [10.0, 11.5]	102, [331]	82″, [8.2″]	4.99	G9V DA	26965, 131063
	Keid. Local, very high CPM, solar type 1+2 triple; YO/− color; dark field. Fascinating system: A is a del Sct type variable; B is a white dwarf, the most easily observed from Earth; its companion C is a flare star (DY Eri). MSC system mass 1.5 M⊙. AB: ps = 550 AU; BC: *P* = 252 y, orbit *r* = 35 AU. (2011)									

Label	Catalog ID	Components	Coordinates (J2000)	Mag.	θ	ρ	Dist.	Spectral type(s)	HD, SAO No.
	β 548		04 16.6 −10 05	7.5, 11.4	345	6.2″	250	A3III	27093, 149503
	h 3642		04 19.0 −33 54	6.5, 8.7	160	6.3″	102	A3V	27490, 194923
	h 3644	(AB) D	04 21.5 −25 44	6.2, 8.2	42	44″	59	F2V	27710, 169368
			Solar type (2)+1 triple, visual binary. MSC system mass 4.1 M$_\odot$. AB: P = 81 y, e = 0.59, apastron 2045. (2011)						
	Σ 536		04 22.2 −04 41	7.9, 8.7	190	1.5″	142	A7IV	27699, 131150
	β 311		04 26.9 −24 05	6.7, 7.1	150	0.4″	98	A3V	28312, 169455
			A type binary; difficult resolution challenge typical of Burnham inventory. AB: P = 596 y, orbit r = 100 AU. (2010)						
	β 184		04 27.9 −21 30	7.4, 7.7	248	1.9″	101	F6V	28396, 169475
	Stone 8	(AB) C	04 28.9 −25 12	7.9, 9.5	352	7.0″	340	K0IV	28521, —
	Σ 560		04 31.4 −13 39	6.3, 9.3	45	30″	113	A2V	28763, 149702
	Σ 570		04 35.2 −09 44	6.7, 7.6	260	13″	198	A1m	29173, 131335
51	Wal 32	A (CaCb)	04 37.6 −02 28	5.2, 10.6	163	67″	29.4	F0V	29391, 131358
	Σ 576	A B	04 38.0 −13 02	7.3, 7.9	172	12″	118	B9.5IV	29482, 149776
53	Kui 18		04 38.2 −14 18	4.0, 7.0	0	1.1″	33.7	K2III	29503, 149781
			Doube with K giant primary; brightness contrast challenge. AB: P = 77 y, ps = 24 AU, near apastron. (2010)						
	β 1236	A B C	04 39.6 −21 15	7.3, 10.3, 9.0	97, 314	1.5″, 40″	87	K1III	29674, 169640
55	Σ 590	(AaAb) (BaBb)	04 43.6 −08 48	6.7, 6.8	318	9.3″	650	G5III	30021, 131443
			High mass, distant (2)+(2) spectroscopic quadruple with G giant; dark field, YW/− color. B (DW Eri) is a del Sct type variable. MSC system mass 7.2 M$_\odot$. AB: ps = 8,150 AU. (2011)						
62	Sh 48	Aa Ab B	04 56.4 −05 10	5.5, 9.6, 8.9	247, 76	0.6″, 67″	230	B6V	31512, 131614
	Σ 631		05 00.7 −13 30	7.5, 8.8	107	5.9″	350	A0	32179, 150076
	Σ 636		05 03.0 −08 40	7.1, 8.5	107	3.5″	188	A0V	32468, 131720
66	Σ 642	(AaAbB) C	05 06.8 −04 39	5.1, 10.8	10	52″	95	B9mn	32964, 131777

Fornax   For   Chart 21

Label	Catalog ID	Components	Coordinates (J2000)	Mag.	θ	ρ	Dist.	Spectral type(s)	HD, SAO No.
	Stone 5		02 17.4 −30 43	8.2, 8.9	199	2.5″	107	F6V	14246, 193623
	β 738		02 23.2 −29 52	7.6, 8.0	212	1.9″	39.0	G1V	14882, 193680
			Nearly matched, solar type binary. AB: P = 560 y, orbit r = 85 AU, widening along edgewise orbit. (2010)						
	Skf 1273		02 24.4 −37 22	7.0, 10.7	195	57″	71	F5V	15048, 193690
	Shy 140		02 28.0 −33 49	5.1, 7.6	218	6.5′	46.6	A2V F7V	15427, 193723
ome	h 3506		02 33.8 −28 14	5.0, 7.7	246	11″	148	B9.5IV	16046, 167882
	h 3509	(AaAb) B	02 34.2 −31 31	7.6, 11.3	59	24″	102	A9V	16087, 193774
	β 261	A B C	02 43.8 −27 54	7.9, 9.2, 10.5	101, 133	3.1″, 69″	270	G5III	17082, 168005
	BrsO 1	(AaAb) B	02 44.2 −25 30	7.0, 8.5	193	13″	41.3	G3V+G0	17134, 168012
			Solar type (2)+1 triple, visual binary; all three stars G type. MSC system mass 2.8 M$_\odot$. Aa,Ab: P = 6.7 y, e = 0.51. (2013)						
	h 3532		02 48.6 −37 24	7.0, 8.1	144	5.3″	57	F3IV	17627, 193926
eta 2	h 3536		02 50.2 −35 51	6.0, 10.0	18	4.9″	138	K0III	17793, 193940
	β 741	A B [(CaCb) D]	02 57.2 −24 58	8.1, 8.2, [7.9, 11.9]	346, 226, [173]	0.8″, 29″, [5.1″]	22.5	K1V	18455, 168181
			Local, solar type 2+(2)+1 quintuple system. MSC system mass 3.4 M$_\odot$. AB: P = 150 y, orbit r = 32 AU, e = 0.60, periastron 2020. (Ca,Cb): P = 1.5 y, e = 0.56. (2013)						
★ alp	h 3555		03 12.1 −28 59	4.0, 7.2	300	5.2″	14.2	F8V	20010, 168373
			Local, high CPM, solar type binary. AB: P = 269 y, orbit r = 57 AU, near cusp of apparent orbit. (2009)						
	LDS 93		03 20.1 −28 51	7.4, 8.5	358	4.2′	35.5	G3V	20782, 168469
	h 3572		03 24.0 −26 13	8.2, 8.5	94	21″	97*	F4V F4V	21145, 168511
★ chi 1	Skf 1280	A (BC) D	03 25.9 −35 55	6.4, 7.3, 10.0	164, 214	2.3′, 85″	108	A1IV+A9V	21423, 194289
★ chi 3	I 58	A B C	03 28.2 −35 51	6.5, 10.1, 9.3	247, 193	6.5″, 3.6′	114	A1V	21635, 194318
			A type 2+1 triple, comoving with A type chi 1 Fornacis (Skf 1280) 34′ s.p. AC: ps = 32,300 AU. chi1/chi3: ps = 1.3 pc. (2010)						

Label	Catalog ID	Components	Coordinates (J2000)	Mag.	θ	ρ	Dist.	Spectral type(s)	HD, SAO No.
	Gemini		Gem						*Chart 8*
	ΟΣ 134	A [B C] D	06 09.3 +24 26	7.6, [9.1, 10.1], 12.6	189, [334], 252	31″, [1.9″], 44″	43*	G0	41996, 78038
7 eta	β 1008	(AaAb) B	06 14.9 +22 30	3.5, 6.2	260	1.7″	118	M3.5I	42995, 78135
			Propus. High mass (2)+1 spectroscopic triple with M type supergiant; faint field. MSC system mass 9.5 M$_\odot$. Aa,Ab: *P* = 8.2 y. AB: *P* = 474 y, orbit *r* = 127 AU, decreasing *θ*. (2012)						
	Σ 899		06 22.8 +17 34	7.4, 8.0	13	2.4″	330	A0V	44496, 95602
	ΟΣ 140	A B	06 26.6 +15 31	6.9, 10.1	119	3.0″	340	B9.5IV	45180, 95684
18 nu	ΟΣΣ 77	(AaAb) (BaBb)	06 29.0 +20 13	4.1, 8.0	330	111″	167	B6IIIe	45542, 78423
			High mass, spectroscopic (2)+(2) quadruple with B giant; Aa,Ab: *P* = 18.8 y, *e* = 0.3, apastron 2020. AB: ps = 25,000 AU. (2013)						
	ΟΣ 145		06 32.3 +15 42	7.3, 9.9	338	1.5″	198	F5	46148, 95800
20	Σ 924	(AaAb) B	06 32.3 +17 47	6.3, 6.9	211	20.2″	80	F8III	46136, 95795
	S 524	A B	06 34.1 +22 07	7.2, 7.4	244	53″	86	A3	46401, 78508
	Σ 932		06 34.4 +14 45	8.3, 8.5	305	1.7″	80	F5	46495, 95847
	ΟΣ 149		06 36.4 +27 17	7.1, 9.0	285	0.7″	33.1	dG2	46780, 78540
			Solar type binary; YW/– color, sparse field. AB: system mass 1.8 M$_\odot$, *P* = 119 y, orbit *r* = 28 AU, *e* = 0.73, periastron 2042. (2009)						
	Σ 957	A B	06 45.2 +30 50	7.5, 9.4	91	3.6″	81	A0	48510, 59444
	ΟΣ 160		06 54.4 +21 10	6.7, 9.9	188	1.3″	118	K1III	50482, 78852
★ 38 e	Σ 982	(AaAb) B	06 54.6 +13 11	4.8, 7.8	147	7.1″	25.6	F0V	50635, 96265
			Solar type (2)+1 triple, visual binary with del Sct type variable; pretty YW/B color. MSC system mass 5.7 M$_\odot$. AB: ps = 250 AU. (295 measures; 2012)						
	Σ 1007	A D	07 00.6 +12 43	7.4, 7.7	28	68″	280	A2V	52155, 96372
	Ho 342		07 02.8 +13 05	8.0, 8.7	88	1.2″	210	F5	52715, —
	Eng 28	A B	07 08.0 +15 32	7.9, 7.7	99	2.9′	46.8	G0V	54046, 96526
	Σ 1035		07 12.0 +22 17	8.1, 8.4	41	8.9″	164	F7IV	55005, 79151
	Wei 14		07 12.8 +15 11	7.8, 8.9	160	2.1″	450*	B9.5IV	55283, 96630
★	Σ 1037	A B	07 12.8 +27 14	7.2, 7.3	307	1.0″	42.5	F8V	55130, 79170
			Solar type matched binary. AB: *P* = 119 y, orbit *r* = 34 AU, *e* = 0.93, closing to periastron 2039. (421 measures; 2012)						
	ΟΣ 167	A B	07 13.5 +32 09	7.4, 10.9	180	4.3″	190	A8V	55225, 59896
★ 55 del	Σ 1066	(AaAb) B	07 20.1 +21 59	3.6, 8.2	230	5.6″	18.5	A9III K3V	56986, 79294
			Wasat. Local (2)+1 spectroscopic triple with A type giant; pretty Y/R color. Aa,Ab: *P* = 6.1 y, *e* = 0.35. *P* = 1,200 y, widening and a good occultation target. (254 measures; 2013)						
	Σ 1081	A B	07 24.1 +21 27	7.7, 8.5	238	1.9″	210	B9	57900, 79361
	Σ 1083		07 25.6 +20 30	7.3, 8.1	47	6.7″	132	A5	58246, 79375
	Ho 346	A B	07 25.9 +18 09	7.0, 11.7	58	14″	320	G5	58338, 96888
	Σ 1090	A B	07 26.5 +18 31	7.3, 8.2	99	60″	112	F2V	58453, 96897
	ΟΣ 171		07 26.7 +31 37	7.4, 9.2	137	1.1″	159	G5	58382, 60080
	Σ 1094		07 27.4 +15 19	7.6, 8.5	96	2.5″	510	A0V	58729, 96914
63	Sh 368	(Aa1Aa2Ab) D B	07 27.7 +21 27	5.3, 9.3, 10.9	98, 324	3.9″, 43″	32.0	F5V	58728, 79403
			Solar type (2+1)+1+1 quintuple with 2+1 spectroscopic and occultation triple; visual triple has Y/B color. MSC system mass 4.7 M$_\odot$. Aa,Ab: *P* = 2.1 y, *e* = 0.42. Aa1,Aa2: *P* = 1.9 d, *e* = 0. AD: ps = 125 AU, est. *P* = 760 y; AB: ps = 1860 AU, est. *P* = 26,000 y. (2006)						
62 rho	Alc 3	(AaAb) B E	07 29.1 +31 47	4.2, 12.5, 7.8	8, 355	2.8″, 12.6′	18.1	F0V	58946, 60118
			Local, solar type (2)+1+1 spectroscopic quadruple, visual binary; bright sparse field. AE: ps = 13,600 AU, near 100% probability pair is physical. (2001)						
	Σ 1102	A B D	07 30.4 +13 52	7.4, 9.2, 8.0	46, 131	7.5″, 112″	45.4	F5	59432, 96964
	Σ 1116		07 34.5 +12 18	7.8, 8.5	96	1.7″	250	B8	60355, 97033

Label	Catalog ID	Components	Coordinates (J2000)	Mag.	θ	ρ	Dist.	Spectral type(s)	HD, SAO No.
★ 66 alp	Σ 1110	(AaAb) (BaBb) (CaCb)	07 34.6 +31 53	1.9, 3.0, 9.8	55, 165	5.0″, 71″	15.6	A1V A2Vm	60179, 60198
		Castor. Local, A type (2)+(2)+(2) spectroscopic sextuple system, closest sextuple to the Sun. One of the grandest doubles in the northern skies. Discovered by Bradley and Pound (1719), one of six cited by W. Herschel (1802) as proof of gravitational attraction (orbital motion) in double stars. MSC system mass 7.6 M$_\odot$. Aa,Ab: P = 9.2 d, ps = 0.12 AU; Ba,Bb: P = 2.9 d, ps = 0.03 AU; AB: P = 467 y, orbit r = 105 AU, e = 0.34, widening; Ca,Cb (eclipsing variable YY Gem): P = 19.5h, ps = 0.03 AU; AB,C: P = 25,800 y, orbit r = 1,490 AU. (1,414 measures; 2014)								
77 kap	OΣ 179		07 44.4 +24 24	3.7, 8.2	242	7.5″	43.4	G8III	62345, 79653
	Σ 1140		07 48.4 +18 20	7.0, 8.7	275	6.2″	210	K0III	63210, 97260
	OΣΣ 89		07 51.0 +31 37	6.8, 7.7	83	77″	200	A6III	63610, 60380

Grus Gru Charts 20, 26

Label	Catalog ID	Components	Coordinates (J2000)	Mag.	θ	ρ	Dist.	Spectral type(s)	HD, SAO No.
	h 5275	(AB) C	21 31.0 −36 33	7.7, 11.3	201	41″	101	F3V	204635, 213058
	h 5319		22 12.0 −38 18	7.7, 7.7	315	2.1″	76	F3V	210571, 213631
	CorO 250		22 13.0 −49 03	7.7, 10.4	353	5.6″	220	G8III	210657, 231031
	I 136		22 25.9 −45 07	7.8, 8.9	278	2.1″	120	G8IV	212538, 231134
sig 2	β 771	A B C	22 37.0 −40 35	5.9, 10.0, 6.3	265, 275	2.7″, 5.6′	66	A1V	214150, 231217
	I 138		22 37.8 −39 51	6.7, 10.7	277	3.4″	90	F7V	214291, 231223
	CorO 252	A B	22 42.6 −47 13	6.0, 11.1	121	8.4″	23.6	F9V M1V	214953, 231257
		Local, unequal solar type double; dark field, 18′ s. of bet Gru, C probably unrelated. AB: ps = 270 AU. (2013)								
	h 5362	(AaAb) B	22 46.7 −46 56	6.6, 9.9	143	11″	78	A9III	215545, 231290
	I 22	A B (CD)	22 55.3 −48 28	7.3, 8.9, 6.7	175, 181	0.6″, 93″	47.7	G3	216655, 231354
		Solar type 2+(2) quadruple, visual triple; faint group 5′ s.f. AB: P = 198 y, orbital r = 46 AU. AB,CD: ps = 5,980 AU, near 100% probability pair is physical. (2013)								
	β 1011		23 02.6 −36 25	6.6, 9.3	293	2.1″	126	K1III	217642, 214261
ups	β 773		23 06.9 −38 54	5.7, 8.2	205	0.9″	87	A1V	218242, 214313
★ the	Jc 20	A B C	23 06.9 −43 31	4.5, 6.6, 7.8	115, 292	1.5″, 2.6′	40.4	F3V F5m	218227, 231444
		Solar type 2+1 triple; sparse field. AC: ps = 8,500 AU, near 100% probability pair is physical. (2013)								
★	Δ 246		23 07.2 −50 41	6.3, 7.1	254	9.1″	45.7	F7IV	218269, 247739
		Solar type double; sparse field. Easy pair in all apertures, characteristic of the Dunlop catalog. AB: ps = 560 AU. (2010)								
	Δ 248	A B	23 20.8 −50 18	6.2, 8.9	232	1.3″	86	A8	220003, 247838
	Δ 249		23 23.9 −53 49	6.1, 7.1	211	27″	119	A4III	220392, 247854

Hercules Her Charts 10, 11

Label	Catalog ID	Components	Coordinates (J2000)	Mag.	θ	ρ	Dist.	Spectral type(s)	HD, SAO No.
	OΣ 307		16 10.5 +47 48	7.7, 10.7	201	18″	240	K0	145768, 45940
	Σ 2024		16 11.8 +42 22	5.9, 10.7	44	23″	194	K4III	145931, 45957
★ 49	Σ 2021	(AaAb) B	16 13.3 +13 32	7.4, 7.5	357	4.1″	23.6	G9V	145958, 102018
		Local, high CPM, solar type (2)+1 triple, visual matched binary. AB: P = 1354 y, orbit r = 120 AU. (406 measures; 2012)								
	Σ 2049		16 27.9 +25 59	7.3, 8.1	195	1.1″	133	A2.5V	148554, 84393
★	Σ 2052	A B	16 28.9 +18 25	7.7, 7.9	119	2.3″	19.7	K1V	148653, 102200
		Local, high CPM, solar type binary, AB: system mass 1.7 M$_\odot$, P = 244 y, orbit r = 44 AU, e = 0.75, apastron 2033. (526 measures; 2013)								
	Σ 2051		16 29.4 +10 36	7.7, 9.4	19	14″	220	G5III	148683, 102204
	Sh 233		16 31.5 +08 18	7.1, 8.3	70	59″	78	G5	148979, 121665
	Σ 2056		16 31.6 +05 26	7.8, 9.2	313	6.8″	118	A3	148980, 121667
	Σ 2063		16 31.8 +45 36	5.7, 8.7	196	17″	69	A2V	149303, 46147
	OΣ 313		16 32.6 +40 07	8.0, 8.3	130	0.9″	460	F9IV	149379, 46152
	Webb 6	(AaAb) B	16 35.4 +17 03	6.4, 7.3	359	2.6′	160	A2V	149632, 102259
	Σ 2079		16 39.6 +23 00	7.6, 8.1	91	17″	410	F0	150340, 84521

Label	Catalog ID	Components	Coordinates (J2000)	Mag.	θ	ρ	Dist.	Spectral type(s)	HD, SAO No.
36,37	Σ I 31	(AaAb) B	16 40.6 +04 13	5.8, 6.9	229	70″	90	A1V+A3IV	150378, 121776
★ 40 zet	Σ 2084		16 41.3 +31 36	3.0, 5.4	149	1.3″	10.7	G1IV	150680, 65485
		Local, high CPM, solar type binary. AB: system mass 2.3 M$_\odot$, P = 34 y, orbit r = 19 AU, e = 0.46, apastron 2019. (834 measures; 2013)							
	Σ 2085		16 42.4 +21 36	7.4, 9.2	311	6.3″	190	A0IV	150781, 84550
	Σ 2094	A B C	16 44.2 +23 31	7.5, 7.9, 11.7	73, 310	1.1″, 25″	156	F5III	151070, 84572
41	OΣ 585	A B	16 45.0 +06 05	6.7, 10.4	191	2.7′	45.9	K0V K3V	151090, 121831
	Σ 2095		16 45.1 +28 21	7.4, 9.2	160	5.3″	220	F7III	151237, 84577
	Σ 2101	A B	16 45.8 +35 38	7.5, 9.4	47	4.0″	59	F6V	151428, 65537
	Σ 2104	A B	16 48.7 +35 55	7.5, 8.8	21	5.9″	173	F2	151878, 65569
52	β 627	A (BC)	16 49.2 +45 59	4.8, 8.5	38	2.0″	55	A1V	152107, 46305
		V637 Her. A type 1+(2) triple, visual binary with alp2 CVn variable. BC: P = 56 y. A,BC: P = 1,977 y, orbit r = 270 AU, widening. (2012)							
	Σ 2107	A B C	16 51.8 +28 40	6.9, 8.5, 11.5	109, 309	1.8″, 83″	58	F5IV	152380, 84655
		Solar type 2+1 triple. MSC system mass 3.7 M$_\odot$. AB: P = 268 y, orbit r = 55 AU, apastron 2029. AC: ps = 6,500 AU. (430 measures; 2012)							
	Σ 2109		16 53.8 +21 10	7.5, 10.3	313	5.7″	189	K0	152629, 84674
	OΣ 318		16 56.7 +14 08	7.0, 9.6	242	2.9″	132	G9III	153064, 102488
	Σ I 32	A B C	16 57.9 +47 22	7.9, 10.9, 8.1	63, 260	4.9″, 112″	18.3	K0	153557, 46409
		Local, high CPM, solar type 2+1 triple; dark field. MSC gives 2.1 M$_\odot$. AB: ps = 120 AU; AC: P = 64,000 y, orbit r = 2,010 AU, near 100% probability pair is physical. (2007)							
	Pry 2	A B	17 04.7 +19 36	6.2, 9.3	227	1.8″	177	A0IV	154441, 102579
63	Shy 713		17 11.1 +24 14	6.2, 7.0	74	3.3′	82	A8V F2V	155514, 84896
	Σ 2142	A B	17 11.7 +49 45	6.2, 9.4	109	5.3″	95	A5III	155860, 46561
★ 64 alp 1,2	Σ 2140	(AaAb) (BaBb)	17 14.6 +14 23	3.5, 5.4	102	4.8″	110	M5II	156014, 102680
		Rasalgethi. High mass (2)+(2) quadruple, visual double with semiregular pulsating M type supergiant; shows YO/YO color at low magnification but Y/B at high magnification. Discovered (with rho Her, below) by C. Mayer (1777). MSC lists a (2)+(2) quadruple, system mass 22.4 M$_\odot$, but there may be a fifth component. Ba,Bb: P = 52 d. AB: P = 3,600 y, orbit r = 520 AU. (471 measures; 2013)							
★ rho	Σ 2161	(AaAb) B	17 23.7 +37 09	4.5, 5.4	321	4.0″	121	B9.5III	157778, 66001
		High mass (2)+1 interferometric triple, visual double. MSC system mass 9.2 M$_\odot$. AB: ps = 650 AU. (375 measures; 2013)							
	OΣ 329		17 24.5 +36 57	6.4, 9.9	12	34″	230	G5III F0V	157910, 66014
	Σ 2160		17 24.6 +15 36	6.4, 9.3	66	3.8″	153	B9V	157741, 102806
	Σ 2189	A B C	17 32.8 +47 53	7.8, 11.2, 8.9	98, 359	21″, 66″	210	A2V	159543, 46791
	Σ 2190	(AaAb) B	17 36.0 +21 00	6.1, 9.5	23	10″	113	A7IV	159834, 85232
	Σ 2194	A B	17 41.1 +24 31	6.5, 9.3	8	16″	162	K1III	160835, 85310
	Σ 2203		17 41.2 +41 39	7.7, 7.8	292	0.8″	158	A4V	161016, 46884
	Σ 2198		17 42.6 +26 33	7.6, 11.2	25	7.7″	177	K0III	161112, 85346
★ 86 mu	Σ 2220	(AaAb) [B C]	17 46.5 +27 43	3.5, [10.2, 10.7]	249, [245]	36″, [1.1″]	8.31	G5IV	161797, 85397
		Local, high CPM, solar type (2)+2 quadruple, visual triple; dominant in dark field. MSC system mass 1.7M$_\odot$. Large amplitude astrometric binary Aa,Ab: P = 65 y, e = 0.32, periastron 2016; BC (AC 7): P = 43.1 y, orbit r = 12 AU, e = 0.18, apastron 2030. (2013)							
	Σ 2215	A B	17 47.1 +17 42	6.0, 6.9	253	0.5″	143	A1V	161833, 103106
		A type spectroscopic binary; 5 stars 6′ f., m.12 binary 4′ n.p. Low quality orbit AB: P = 1,062 y, orbit r = 120 AU. (220 measures; 2013)							
	Σ 2232		17 50.3 +25 17	6.7, 8.9	137	6.3″	147	A1V	162485, 85459
	Σ 2242		17 51.2 +44 54	8.1, 8.3	326	3.5″	118	F0	162880, 47012
	OΣ 338	A B	17 52.0 +15 20	7.2, 7.4	163	0.8″	230	G8III	162734, 103161
		Binary with G giant. AB: P = 1,277 y, orbit r = 260 AU. (320 measures; 2013)							
90	β 130		17 53.3 +40 00	5.3, 8.8	110	1.6″	108	K3III	163217, 47037

Label	Catalog ID	Components	Coordinates (J2000)	Mag.	θ	ρ	Dist.	Spectral type(s)	HD, SAO No.
	Σ 2245	(AaAb) B	17 56.4 +18 20	7.4, 7.6	290	2.6″	240	A0III	163640, 103227
	Hu 235		17 57.1 +45 51	6.9, 9.0	283	1.6″	70	F7IV	164059, 47084
	LDS 6413		18 00.6 +29 34	7.1, 13.1	104	88″	28.4	G2V+M	164595, 85632
	Σ 3129	(AaAb) B	18 01.1 +45 21	7.6, 10.6	168	31″	132	B9	164898, 47139
★ 95	Σ 2264		18 01.5 +21 36	4.9, 5.2	257	6.5″	128	A5IIIn	164669, 85648
		Double with A type giant; faint field. AB: ps = 1,100 AU. (262 measures; 2013)							
	β 1127	A B	18 02.5 +44 14	7.3, 9.2	50	0.7″	81	F5V	165170, 47163
V772	OΣ 341	(AaAb) B	18 05.8 +21 27	7.4, 8.8	92	0.5″	39.5	G0V+G5V	165590, 85723
		Solar type (2)+1 spectroscopic triple; primary is an eclipsing type variable (P = 0.9 d). AB: system mass 2.3 M$_\odot$, P = 20 y, orbital r =10 AU, e = 0.96, periastron 2018, returns to view 2025. (266 measures; 2010)							
	Σ 2282	A (BC)	18 06.5 +40 22	7.9, 8.7	80	2.7″	510	A1V	165941, 47220
99 b	AC 15	A B	18 07.0 +30 34	5.1, 9.0	317	1.2″	15.6	F7V	165908, 66648
		Local, solar type binary; faint field. AB: system mass 1.8 M$_\odot$, P = 56 y, orbit r = 17 AU, e = 0.77, apastron 2025. (212 measures; 2011)							
	OΣ 344	(AaAb) B	18 07.1 +49 43	6.5, 10.3	140	2.3″	220	A2V	166228, 47233
	Hu 674		18 09.7 +50 24	7.7, 8.6	212	0.7″	141	A3V	166820, 30776
	Σ 2289		18 10.1 +16 29	6.7, 7.2	222	1.2″	260	A0V G0III	166479, 103443
		A type binary with G giant; sparse field. Low quality orbit AB: P = 3,040 y, orbit r = 390 AU. (327 measures; 2013)							
	H V 93		18 13.0 +28 15	8.2, 8.3	136	55″	82	F8	167215, 85832
	Σ 2315	(AaAb) B	18 25.0 +27 24	6.6, 7.8	118	0.6″	118	A0V A4V	169718, 86019
		A type (2)+1 spectroscopic triple, visual double; three stars n. unrelated; galaxy NGC 6632 10′ n. AB: P = 2,094 y, orbit r = 240 AU. (284 measures; 2011)							
	β 1326	(AaAb) B C	18 26.7 +26 27	6.5, 12.1, 9.6	104, 60	4.7″, 62″	320	B3V	170111, 86060
	Σ 2319	A B	18 27.7 +19 18	8.4, 8.2	190	5.4″	161	F5	170267, 103740
	Σ 2320	(AaAb) (BaBb)	18 27.8 +24 42	7.1, 8.9	358	1.1″	260	B9V	170314, 86083
★	Σ 2339	(AB) [C D]	18 33.8 +17 44	7.5, [9.3, 9.6]	273, [262]	2.0″, [0.4″]	183	F6V	171365, 103853
		High mass, solar type (2)+2 quadruple, miniature visual 1+2 triple; faint field, CD is a large aperture detection test. AB: P = 42 y, orbit r = 27 AU, apastron 2030; AC: ps = 490 AU. (2011)							
	OΣ 359		18 35.5 +23 36	6.4, 6.6	4	0.8″	144	G9III +G7III	171745, 86224
		Binary with G giant; faint rich field. AB: system mass 4.5 M$_\odot$, P = 219 y, orbit r = 60 AU, e = 0.84, widening to apastron 2038. (265 measures; 2013)							
★	OΣ 358	A B E	18 35.9 +16 59	6.9, 7.1, 12.0	149, 349	1.7″, 35″	32.9	F8V	171746, 103886
		Solar type binary or 2+1 triple (E status is uncertain); pretty field. AB: P = 380 y, orbit r = 60 AU. (503 measures; 2013)							
	Σ 2360		18 39.3 +20 56	8.0, 9.2	358	2.4″	820*	B5IV	172421, 86288
	Σ 2401	A B	18 49.0 +21 10	7.3, 9.3	38	4.3″	620*	B3V	174261, 86458
	Σ 2411	A (BaBb) C D	18 52.3 +14 32	6.6, 9.6, 11.0, 11.2	95, 96, 134	13″, 66″, 113″	111	G9III	174897, 104203
	Σ 2415		18 54.5 +20 37	7.1, 8.7	289	2.0″	195	A0IV	175427, 86563

Horologium — Hor — *Chart 27*

Label	Catalog ID	Components	Coordinates (J2000)	Mag.	θ	ρ	Dist.	Spectral type(s)	HD, SAO No.
	Δ 1	A B C	02 28.0 −58 08	8.0, 8.5, 9.6	32, 301	1.2″, 18″	90	F8	15546, 232780
★	CorO 14	A B C	02 38.7 −52 57	7.9, 8.5, 6.8	129, 75	8.9″, 3.6′	60	F1III F8V G8V	16699, 232841
		Solar type 2+1 triple with F giant; most splendid in a dark sky. AC: ps = 17500 AU. (2013)							
	h 3520		02 38.9 −54 50	7.7, 8.6	205	21″	129	F0V	16744, 232845
	Δ 7	A [B C]	02 39.7 −59 34	7.7, [8.0, 8.9]	97, [9]	37″, [0.4″]	185	K0III	16852, 232851
	φ 333		02 43.4 −66 43	6.5, 8.2	215	0.5″	54	F5V	17326, 248632
		Solar type binary. AB: system mass 2.7 M$_\odot$, P = 33 y, orbit r = 14 AU, e = 0.96, closing to periastron 2027. Large aperture challenge. (2013)							

Label	Catalog ID	Components	Coordinates (J2000)	Mag.	θ	ρ	Dist.	Spectral type(s)	HD, SAO No.
	Δ 10		03 04.6 −51 19	7.6, 8.5	70	38″	53	G1V	19330, 232983
	h 3559	(AaAb) B	03 10.1 −63 55	6.7, 10.1	41	43″	260	A3III	20060, 248748
	h 3576		03 24.6 −45 40	7.3, 8.8	341	2.9″	138	A2V+	21319, 216302
	Skf 949		03 53.6 −46 54	6.1, 8.5	4	77″	108	K2III+G	24706, 216540
	Shy 458	C (AaAb)	04 03.8 −44 29	8.2, 8.6	123	10′	69	G5V	25842, 216623
	h 3643		04 19.3 −44 16	5.5, 8.6	115	70″	77	K2III	27588, 216749

Hydra — Hya — Charts 15, 23

Label	Catalog ID	Components	Coordinates (J2000)	Mag.	θ	ρ	Dist.	Spectral type(s)	HD, SAO No.
	β 102		08 16.8 −09 01	7.1, 10.3	117	3.2″	144	A0V	69460, 135717
	Σ 1216		08 21.3 −01 36	6.9, 7.9	305	0.5″	156	A2Vn	70340, 135804

A type binary; double Wz 12 (m.10, 8″) 8′ f. in sparse field. AB: system mass 6.4 M⊙, P = 402 y, orbit r = 90 AU, e = 0.11, apastron 2121. (2013)

Label	Catalog ID	Components	Coordinates (J2000)	Mag.	θ	ρ	Dist.	Spectral type(s)	HD, SAO No.
	Σ 1243		08 33.9 +01 35	7.9, 9.4	233	1.7″	270	A0	72605, 72605
	Σ 1255	A B	08 39.7 +05 46	7.3, 8.6	33	26″	36.3	G1V	73668, 117000
	Σ 1261		08 40.7 −11 56	7.7, 9.6	303	30″	78	G5	73940, 154533
	Σ 1260		08 40.7 −12 10	7.9, 8.1	302	4.9″	139	A2	73941, 154531
	S 579	A B	08 43.7 −07 14	4.7, 8.2	310	79″	240	G1Ib	74395, 136221
	Σ 1270		08 45.3 −02 36	6.9, 7.5	265	4.5″	62	F2IV	74688, 136243
★ 11 eps	Σ 1273	(AB) (CaCb) D	08 46.8 +06 25	3.5, 6.7, 12.5	307, 200	2.9″, 18″	39.6	F8V	74874, 117112

Ashlesha. Solar type (2)+(2)+1 spectroscopic quintuple; visual triple with BY Dra variable; dark field. MSC system mass 7.0 M⊙. AB: P = 15.1y, e = 0.66. (265 measures); Ca,Cb: P = 9.9 d. AB,C: P = 589 y, orbit r = 135 AU. AD: ps = 710 AU. (426 measures; 2012)

Label	Catalog ID	Components	Coordinates (J2000)	Mag.	θ	ρ	Dist.	Spectral type(s)	HD, SAO No.
	β 335		08 48.2 +02 35	7.5, 9.4	265	2.6″	153	F5	75121, 117140
15	β 587	(AaAb) B	08 51.6 −07 11	5.8, 7.4	121	1.2″	137	A4m	75737, 136345
	β 407	A B C	08 51.7 −06 47	7.8, 10.3, 11.4	167, 165	5.9″, 92″	230	A0	75770, —
	Σ 1290		08 52.1 +04 28	7.4, 9.2	325	2.8″	200	A2	75768, 117208
	β 24		08 54.2 −08 46	8.0, 8.6	176	1.2″	260	F0 A2	76174, 136388
	β 103		08 54.9 −07 49	7.7, 9.8	72	3.1″	130*	A5	76274, 136400
17	Σ 1295		08 55.5 −07 58	6.7, 6.9	4	3.9″	89	A2m A7m	76370, 136409
	β 409		09 00.8 −09 11	7.3, 10.1	187	10″	167	A0	77196, 136489
NP	β 211		09 02.0 +02 40	7.3, 8.8	268	1.1″	147	A2	77314, 117363
	Σ 1309		09 06.6 +02 49	8.5, 8.4	274	12″	98	F5	78126, 117428
	★Σ 197		09 09.5 +02 56	7.9, 9.0	66	1.4″	165	F1V	78637, —
	β 104		09 11.5 +00 17	7.0, 10.6	105	3.3″	230	K2III	79011, 117483
	Ho 363		09 15.0 −20 07	7.5, 9.3	183	2.5″	157	A1V	79709, —
	β 212		09 16.1 −08 21	7.8, 8.3	198	1.6″	131	A5	79825, —
	Σ 1336		09 17.5 +00 33	7.0, 11.1	181	41″	98	A0	80046, 117569
27	Sh 105	(AaAb) [B C]	09 20.5 −09 33	4.9, [7.0, 11.0]	211, [198]	3.8′, [9.1″]	68	G8III F4V	80586, 136768
	Bvd 145		09 21.5 −17 45	8.2, 9.0	317	46″	90*	F5V F7V	80828, 155124
	β 337		09 22.5 −17 54	7.0, 10.2	351	9.6″	102	F4V	80971, 155129
	Σ 1347		09 23.3 +03 30	7.3, 8.3	312	21″	109	F0	81029, 117641
★	Σ 1348	A B	09 24.5 +06 21	7.5, 7.6	316	2.1″	71	F7V	81212, 117661

Exquisite matched double; dark field, CPM component C (m.14) 2′ n.p. AB: ps = 200 AU. (245 measures; 2012)

Label	Catalog ID	Components	Coordinates (J2000)	Mag.	θ	ρ	Dist.	Spectral type(s)	HD, SAO No.
29	β 590	A B C	09 27.2 −09 13	7.8, 7.0, 11.3	196, 169	0.4″, 11″	230	A2V	81728, 136861
★	Σ 1355		09 27.3 +06 14	7.7, 7.8	353	1.8″	54	F7V	81670, 117704

Matched solar type binary; wandering in a lonely field. AB: P = 591 y, orbit r = 95 AU. (2013)

Label	Catalog ID	Components	Coordinates (J2000)	Mag.	θ	ρ	Dist.	Spectral type(s)	HD, SAO No.
	Σ 1357	A B	09 28.3 −09 59	6.9, 9.9	55	7.6″	240	K0	81902, 136883
31 tau 1	h 1167	(AaAb) B	09 29.1 −02 46	4.6, 7.3	4	68″	17.3	F6V K0	81997, 136895

Ukdah. Local, solar type (2)+1 triple; sparse field. MSC system mass 2.3 M⊙. (A)B: ps = 1590 AU. (2012)

Label	Catalog ID	Components	Coordinates (J2000)	Mag.	θ	ρ	Dist.	Spectral type(s)	HD, SAO No.
	β 591		09 29.6 −03 07	7.8, 8.9	32	0.9″	470	F5 A3	82072, 136903

Label	Catalog ID	Components	Coordinates (J2000)	Mag.	θ	ρ	Dist.	Spectral type(s)	HD, SAO No.
	Σ 1365		09 31.5 +01 28	7.4, 8.0	157	3.4″	96	F9III	82355, 117747
	β 910	A B	09 32.9 −14 00	7.3, 10.2	306	7.0″	146	K0IV	82661, 155280
	Cbl 131		09 33.3 −07 11	6.4, 11.0	20	57″	125	K0	82674, 136951
	Σ 1371	A B	09 35.4 +03 54	7.9, 10.2	276	7.7″	47.0	G0	82994, 117792
	S 604		09 35.6 −19 35	6.3, 9.4	90	52″	88	A2V	83104, 155323
	Cbl 132		09 51.1 −18 40	7.4, 11.2	116	51″	120	F7V	85402, 155533
	h 4261		09 53.6 −19 29	7.8, 9.4	82	8.4″	250	K0III	85752, —
	β 217		10 06.8 −24 43	7.9, 8.0	132	2.1″	88	F8IV	87793, 178425
★	β 218		10 07.4 −19 43	8.2, 8.2	138	0.5″	210	A2IV	87840, 155744
			A type double; sparse or dark field, m.12 star 1′ n.p. is unrelated. AB: ps = 140 AU. Neglected. (1998)						
	β 911	A B	10 08.4 −19 45	7.4, 11.2	313	4.1″	37.6	G0V	87998, 155757
	Σ 1416		10 12.3 −16 05	7.7, 9.3	277	12″	240	A3III	88536, 155811
	β 219		10 21.6 −22 32	6.7, 8.5	186	1.8″	165	A1V	89828, 178723
	h 4311		10 23.3 −13 23	6.7, 10.3	121	5.1″	53	F5V	90045, 155947
	UC 1941		10 29.7 −22 03	7.2, 10.6	216	38″	164	F0V	90954, 178895
	UC 1946		10 30.9 −26 29	6.6, 10.2	29	29″	86	F7III	91135, 178917
	β 411		10 36.1 −26 41	6.7, 7.8	306	1.3″	41.3	F6V	91881, 179014
			Solar type binary; brightest star in field. System mass 2.1 M⊙. AB: P = 159 y, orbit r = 35 AU, e = 0.76, apastron 2028. (2013)						
	Σ 1474	A B	10 47.6 −15 16	6.7, 7.0	28	68″	410	B9IV	93526, 156235
	Σ 1473	A B	10 47.6 −15 38	7.7, 8.9	10	30″	83	F7II	93527, 156233
	I 503		10 49.3 −26 49	7.8, 8.8	119	1.2″	136	F3V	93785, —
	I 211		10 59.2 −33 44	5.8, 9.5	220	1.9″	48.5	F2V	95221, 201976
★ N	H III 96		11 32.3 −29 16	5.6, 5.7	210	9.4″	26.3	F8V F8V	100286, 179968
			Matched solar type double; s.p. of two stars, beautifully matched in sparse faint field. AB: ps = 335 AU. (2007)						
	h 4455	A B	11 36.6 −33 34	6.0, 7.8	241	3.4″	112	K0III	100893, 202622
	I 232		11 40.0 −33 27	7.0, 10.1	161	2.2″	111	K1III	101387, 202686
	h 4465	A C	11 41.7 −32 30	5.4, 8.3	43	67″	124	K5III F7V	101666, 202717
★ bet	h 4478		11 52.9 −33 54	4.7, 5.5	37	0.7″	95	B9III	103192, 202901
			High mass double with B giant, alp2 CVn type variable; sparse field. AB; ps = 90 AU. Neglected. (1998)						
	Δ 116	A B	11 56.7 −32 16	7.7, 7.8	82	19″	47.3	G8V G8V	103743, 202965
	Jc 17	A B	12 10.0 −34 42	6.4, 8.0	17	3.2″	99	A0V	105686, 203183
	h 4505	A B	12 11.7 −30 36	7.8, 10.8	274	10″	410	G8III	105953, 203219
	Howe 72		12 13.6 −33 48	6.5, 8.6	164	1.3″	128	A0V	106257, 203252
	h 4556		12 54.3 −27 58	7.7, 8.8	82	5.9″	101	F9V	112086, 181664
	Stone 28	(AB) C	13 14.5 −24 17	6.7, 11.5	331	12″	100	F0V	114993, 181476
			Solar type (2)+1 triple, visual binary; faint field. MSC gives 4.0 M⊙. AB: P = 61 y, e = 0.69, periastron 2030; AC: ps = 1,600 AU. (2013)						
★	H N 69	A B	13 36.8 −26 30	5.7, 6.6	189	10″	82	A7III A7IV	118349, 181790
			Matched A type double with A giant; dark field, m.11 double 5′ p. AB: ps = 1,105 AU. (2009)						
	β 938		14 06.3 −26 35	8.1, 8.2	125	0.4″	111	A8V	123107, 182242
V353	UC 2694		14 10.2 −25 24	7.5, 10.9	115	43″	132	F5V	123767, 182302
	β 345	A B	14 41.8 −29 42	7.6, 8.1	287	1.0″	88	A5III	129160, 182772
★ 54 m	H III 97		14 46.0 −25 27	5.1, 7.3	121	8.3″	30.3	F2V	129926, 182855
			Solar type double; pretty brightness contrast, dark field, m.13 double 4′ n. AB: ps = 340 AU. (2009)						
59	β 239		14 58.7 −27 39	6.2, 6.8	8	0.5″	113	A4V+A6V	132219, 183058
			Binary with A type giant; brighter of two stars in field. AB: P = 429 y, orbit r = 95 AU. (2010)						

Hydrus Hyi Charts 26, 30

Label	Catalog ID	Components	Coordinates (J2000)	Mag.	θ	ρ	Dist.	Spectral type(s)	HD, SAO No.
	Gli 8		00 59.6 −75 49	8.1, 8.9	84	27″	102	G0V	6058, 255726
	h 3435		01 25.3 −59 30	7.1, 9.4	1	25″	124	F2IV	8787, 232418

Label	Catalog ID	Components	Coordinates (J2000)	Mag.	θ	ρ	Dist.	Spectral type(s)	HD, SAO No.
	h 3475		01 55.3 −60 19	7.2, 7.2	77	2.5″	56	F2V	11944, 248461
	h 3484		02 07.4 −59 41	7.6, 10.5	62	52″	44.2	F8V	13246, 232642
	h 3568	(AaAb) B	03 07.5 −78 59	5.7, 7.7	226	15″	83	F2II	20313, 255962
★	Lfr 1		03 44.8 −70 02	7.4, 7.6	83	76″	55	G0V G0V	24062, 256022
			Matched solar type double; distinctive in sparse field. AB: ps = 5,640 AU, near 100% probability the pair is physical. (2000)						

Indus   Ind   *Charts 26, 29*

Label	Catalog ID	Components	Coordinates (J2000)	Mag.	θ	ρ	Dist.	Spectral type(s)	HD, SAO No.
	I 41		20 36.5 −45 33	7.7, 8.5	356	2.2″	125	A7III	196014, 230292
	I 17	A B C	20 45.0 −50 29	8.0, 8.0, 7.5	35, 122	1.0″, 2.1′	330	A0IV K0III	197322, 246715
			Pretty A type 2+1 wide matched triple. AB: ps = 450 AU, neglected (1992); AC (Δ 235): ps = 56,100 AU, also neglected. (1999)						
	Skf 1168		20 53.1 −46 37	7.6, 11.6	231	17″	220	A3III	198592, 230414
	I 1429		20 54.8 −46 36	7.8, 8.8	143	0.9″	77	F7V	198828, 230431
	I 130		21 04.1 −47 58	7.2, 9.9	319	3.3″	270	K2III	200248, 230501
	h 5246		21 10.4 −54 34	7.8, 8.0	131	4.1″	34.8	K1V F	201247, 246894
BR	Hu 1626	(AaAb) B	21 11.4 −52 20	7.3, 8.8	117	1.1″	48.4	F8V	201427, 246896
	h 5267	A D	21 26.6 −46 04	7.3, 10.0	182	44″	63	F7V	203934, 230687
			Solar type CPM double; stars n. unrelated, component B not seen since 1834. AD: ps = 3,740 AU, neglected. (1999)						
★	Jc 25	A B	21 44.0 −57 20	6.5, 6.9	4	2.5′	45.5	F5V	206429, 247151
			Wide solar type CPM double; dark field, Y/YO color. Despite similarity, C (3′ s.p.) is unrelated. AB: ps = 9,210 AU. Neglected. (1999)						
	I 19	A B	21 48.7 −65 30	7.3, 8.7	309	1.3″	101	F3V	207015, 255076

Lacerta   Lac   *Charts 2, 6*

Label	Catalog ID	Components	Coordinates (J2000)	Mag.	θ	ρ	Dist.	Spectral type(s)	HD, SAO No.
	Sei 1549	(AB) C	22 01.1 +39 15	7.1, 10.3	20	38″	186	A0	209260, 71949
	h 1741	A D	22 11.2 +50 49	5.4, 10.0	271	74″	56	A5V	210715, 34143
	Σ 2894	A B	22 18.9 +37 46	6.2, 8.9	189	16″	74	A8III	211797, 72228
	h 1756	A B	22 21.9 +40 40	6.7, 10.5	286	22″	165	K3.5III	212212, 51919
	Σ 2902	A B	22 23.6 +45 21	7.6, 8.2	88	6.4″	200	G5	212468, 51957
			Solar type double; CD (m.13, 9″) 3′ n.f. is unrelated. AB: ps = 1,730 AU. (2004)						
	Σ 2906		22 26.8 +37 27	6.5, 9.6	1	4.2″	460	B2V	212883, 72344
	Σ 2917	A B	22 30.6 +53 32	8.3, 8.6	71	4.6″	370	F0IV	213495, 34534
	Σ 2918		22 31.3 +50 52	8.0, 9.4	237	1.6″	230	A1V	213557, 34541
	h 1791	A B	22 35.7 +56 52	7.7, 9.7	59	17″	46*	G0	214238, 34602
			Solar type double; rich field, Sti 2828 (m.11, 13″) 7′ n.p. AB: ps = 1,060 AU. (2006)						
	Es 1028		22 42.4 +54 15	7.6, 10.6	243	5.8″	210	A0V	215178, 34718
	OΣ 476	A (BC)	22 43.1 +47 10	7.4, 7.1	301	0.5″	780	A1V G	215242, 52296
			Distant, matched A type 1+(2) triple, visual double; rich field. MSC 8.5 M☉. Low quality orbit BC: P = 240 y, orbit r = 110 AU. (2007)						
	Σ 2942	A B C	22 44.1 +39 28	6.2, 8.9, 11.7	277, 247	3.2″, 9.0″	210	K5III	215359, 72675
	Σ 2946		22 49.7 +40 31	8.1, 8.3	261	5.2″	91	F8 F8	216122, 52401
	β 382	(AaAb) B	22 53.7 +44 45	6.0, 7.8	237	0.8″	57	A3m F6V	216608, 52465
			A type (2)+1 spectroscopic triple; AB: P = 105 y, orbit r = 36 AU, e = 0.54, closing to periastron 2044. (2011)						

Leo   Leo   *Charts 9, 15, 16*

Label	Catalog ID	Components	Coordinates (J2000)	Mag.	θ	ρ	Dist.	Spectral type(s)	HD, SAO No.
1 kap	β 105	A B	09 24.7 +26 11	4.6, 9.7	211	2.4″	62	K3III	81146, 80807
			Al Minliar al Asad. Solar type double with K giant; dark field. AB: ps = 200 AU. Neglected. (1975)						
3	H IV 47		09 28.5 +08 11	5.8, 11.1	80	25″	124	G9III	81873, 117718

Label	Catalog ID	Components	Coordinates (J2000)	Mag.	θ	ρ	Dist.	Spectral type(s)	HD, SAO No.
2 ome	Σ 1356		09 28.5 +09 03	5.7, 7.3	111	0.9″	33.2	F9IV	81858, 117717
			Solar type spectroscopic binary; dark field. AB: *P* = 118 y, orbit *r* = 29 AU, *e* = 0.56, widening. (663 measures; 2013)						
6	Sh 107	(AaAb) B	09 32.0 +09 43	5.2, 9.3	75	37″	153	K3III	82381, 117751
7	H V 58		09 35.9 +14 23	6.3, 9.4	80	41″	210	A1V	83023, 98662
	Wal 56	A B C	09 38.8 +10 47	6.7, 11.6, 10.7	99, 82	7.8″, 85″	94	A5III	83452, 98690
	Σ 1379		09 45.3 +08 53	7.2, 10.8	176	10″	63	F5	84407, 117890
	Hjl 115	A [B C]	09 47.3 +08 34	7.1, [9.7, 12.0]	190, [281]	59″, [1.8″]	82	F0	84701, 117908
	Σ 1399		09 57.0 +19 46	7.7, 8.4	176	31″	41.8	G0	86133, 81101
32 alp	Σ II 6	A (BC)	10 08.4 +11 58	1.4, 8.2	307	2.9′	24.3	B7V	87901, 98967
			Regulus. Local, high mass 1+(2) triple, visual double; featureless field. MSC system mass 6.2 M$_\odot$. AB: ps = 5,700 AU. (2012)						
	h 476		10 12.0 +20 07	7.4, 11.7	49	24″	440	G5	88403, 81225
	OΣ 215		10 16.3 +17 44	7.3, 7.5	176	1.4″	115	A9IV	88987, 99032
			High mass A type binary; very dark field. AB: *P* = 670y, orbit *r* = 155 AU, widening. (450 measures; 2013)						
	Bvd 81		10 16.7 +25 22	5.8, 10.0	28	79″	109	K2III G6V	89024, 81264
39	OΣ 523		10 17.2 +23 06	5.8, 11.4	299	7.7″	22.8	F7V dM1	89125, 81270
			Local, solar type double with low mass component; dark field. AB: ps = 235 AU. (2005)						
★ 41 gam	Σ 1424	A B	10 20.0 +19 51	2.4, 3.6	126	4.6″	39.9	K0III	89484, 81298
			Algieba. High CPM binary with K giant; m.10 stars n.p. C, D are unrelated. AB: *P* = 510 y, ps = 170 AU. (823 measures; 2013)						
★	Σ 1426	A B C	10 20.5 +06 26	8.0, 8.3, 9.4	311, 9	0.9″, 7.8″	150	F5	89619, 118241
			Probable 2+1 solar type triple; difficult, compact object in very dark field. AC: ps = 1,580 AU. (2011)						
	OΣ 216		10 22.7 +15 21	7.4, 10.3	232	2.3″	28.8	G5	89906, 99091
			Solar type binary; dark field, uncatalogued m.13 star 2′ n.p. AB: *P* = 315 y, orbit *r* = 55 AU, widening. (2013)						
	Σ 1431		10 25.6 +08 47	7.8, 9.1	74	3.6″	152	A8III	90303, 118292
	OΣ 217		10 26.9 +17 13	7.9, 8.6	148	0.8″	85	F6V	90444, 99127
			Solar type binary; dark field, 7′ f. m.9 star. AB: system mass 2.7 M$_\odot$, *P* = 140 y, orbit *r* = 40 AU, *e* = 0.98, widening to apastron 2036. (2012)						
	OΣ 220		10 29.2 +10 09	7.5, 8.6	98	0.6″	50*	F8	90791, 99153
	Σ 1448	A C	10 34.4 +21 36	7.5, 9.6	259	11″	530	K0	91527, 81420
★ 49	Σ 1450	(AaAb) B	10 35.0 +08 39	5.8, 7.9	157	2.1″	131	A2V	91636, 118380
			A = TX Leo. A type (2)+1 triple with Algol type eclipsing binary (*P* = 2.5 d); very dark field. AB: ps = 370 AU. (2009)						
	S 617	(AaAb) B	10 53.4 −02 15	6.2, 8.7	178	35″	79	G9IV	94363, 137863
			Solar type (2)+1 spectroscopic triple; dark field, 7′ s.p. p1 Leo. MSC system mass 5.7 M$_\odot$. Aa,Ab: *P* = 3.2 y, *e* = 0.38. (2012)						
★ 54	Σ 1487	A (BaBb)	10 55.6 +24 45	4.5, 6.3	113	6.4″	88	A1V A2Vn	94601, 81583
			Pretty A type 1+(2) spectroscopic triple; dark field, Algieba twin. AB: ps = 760 AU. (246 measures; 2013)						
55	β 1076		10 55.7 +00 44	6.0, 9.0	53	1.1″	44.9	F4V	94672, 118574
			Solar type binary; dark field. AB: *P* = 138 y, orbit *r* = 34 AU, at cusp of orbit. (2013)						
	Σ 1500		11 00.0 −03 28	7.9, 8.3	300	1.4″	54	F8V	95280, 137933
	Σ 1504		11 04.0 +03 38	7.9, 8.1	122	1.3″	220	F0	95899, 118638
	Σ 1506	A (BC)	11 04.7 −04 13	7.7, 10.2	224	12″	26.3	G5 M5	96064, 137978
			Solar type 1+(2) triple, visual binary with BY Dra variable; several nearby faint stars in dark field. MSC system mass 1.8 M$_\odot$. AB: ps = 430 AU. BC: *P* = 23 y, *e* = 0.12, periastron 2018. (2012)						
65 p4	β 599		11 06.9 +01 57	5.7, 9.7	104	2.7″	63	G9IIICN	96436, 118668
			High CPM double with R Leporis type giant carbon star; YO/− color, unrelated m.7 star 7′ s.p. AB: ps = 230 AU. Neglected. (1991)						
	Σ 1517	A B	11 13.7 +20 08	7.5, 8.0	320	0.7″	55	G5III	97561, 81725
			High CPM binary with G type giant; dark field. AB: *P* = 924 y, orbit *r* = 125 AU. (224 measures; 2010)						
	Σ 1521		11 15.3 +27 34	7.7, 8.1	98	3.5″	116	A5	97799, 81740
	Σ 1529		11 19.4 −01 39	7.1, 7.9	255	8.9″	48.1	F6IV dG3	98427, 138130
★ 78 iot	Σ 1536	(AaAb) B C	11 23.9 +10 32	4.1, 6.7, 11.1	98, 346	2.1″, 5.5′	23.7	F4IV	99028, 99587
			Local, solar type (2)+1+1 quadruple with del Sct type variable; dark field. AB: system mass 2.5 M$_\odot$, *P* = 186 y, orbit *r* = 45 AU, *e* = 0.53, apastron 2041. (536 measures; 2013)						

Label	Catalog ID	Components	Coordinates (J2000)	Mag.	θ	ρ	Dist.	Spectral type(s)	HD, SAO No.
83	Σ 1540	A B	11 26.8 +03 01	6.6, 7.5	150	28″	17.8	G7V	99491, 118864
		Local, high CPM, solar type double; dark field. AB: ps = 670 AU. (2013)							
88	Σ 1547	(AaAb) B	11 31.7 +14 22	6.3, 9.1	331	15″	23.3	G0IV	100180, 99648
		Local, solar type (2)+1 triple, visual binary; m.10 double 6′ s.f. MSC system mass 2.4 M☉. AB: ps = 470 AU. (2013)							
90	Σ 1552	A (BaBb) C	11 34.7 +16 48	6.3, 7.3, 9.8	208, 235	3.5″, 64″	580	B4V B9V F5	100600, 99673
		Distant, high mass 1+(2)+1 quadruple; dark field. MSC gives 16.5 M☉. AB: ps = 2,740 AU. AC: ps = 50,000 AU. (2013)							
	Σ 1565		11 39.6 +19 00	7.3, 8.4	305	22″	670	F4IV	101302, 99718
93	Σ II 7	(AaAb) (BaBb)	11 48.0 +20 13	4.6, 9.0	355	75″	71	A7V G5III G5V	102509, 81998
		Solar type (2)+(2) quadruple with G giant; Aa is RS CVn type variable. MSC gives 5.2 M☉. Aa,Ab: P = 72 d, e = 0. Ba,Bb: P = 110 y. (2012)							
	β 603		11 48.6 +14 17	6.0, 8.5	334	1.0″	60	A8III	102590, 99800
		Binary with A type giant; dark field, 20′ s.p. bet Leo. AB: P = 136 y, orbit r = 42 AU, decreasing θ. (2011)							
	Sh 132	A B	11 52.8 +15 26	6.9, 10.2	14	39″	93	A2	103152, 99840
	OΣΣ 112	A B	11 54.5 +19 25	8.3, 8.5	35	73″	40.5	G0	103432, 99858

Leo Minor   LMi   *Chart 9*

Label	Catalog ID	Components	Coordinates (J2000)	Mag.	θ	ρ	Dist.	Spectral type(s)	HD, SAO No.
11	Hu 1128		09 35.7 +35 49	5.3, 12.5	61	5.7″	11.4	G8V M5V	82885, 61586
		SV LMi. Local, high CPM, solar type double with RS CVn type variable. AB: ps = 90 AU. (2006)							
	Shy 212		09 40.0 +35 20	7.0, 7.2	228	8.2′	49.0	F5 F5	83525, 61620
	Σ 1374	A B	09 41.4 +38 57	7.3, 8.7	313	2.8″	51	G3IV	83698, 61629
	A 2142		10 05.7 +41 03	8.0, 8.8	295	1.1″	138	F0	87411, 43178
	Σ 1405		10 05.9 +39 35	7.3, 11.8	252	22″	106	A9V	87442, 61855
	Shy 552		10 16.6 +41 17	7.4, 8.8	6	107″	56	F5 F8	88959, 43264
	Σ 1432		10 27.0 +29 41	7.8, 10.3	121	29″	100	F2	90441, 81347
★ 31 bet	Hu 879	(AaAb) B	10 27.9 +36 42	4.6, 6.0	225	0.5″	47.2	G8III	90537, 62053
		Spectroscopic (2)+1 triple with G giant; very dark field. AB: system mass 3.0 M☉, P = 39 y, orbit r = 17 AU, e = 0.67, apastron 2018.5 (2012)							
	Cou 961	(AaAb) B	11 01.8 +29 52	7.2, 9.4	327	1.2″	132	K0	95515, 62361

Lepus   Lep   *Chart 21*

Label	Catalog ID	Components	Coordinates (J2000)	Mag.	θ	ρ	Dist.	Spectral type(s)	HD, SAO No.
	β 314	A B	04 59.0 −16 23	5.9, 7.5	320	0.8″	40.5	F3V F9V	31925, 150052
		Solar type binary. AB: system mass 2.3 M☉, P = 55y, orbit r = 18 AU, e = 0.88, periastron 2034. (2012)							
★ kap	Σ 661		05 13.2 −12 56	4.4, 6.8	357	2.2″	220	B9V	33949, 150239
		High mass double; bright mixed field, stars n. unrelated. AB: ps = 655 AU. (2008)							
	β 317	A B	05 13.9 −22 59	7.5, 10.3	13	8.6″	240	K0III	34087, 170216
	Bvd 52		05 18.1 −16 11	7.7, 9.1	226	90″	57	KIV G3V	34616, 150319
	S 476	A B	05 19.3 −18 31	6.3, 6.5	20	38″	260	B3V A	34798, 150335
	h 3750		05 20.4 −21 14	4.7, 8.5	279	4.1″	128	A0V	34968, 170327
★	h 3752	A B	05 21.8 −24 46	5.4, 6.6	99	3.5″	108	G0	35162, 170351
		Solar type double; sparse field, globular cluster NGC 1904 1/2º n.f. AB: ps = 510 AU. (2009)							
	CorO 31		05 23.6 −22 18	7.5, 10.2	284	17″	185	A7V	35430, 170383
★ 9 bet	β 320	A B E	05 28.2 −20 46	2.9, 7.5, 12.2	3, 59	2.6″, 4.0′	49.2	G5II	36079, 170457
		Nihal. 2+1 triple with G type supergiant; other nearby stars are unrelated. AB: ps = 170 AU; AE: ps = 15,900 AU. (2008)							
	h 3770		05 33.5 −24 20	7.8, 11.4	23	3.8″	51	F7V	36846, 170549
	β 321		05 39.3 −17 51	6.7, 7.8	160	0.5″	470	B7V	37643, 150652
	UC 1344		05 40.1 −23 43	7.3, 10.0	269	21″	164	K2III	37778, 170663

Label	Catalog ID	Components	Coordinates (J2000)	Mag.	θ	ρ	Dist.	Spectral type(s)	HD, SAO No.
	h 3788		05 41.3 −26 21	7.6, 9.2	155	26″	68	F3V	37987, 170697
★ 13 gam	H VI 40	A B	05 44.5 −22 27	3.6, 6.3	350	95″	8.93	F6V K2V	38393, 170759
		Local, high CPM, solar type double; dramatic in sparse field. AB: ps = 1,100 AU. (2012)							
	h 3798		05 47.2 −24 30	8.2, 8.8	70	16″	110*	F2	38788, 170838
	Σ 801	A (BC)	05 48.5 −13 21	7.5, 10.5	328	27″	320	K0III	38904, 150823
	β 94		05 49.6 −14 29	5.7, 8.2	165	2.3″	82	G8III	39070, 150845
	Gan 2		05 54.5 −19 42	7.6, 10.6	19	11″	23.6	G6V	39855, 150931
		Local, solar type double; f. of two stars in sparse field. AB: ps = 350 AU. (2010)							
	Arg 12		06 05.3 −25 02	8.4, 8.5	296	4.6″	270	A2V	41628, —
	β 17	A B C	06 08.4 −11 09	6.9, 9.3, 10.8	189, 257	3.7″, 10″	105	A5IV	42116, 151154

Libra   Lib   *Charts 17, 24*

Label	Catalog ID	Components	Coordinates (J2000)	Mag.	θ	ρ	Dist.	Spectral type(s)	HD, SAO No.
	Σ 1837		14 24.7 −11 40	6.9, 7.9	274	1.2″	91	F5III	126251, 158550
	Sh 179	(AaAb) [B C]	14 25.5 −19 58	6.6, [7.2, 8.4]	296, [91]	35″, [1.2″]	128	A2V	126367, 158558
		A type (2)+2 quadruple, visual triple; impressive in faint field. MSC system mass 6.8M☉. AB: ps = 6,040 AU. (2012)							
	Bvd 108		14 38.1 −09 44	7.7, 10.1	279	46″	1180	M1III	128595, 140042
	β 226		14 38.9 −22 20	7.9, 8.4	117	0.8″	195	A2V	128687, 182731
5	Hld 20		14 46.0 −15 28	6.4, 10.1	249	4.7″	440	K1III	129978, 158788
7 mu	β 106	A B E	14 49.3 −14 09	5.6, 6.6, 12.6	2, 231	1.9″, 27″	73	A1	130559, 158821
★ 9 alp 1,2	Sh 186	(AaAb) (BaBb) C	14 50.9 −16 02	2.7, 5.2, 12.5	314, 291	3.9′, 4.6′	23.2	A3IV F4IV	130841, 158840
		Zubeneschamali. Local, A type (2)+(2)+1 spectroscopic quintuple, comoving with BY Dra type variable D (KU Lib, HD 128987) 2.6° preceding. MSC system mass 5.1 M☉. AB: ps = 7,300 AU. AD (Cab 1): ps = 1.0 pc. (2012)							
	H N 28	A (BaBb)	14 57.5 −21 25	5.9, 8.2	305	26″	5.84	K5Ve M2V	131977, 183040
		Local, high CPM, low mass 1+(2) triple. MSC system mass 1.7 M☉. Ba,Bb: P = 308 d. AB: ps = 205 AU, but may be optical. (2011)							
	h 2757		14 58.7 −22 24	8.0, 9.6	95	12″	117	F6V	132234, 183060
	Σ 1899		15 01.6 −03 10	6.7, 10.2	67	28″	109	K2IV K0V	132883, 140278
	h 4727	A B	15 03.6 −27 50	8.5, 8.5	220	7.5″	44.0	G2V G2V	133131, 183129
	S 665		15 04.5 −17 54	8.1, 8.9	90	25″	—	K2IIICN	133353, 159004
	β 119	A B	15 05.5 −07 01	8.1, 8.8	273	2.1″	47.6	G0	133584, —
	Sh 195		15 14.5 −18 26	6.8, 8.3	140	47″	70	F3V	135208, 159118
	Σ 3091	A B	15 16.0 −04 54	7.7, 8.5	225	0.6″	63	F8V	135578, 140421
		Solar type binary; sparse field. AB: P = 148 y, orbit r = 45 AU, e = 0.82, periastron 2032. (2010)							
	β 227	A B C	15 19.2 −24 16	7.5, 8.6, 11.2	161, 84	1.8″, 46″	70	K1III	136032, 183368
	h 4756		15 19.7 −24 16	7.9, 8.3	247	0.6″	70	F4V	136121, 183377
	Lv 6		15 20.6 −27 01	7.9, 10.0	29	17″	650	K2III	136259, 183391
	h 4769	A B	15 25.4 −21 56	7.9, 9.7	192	9.8″	210	K0III	137210, 184077
	Sh 202	(AaAb) B	15 28.2 −09 21	7.0, 7.6	133	52″	20.6	K1V K2	137763, 140550
		Local, high CPM, solar type (2)+1 triple. MSC system mass 2.0 M☉. Aa,Ab: P = 2.4 y, orbit r = 3 AU, e = 0.97 (!). (2012)							
	S 672	(AaAb) B	15 31.7 −20 10	6.3, 8.9	281	11″	87	A8V	138268, 159317
	Lal 123	(AaAb) (BaBb)	15 33.2 −24 29	6.9, 7.0	300	9.2″	97	A3+F0V	138488, 183567
		Matched (2)+(2) quadruple, visual binary; sparse field. MSC gives 6.9 M☉. Ba,Bb: P = 60.44 y, e = 0.64, apastron 2028. (2013)							
	Σ 1962	(AaAb) B	15 38.7 −08 47	6.4, 6.5	190	12″	26.6	F8V F8V	139461, 140672
	β 121		15 39.6 −27 39	8.4, 8.5	282	1.8″	160	A2V	139519, 183668
	β 122		15 39.9 −19 46	7.7, 7.7	228	1.9″	57	F5V	139628, 159421
	β 35	A B	15 42.8 −16 01	7.3, 8.7	109	1.7″	64	F7IV	140164, 159453
	β 354		15 43.2 −25 25	7.3, 9.3	288	5.7″	126	F1V	140192, 183731

Label	Catalog ID	Components	Coordinates (J2000)	Mag.	θ	ρ	Dist.	Spectral type(s)	HD, SAO No.
	β 620	(AaAb) B (CaCb)	15 46.2 −28 04	7.6, 7.0, 9.0	174, 213	0.6″, 51″	71	F2IV	140722, 183772
		Solar type (2)+1+(2) quintuple, visual triple; rich field. MSC system mass 6.0 M$_\odot$. AB: ps = 60 AU. AC: ps = 4,890 AU. (2013)							
	Sh 213		15 59.1 −19 56	8.1, 8.5	318	18″	57	A5V	143094, 159618

Lupus   Lup   *Charts 24, 28*

Label	Catalog ID	Components	Coordinates (J2000)	Mag.	θ	ρ	Dist.	Spectral type(s)	HD, SAO No.
	h 4672		14 20.2 −43 04	5.8, 7.9	301	3.5″	113	G8III	125383, 224838
	R 244		14 22.6 −48 19	6.1, 9.5	122	4.5″	670	B1III	125721, 224870
	h 4706		14 51.3 −47 24	7.7, 9.0	219	6.8″	320	K0	130717, 225239
	h 4715	(AaAb) B	14 56.5 −47 53	6.0, 6.8	278	2.1″	169	B9V	131657, 225306
	h 4723	A B	15 01.9 −51 55	7.6, 10.0	170	5.2″	220	K0III	132606, 242180
★ pi	h 4728	(AaAb) (BaBb)	15 05.1 −47 03	4.6, 4.6	65	1.7″	136	B5V	133242, 225426
		High mass (2)+(2) spectroscopic quadruple, visual CPM double; dominant in faint rich field. AB: ps = 310 AU. (2010)							
	Hd 242	A C	15 05.3 −41 04	5.2, 11.1	182	49″	108	G8III	133340, 225435
	CapO 415		15 10.7 −43 44	7.1, 7.7	20	50″	30.4	G5V	134331, 225508
	Δ 178		15 11.6 −45 17	6.5, 9.6	314	1.2″	156	K1III	134444, 225517
★ kap 1,2	Δ 177		15 11.9 −48 44	3.8, 5.5	143	27″	55	B9.5V A5V	134481, 225525
		High mass double; brilliant pair in faint rich field. AB: ps = 2,000 AU. (2010)							
zet	Δ 176		15 12.3 −52 06	3.5, 6.7	249	71″	36.0	K0III F8V	134505, 242304
	Hld 121		15 12.8 −52 02	7.7, 8.9	216	2.1″	510	B6III	134598, 242314
	CorO 179		15 13.0 −37 15	8.0, 8.1	228	6.5″	117	A3V	134799, 206805
V348	I 228		15 14.0 −43 48	8.0, 8.2	16	1.3″	550	A4V	134930, 225554
		A type double; primary is bet Cep type variable; double I 565 4′ n.f. AB: ps = 550 AU. Neglected. (1991)							
★ mu	h 4753	A B C	15 18.5 −47 53	4.9, 5.0, 6.3	302, 128	0.9″, 23″	103	B8V	135734, 225638
		High mass 2+1 triple; dominant in rich field, stars s.p. unrelated. MSC system mass 11.0 M$_\odot$. AC: ps = 3,200 AU. (2013)							
	Howe 76		15 21.5 −38 13	6.6, 9.3	121	5.7″	143	A0	136347, 206543
	h 4776		15 30.4 −41 55	6.3, 8.4	229	5.6″	126	B9.5V	137919, 225846
	B 2036	A B C	15 31.3 −33 49	7.7, 7.9, 9.1	2, 120	0.4″, 1.5″	103	A2V	138138, 206720
★ gam	h 4786	A B	15 35.1 −41 10	3.0, 4.5	275	0.8″	129	B2IV	138690, 225938
		High mass binary with ellipsoidal variable (RV Lup); faint field. AB: *P* = 190 y, orbit *r* = 50 AU, *e* = 0.51, periastron 2075. (2013)							
	h 4788	(AaAb) B	15 35.9 −44 58	4.7, 6.5	10	2.1″	131	B3V	138769, 225950
	I 89		15 41.1 −39 59	6.8, 8.1	163	1.4″	117	F5V	139677, 206867
	Arg 28	A B	15 41.9 −30 09	7.9, 10.1	24	35″	260	K3III	139911, 183710
	Skf 2047	A C	15 43.8 −39 28	7.6, 8.8	132	114″	140*	A2V	140197, 206910
	Howe 79		15 44.4 −41 49	6.1, 7.9	338	3.2″	144	A0V B	140285, 226132
	Δ 192	A B C	15 47.1 −35 31	7.1, 9.0, 7.3	67, 143	0.7″, 35″	147	B9.5V	140817, 206968
xi 1,2	Pz 4		15 56.9 −33 58	5.1, 5.6	49	10″	42.4	A3V B9V	142629, 207144
★ eta	Rmk 21	A B C	16 00.1 −38 24	3.4, 7.5, 9.4	19, 248	15″, 116″	136	B2.5IV	143118, 207208
		High mass 2+1 triple; faint rich field, D 2′ n.p. possibly related. AB: ps = 2,750 AU. (2013)							
	Howe 82		16 03.8 −33 04	7.7, 7.9	346	2.4″	320	F3V	143823, 207285
V856	Δ 199	A B (CaCbD)	16 08.6 −39 06	6.6, 13.0, 7.1	297, 184	16″, 44″	171	A7IVe	144668, 207368
		A type 2+(2+1) quintuple with del Sct type variable; sparse field. MSC system mass 6.1 M$_\odot$. AC: ps = 10,200 AU. (2000)							

Lynx   Lyn   *Charts 3, 8*

Label	Catalog ID	Components	Coordinates (J2000)	Mag.	θ	ρ	Dist.	Spectral type(s)	HD, SAO No.
4	Σ 881	(AaAb) B	06 22.1 +59 22	6.1, 7.7	149	0.6″	152	A3V	43812, 25678
		A type (2)+1 triple, visual binary; sparse field. Aa,Ab: *P* = 503 y. (2013)							
	OΣΣ 72	A B C	06 24.7 +59 40	7.6, 10.8, 7.6	305, 323	46″, 2.2′	212	K0III A3	44271, 25709
	Σ 936		06 39.7 +58 06	7.3, 9.0	283	1.2″	250	G5	46963, 25861

Label	Catalog ID	Components	Coordinates (J2000)	Mag.	θ	ρ	Dist.	Spectral type(s)	HD, SAO No.
★	Σ 946		06 44.9 +59 27	7.3, 9.1	130	3.9″	101	F5	47977, 25925
			Solar type double; lovely W/B color, group of stars p. and n.p., including I 757 (m.9, 1.6″). AB: ps = 530 AU. (2012)						
★ 12	Σ 948	A B C	06 46.2 +59 27	5.4, 6.0, 7.1	69, 311	1.9″, 8.5″	66	A3V	48250, 25939
			A type, compact 2+1 triple. MSC system mass 5.9 M$_\odot$. AB: P = 903 y, orbit r = 150 AU, decreasing θ (394 measures; 2012)						
★	Σ 958	A (BaBb)	06 48.2 +55 42	6.3, 6.3	257	4.5″	44.3	dF5 dF6	48766, 25962
			Solar type 1+(2) spectroscopic triple, visual double; sparse or dark field. AB: P = 2,200 y, orbit r = 215 AU. (2012)						
	Σ 960		06 49.6 +53 02	7.9, 9.9	69	22″	270	F0	49082, 25973
★ 15	OΣ 159	A B C	06 57.3 +58 25	4.5, 5.5, 12.2	232, 351	0.7″, 40″	55	G8III F8V	50522, 26051
			Solar type 2+1 triple with G giant; WY/W color. AB: system mass 3.9 M$_\odot$, P = 262 y, orbit r = 65 AU, e = 0.74, widening. (2012)						
	Σ 1032	A B	07 13.9 +48 30	7.3, 10.3	113	2.7″	149	A2V	55078, 41630
	Σ 1033	A B	07 14.8 +52 33	7.8, 8.4	275	1.6″	270	F0	55199, 26240
	Σ 1050	A B	07 19.9 +54 55	8.1, 8.8	24	19″	141	A0V	56385, 26279
★ 20	Σ 1065		07 22.3 +50 09	7.5, 7.7	255	15″	95*	F0	57066, 26306
			Solar type spectroscopic binary, visual CPM double; radiant W/W color in dark field. AB: ps = 1,920 AU. (2005)						
★ 19	Σ 1062	A B C D	07 22.9 +55 17	5.8, 6.7, 12.8, 7.6	316, 293, 6	15″, 87″, 3.6′	140*	B8V B9V	57103, 26312
			Exquisite, high mass 2+1+1 quadruple, possible spectroscopic (2)+(2)+1+1 sextuple. MSC gives 10.7 M$_\odot$. AB: ps = 2,830 AU; AD: ps = 41,800 AU. (2012)						
	β 758	A B C	07 28.9 +48 11	5.7, 9.9, 10.2	94, 93	1.1″, 17″	153	B9	58661, 41797
CC	OΣ 174	(AaAb) B	07 35.9 +43 02	6.6, 8.3	88	2.2″	141	F0	60335, 41877
			Solar type (2)+1 triple: primary is W UMa type contact binary; pretty field, YW/– color, MAD 3 (m.10, 4″) 2′ p. AB: ps = 420 AU. (2012)						
	Σ 1172		08 04.6 +54 45	7.9, 9.6	244	1.7″	390	A0IV	66067, 26648
	h 2424		08 06.0 +59 15	6.7, 10.8	150	43″	230	A0	66286, 26662
	OΣ 189		08 14.8 +43 02	6.9, 10.7	295	4.3″	200	A2IV	68562, 42242
	OΣΣ 92	A B	08 23.8 +57 25	7.6, 9.3	181	58″	70	F2	70253, 26815
	Σ 1217		08 24.3 +44 57	7.8, 9.4	242	29″	37.8	G0	70516, 42337
	OΣ 193		08 28.1 +33 32	7.7, 11.7	297	14″	23*	K0	71354, 60832
	Σ 1274		08 49.0 +38 21	7.4, 9.3	43	9.4″	240	A2	75052, 61060
	Σ 1282	A B C	08 50.7 +35 04	7.6, 7.8, 12.4	277, 10	3.5″, 49″	52	F8	75353, 61077
	Σ 1289		08 54.7 +43 35	8.2, 8.9	7	3.7″	58	G2V	75949, 42594
★	Σ 1333		09 18.4 +35 22	6.6, 6.7	51	1.8″	87	A8V	80024, 61387
			A type double; very dark field. AB: ps = 210 AU. (2013)						
★ 38	Σ 1334	(AaAb) (BaBb)	09 18.8 +36 48	3.9, 6.1	224	2.6″	38.3	A3V	80081, 61391
			A type (2)+(2) spectroscopic quadruple; a pure white "diamond ring." MSC gives 6.2 M$_\odot$. AB: ps = 135 AU. (220 measures; 2013)						
★	Σ 1338	A B	09 21.0 +38 11	6.7, 7.1	308	1.1″	42.7	F2V F4V	80441, 61411
			Matched solar type binary; isolated in field. AB: system mass 2.3 M$_\odot$, P = 303 y, orbit r = 57 AU, e = 0.25. (432 measures; 2013)						
DI	Σ 1369	(AaAb) (BaBb)	09 35.4 +39 58	7.0, 8.0	151	25″	85	F2V	82780, 42931

Lyra   Lyr   *Chart 11*

Label	Catalog ID	Components	Coordinates (J2000)	Mag.	θ	ρ	Dist.	Spectral type(s)	HD, SAO No.
	OΣ 352		18 26.4 +46 49	7.9, 9.4	221	24″	100	F2	170313, 47497
	Σ 2333	A B	18 31.1 +32 15	7.8, 8.6	333	6.4″	220	B9IV	171026, 67059
	Σ 2351		18 36.2 +41 17	7.6, 7.6	160	5.1″	210	A1V A0V	172068, 47639
	Σ 2349	(AaAb) B C	18 36.6 +33 28	5.4, 9.4, 12.1	204, 314	7.4″, 33″	155	B8II	172044, 67164
	Σ 2362		18 38.4 +36 03	7.5, 8.7	187	4.4″	105	A5	172394, 67198

Label	Catalog ID	Components	Coordinates (J2000)	Mag.	θ	ρ	Dist.	Spectral type(s)	HD, SAO No.
★	Σ 2367	A B C	18 41.3 +30 18	7.7, 8.0, 8.8	75, 192	0.4″, 14″	65*	G5III F8IV F7V	172865, 67250
		Solar type 2+1 spectroscopic triple with G giant; YO/YW color. MSC 5.6 M$_\odot$. AB: P = 92 y, orbit r = 16 AU, e = 0.91, apastron 2026. (2012)							
	Σ 2380		18 42.9 +44 56	7.3, 8.7	8	26″	138	G8III	173399, 47727
★ 4,5 eps 1,2	Σ I 37	A B [(CaCb) D]	18 44.3 +39 40	5.2, 6.1, [5.3, 5.4]	344, 171, [79]	2.3″, 3.5′, [2.5″]	49.8	A4V F1V	173582, 67310
		A type 2+[(2)+1] quintuple, visual quadruple; the famous "Double Double". C. Mayer (1777) noted the wide pair of binaries but W. Herschel (1779) first resolved them. MSC system mass 10.3 M$_\odot$. AB (Σ 2382): P = 1,804 y, orbit r = 235 AU, closing. (585 measures). CD (Σ 2383): P = 724 y, orbit r = 44 AU, decreasing θ. (616 measures) Ca,Cb pair uncertain. AB,CD: ps = 14,100 AU. (2013)							
	OΣΣ 172		18 44.5 +34 00	7.9, 8.7	6	61″	54	F8	173548, 67311
★ 6,7 zet 1,2	Σ I 38	(AaAb) D	18 44.8 +37 36	4.3, 5.6	150	44″	47.9	F0IV	173648, 67321
		Solar type (2)+1 spectroscopic triple, visual binary; rich field. MSC system mass 4.9 M$_\odot$. AD: ps = 2,840 AU. (2013)							
	Σ 2390		18 45.8 +34 31	7.4, 8.6	155	4.2″	360	A7V	173815, 67350
	Σ 2397		18 47.2 +31 24	7.5, 9.1	266	3.9″	270	G3III	174022, 67378
	Σ 2406		18 49.9 +26 26	7.1, 11.2	5	4.8″	118	A3V	174549, 86481
★ 10 bet	Σ I 39	(Aa1Aa2) Ab B F	18 50.1 +33 22	3.6, 8.2, 6.7, 10.6	176, 147, 19	0.5″, 45″, 87″	296	B7IIe ~B2V	174638, 67451
		Sheliak. High-mass (2)+1+1+1 quintuple, visual quadruple; in a beautiful field with unidentified m.10 double 4′ s. The eponymous eclipsing, semi-detached binary: variability detected by J. Goodricke (1784), the Aa1,Aa2 mass transfer binary imaged (2008) with terrestrial CHARA interferometer (USA). Aa1,Aa2: P = 12.9 d, orbit r = ~0.25 AU, e = 0.0 (circular). Estimated system mass 15.6 M$_\odot$ (supergiant donor star 2.8 M$_\odot$, ellipsoidal main sequence gainer star 12.8 M$_\odot$); the nearly edge-on orbit brackets a thick accretion disk with bipolar plasma jets. An astonishing astrophysical system. (2012)							
	Sh 282	(AaAb) B C	18 54.9 +33 58	6.1, 9.1, 7.6	129, 350	1.8″, 45″	360	A8V G5III	175635, 67566
	Σ 3130	A B C D	18 56.0 +44 14	7.2, 8.3, 10.5, 12.0	288, 258, 310	0.4″, 2.7″, 3.0′	187	A2	176003, 47928
	β 648	A (BaBb)	18 57.0 +32 54	5.3, 8.0	251	1.2″	14.9	G0V	176051, 67612
		Local, solar type 1+(2) triple, visual binary; faint rich field. AB: P = 61 y, orbit r = 19 AU, e = 0.27. (411 measures; 2011)							
	Dawes 9		19 04.3 +43 53	7.1, 10.2	172	2.0″	230	A0V	177829, 48054
17	Σ 2461	(AaAb) B	19 07.4 +32 30	5.3, 9.1	290	3.7″	41.6	F0V	178449, 67835
		Very high CPM, solar type (2)+1 spectroscopic triple, visual double; group of stars n. in rich field. AB: ps = 210 AU. (2001)							
	Σ 2469	(AaAb) B C	19 07.8 +38 56	7.9, 9.1, 12.4	125, 163	1.3″, 37″	139	A3	178661, 67846
★	Σ 2474	(AaAb) B	19 09.1 +34 36	6.8, 7.9	263	16″	52	G1V	178911, 67879
		Solar type (2)+1 spectroscopic triple, visual double; a charming "reflection" of optical pair Σ 2470 (m.7, 14″) 10′ n.p. in a rich field. MSC system mass 2.8 M$_\odot$. Aa,Ab: P = 3.6 y, orbit r = 5 AU, e = 0.60. AB: ps = 1,120 AU. (2013)							
	Σ 2481	(AaAb) (BCaCb)	19 11.1 +38 47	8.4, 8.3	20	4.6″	53	G6V	179484, 67936
		Solar type (2)+(1+2) spectroscopic quintuple. BC: P = 63 y, e = 0.50, periastron 2019. Ba,Bb: P = 0.6 d. AB: ps = 330 AU. (250 measures; 2012)							
	Σ 2483	A B	19 12.4 +30 21	8.0, 9.1	318	10″	520	B9IV	179709, 67963
	Sh 289		19 13.5 +39 02	8.0, 8.7	55	39″	—	A	180077, 68003
★	OΣ 367	A [B C]	19 14.5 +34 34	7.3, [9.7, 9.8]	226, [262]	35″, [1.0″]	67	F5IV	180286, 68022
		Solar type 1+2 visual triple; rich field, faint doubles nearby, BC a resolution challenge. AB: ps = 3,160 AU; BC: ps = 90 AU. (2013)							
	OΣ 371	A B	19 15.9 +27 27	7.0, 7.6	160	0.9″	280	B8V	180553, 87005
		Distant, high mass, spectroscopic binary; rich field. MSC gives 13.3 M$_\odot$. AB: ps = 340 AU. AC: ps = 17,800 AU. (229 measures; 2013)							

Label	Catalog ID	Components	Coordinates (J2000)	Mag.	θ	ρ	Dist.	Spectral type(s)	HD, SAO No.
	Mensa		Men						*Chart 30*
	h 3673		04 24.9 −77 41	8.1, 8.3	67	10″	71	F7V	29058, 256092
	Hrg 2		05 01.2 −74 20	7.4, 8.0	169	0.8″	270	B9.5V	33244, 256152
	h 3888		06 28.6 −78 54	7.6, 10.3	125	35″	188	F0IV	47674, 256310
	h 3996		07 11.6 −84 28	7.5, 11.9	260	16″	210	B9.5V	60102, 258468
	Microscopium		Mic						*Chart 25*
	Gli 259	A B	20 31.9 −40 54	8.4, 8.4	158	4.2″	43.8	K1	195284, 230257
★	Jc 18	A B C	20 33.8 −40 33	7.8, 8.5, 10.9	223, 145	4.4″, 108″	153	A1V	195599, 230275
			A type 2+1 triple; dark field. AB: ps = 910 AU. Like many southern hemisphere systems, this one is neglected. (1998)						
	h 5211	A B	20 40.9 −42 24	6.3, 10.1	298	20″	164	G6III	196748, 230321
	h 5228		20 51.7 −40 54	7.4, 9.3	104	32″	520	K0IIICN	198433, 230405
	Skf 2097		21 01.1 −29 44	7.0, 8.1	152	88″	186	A2IV+F0V	199917, 189930
	Δ 236		21 02.2 −43 00	6.7, 7.0	72	58″	82	G3IV K0IV	200011, 230492
	UC 4365		21 03.0 −38 38	5.4, 10.7	207	100″	35.4	F5V	200163, 212666
	β 251		21 12.1 −30 35	7.4, 9.4	232	2.1″	105	F5V	201695, 212786
	Skf 2098		21 19.1 −32 21	7.1, 11.0	27	101″	330	K3III	202774, 212895
	Skf 1173		21 23.7 −39 47	7.5, 9.0	153	94″	105	F3IV+F6V	203493, 212968
	MlbO 6	A B C	21 27.0 −42 33	5.6, 8.2, 11.9	150, 75	2.9″, 4.9′	56	Am	204018, 230692
	Monoceros		Mon						*Chart 14*
3	β 16	(AaAb) B	06 01.8 −10 36	5.0, 8.0	355	1.9″	240	B5III	40967, 151037
	AC 3		06 11.7 −04 40	6.3, 8.2	203	0.7″	176	B9	42657, 132941
	β 567		06 15.5 −04 55	6.0, 10.0	240	4.2″	65	A5IV	43319, 133027
	β 569	A B	06 25.2 −10 56	7.9, 9.7	117	1.7″	186	K0	45140, 151478
	Σ 914		06 26.7 −07 31	6.3, 9.3	299	21″	75	A0Vn	45380, 133263
	Σ 910	A [B C]	06 26.7 +00 27	7.0, [9.2, 8.5]	152, [151]	66″, [0.5″]	210	G5	45317, 113892
★ 11 bet	Σ 919	A B C	06 28.8 −07 02	4.6, 5.0, 5.4	133, 125	7.1″, 9.8″	210	B3Ve B3n	45725, 133316
			High mass, matched triple; discovered independently by B. Castelli (1617) and W. Herschel (1779), who declared it "one of the most beautiful sights in the heavens." B possibly optical (divergent CPM); faint pair 25″ n.f. AC: ps = 2,780 AU. (2012)						
	OΣ 142	(AaAb) B	06 29.9 +07 07	7.1, 10.6	354	8.4″	410	B2.5V	45789, 113953
	Σ 926	Aa Ab B	06 31.7 +05 46	7.2, 9.6, 8.6	18, 288	0.9″, 11″	130*	A1	46105, 114003
	A 509	A B	06 37.9 −08 47	7.7, 9.7	137	1.4″	340	K0	47364, 133500
★ 15	Σ 950	(AaAb) B C	06 41.0 +09 54	4.7, 7.8, 9.9	213, 16	2.9″, 16″	280	O7Ve	47839, 114258
			S Mon. Young cluster of high mass OB stars (Σ 952); proper motions are complex, but suggest Aa, Ab, B and C are related as (2)+1+1. Aa,Ab: P = 74 y, widening. The three wide pairs s. are apparently unrelated. Tiny double D 11 (m.9, 3.5″) 73″ s.f. seems unrelated. (2012)						
	Σ 953		06 41.2 +08 59	7.1, 7.7	331	7.3″	500	F5	47888, 114266
	Σ 987		06 54.1 −05 51	7.1, 7.2	176	1.3″	164	A6Vn	50700, 133855
★	β 327	A B C	06 58.5 −03 01	7.8, 8.2, 11.1	102, 100	0.7″, 13″	3020*	B0.5IV	51756, 133972
			Distant, high mass 2+1 triple; in faint cluster in a splendid field. AB: ps = 2,850 AU; AC: ps = 53,000 AU. Neglected. (1997)						
	β 573		07 01.8 −10 53	7.2, 7.8	301	0.8″	118	F8	52694, 152256
V752	Σ 1029		07 07.9 −04 41	7.5, 8.0	26	1.6″	230	A9V	54250, 134234
	β 197	A B	07 12.8 −07 09	7.8, 9.3	148	2.2″	700	F2	55489, 134360
	A 524		07 14.2 −03 54	6.1, 10.2	152	2.7″	400	M1III	55775, 134391
	Σ 1056		07 15.6 −01 52	8.0, 8.9	300	4.0″	250	G0	56083, 134420
	Ho 245	A B	07 38.7 −01 27	7.9, 8.7	186	0.7″	380*	A3III+	61343, 134933

Label	Catalog ID	Components	Coordinates (J2000)	Mag.	θ	ρ	Dist.	Spectral type(s)	HD, SAO No.
	Σ 1128		07 39.8 −06 15	7.9, 10.0	169	16″	101	G5	61583, 134950
	Bgh 3		07 40.0 −03 36	7.3, 9.0	113	58″	14.2	K2V	61606, 134954
			Local, solar type double with BY Dra type variable; bright rich field. AB: ps = 1,110 AU. (2003)						
	Σ 1154		07 52.1 −03 03	7.1, 9.3	354	2.8″	112	A5	64110, 135190
	φ 325		07 52.8 −05 26	7.3, 6.2	358	0.4″	39.7	F5IV	64235, 135205
			Solar type binary; sparkling field. System mass 2.2 M$_\odot$. AB: P = 31 y, orbit r = 13 AU, e = 0.65, closing, but reappears around 2040. (2013)						
★	Σ 1157		07 54.6 −02 48	7.9, 7.9	177	0.7″	115*	F0	64607, 135238
			Matched solar type double, in glorious field; fine subarcsecond challenge. AB: ps = 110 AU. (2013)						
	Σ 1189		08 07.9 −01 21	7.8, 10.1	334	8.9″	180*	A1V	67452, —

Musca   Mus   *Chart 28*

Label	Catalog ID	Components	Coordinates (J2000)	Mag.	θ	ρ	Dist.	Spectral type(s)	HD, SAO No.
★	h 4432		11 23.4 −64 57	5.4, 6.6	308	2.4″	124	B5V	99104, 251382
			High mass double; pretty brightness contrast, faint field. AB: ps = 400 AU. Neglected. (1991)						
	CorO 130		11 51.9 −65 12	5.0, 7.3	159	1.6″	111	B4V	103079, 251617
	Shy 230		11 54.7 −66 23	6.4, 8.5	123	3.7′	43.5	F2IV G3V	103482, 251637
	h 4498	(AaAb) B	12 06.4 −65 43	6.1, 7.7	60	8.9″	126	K0III	105151, 251738
			(2)+1 triple with K giant, visual binary; rich field. Low quality orbit Aa,Ab: P = 48 y, widening. (2000)						
	h 4522	A B	12 25.5 −69 29	7.9, 8.7	66	13″	260	B8	108073, 251894
	I 296		12 39.2 −75 22	6.6, 9.1	271	2.0″	175	B7Vn +A2V	109857, 256967
★ bet	R 207	(AaAb) B (C)	12 46.3 −68 06	3.5, 4.0	50	1.0″	105	B2.5V	110879, 252019
			High mass (2)+1(+1) quadruple (C is m.16); faint rich field. System mass 20.9 M$_\odot$; AB: ps = 140 AU. (2013)						
	h 4550		12 48.3 −67 08	7.6, 8.7	98	12″	123	A3III	111161, 252034
	Gli 185		12 49.0 −65 36	7.3, 9.7	8	8.7″	910	B6IV	111283, 252038
★ eta	Δ 131	(AaAb) B C	13 15.2 −67 54	4.8, 10.0, 7.2	125, 332	2.7″, 58″	117	B8V	114911, 252224
			High mass (2)+1+1 spectroscopic quadruple; nicely displayed in faint field. MSC system mass 10.2 M$_\odot$. AC: ps = 9,150 AU. (2002)						
	Hd 224		13 17.9 −68 30	7.0, 8.1	208	0.7″	220	B9IV	115286, 252245
	h 4586		13 28.4 −67 52	7.3, 9.1	141	2.9″	142	A5III	116865, 252318
	I 298		13 32.5 −69 14	7.4, 8.5	156	0.5″	126	F5II	117445, 252353
	h 4596		13 37.4 −64 56	8.2, 8.4	280	1.4″	390	A	118242, 252388

Norma   Nor   *Chart 29*

Label	Catalog ID	Components	Coordinates (J2000)	Mag.	θ	ρ	Dist.	Spectral type(s)	HD, SAO No.
	h 4777		15 32.7 −57 24	7.5, 9.1	295	5.6″	133	F2IV	138109, 242668
	Hld 124		15 45.0 −50 47	6.6, 8.5	195	2.2″	122	A3V	140274, 242915
	h 4795	A B	15 45.0 −59 07	7.6, 10.4	222	7.6″	129	A2III	140111, 242901
	Δ 191	(AB) C	15 45.3 −58 41	7.8, 8.1	296	33″	590	A5V	140178, 242913
★ V360	Δ 193	(AaAb) B	15 51.1 −55 03	5.8, 9.1	10	16″	660	B2II	141318, 243044
			Distant, high mass (2)+1 spectroscopic triple with pulsating variable star; sparse field. AB: ps = 14,200 AU. (2013)						
	Δ 195	A B	15 54.8 −50 20	6.8, 7.5	9	12″	95	A2	142080, 243110
	h 4813		15 55.5 −60 11	5.9, 8.4	100	4.4″	51	G5II	142049, 253349
	Spm 33		15 57.1 −48 10	6.3, 11.5	314	34″	48.5	F2V	142529, 226392
iot 1	h 4825	A B C	16 03.5 −57 46	5.2, 5.8, 8.0	228, 241	0.4″, 11″	39.4	A7IV	143474, 243279
			A type 2+1 triple; rich field. MSC gives 4.5 M$_\odot$. AB: P = 27 y, orbit r = 13 AU, e = 0.52, periastron 2017, reappears in 2025. (2010)						
	Rss 30		16 21.5 −53 06	8.3, 8.4	238	47″	360	B9III	146921, 243684
	CorO 197	(AaAb) (BaBb)	16 25.3 −49 09	8.1, 8.2	96	2.3″	47.0	K1V	147633, 226738
★ eps	h 4853	(AaAb) B	16 27.2 −47 33	4.5, 6.1	334	23″	163	B2V	147971, 226773
			High mass (2)+1 spectroscopic triple, visual binary; easy pair in faint rich field. AB: ps = 5,060 AU. (2013)						

Label	Catalog ID	Components	Coordinates (J2000)	Mag.	θ	ρ	Dist.	Spectral type(s)	HD, SAO No.
	Octans		Oct					*Chart 30*	
	Gli 14		01 37.4 −82 17	7.6, 8.4	54	5.4″	123	K1	10693, 258270
	R 38		03 42.5 −85 16	6.6, 8.1	252	1.9″	153	B9.5IV	25887, 258356
	Δ 82		09 33.3 −86 01	7.1, 7.6	276	16″	48.6	F5IV	85300, 258542
	h 4310		10 05.6 −84 05	7.7, 8.4	262	4.3″	64	F8V	88948, 258564
	h 4490		12 02.3 −85 38	6.2, 9.0	146	25″	114	K3III	104555, 258632
iot	Rst 2819	A B	12 55.0 −85 07	5.9, 6.9	240	0.7″	108	K0III	111482, 258654
	Grv 1247		15 43.3 −84 28	5.6, 11.6	333	66″	66	A2V K0	137333, 258731
	h 4798		16 10.3 −84 14	7.7, 11.2	130	20″	34.3	K0	142022, 258738
mu 2	Δ 232		20 41.7 −75 21	6.5, 7.1	19	17″	44.3	G5III	196067, 257836
	Gli 263		21 06.7 −80 42	7.3, 9.6	245	4.6″	79	F0IV	199391, 258879
	h 5235		21 21.3 −84 19	8.2, 8.4	80	3.1″	190	F5II	200816, 258888
	h 5262		21 33.4 −80 02	6.5, 10.4	92	24″	117	A0V	203955, 258904
★ lam	h 5278		21 50.9 −82 43	5.6, 7.3	63	3.5″	125	G9III	206240, 258914
	Probable A type double with G giant; YO/W color, sparse or dark field. AB: ps = 590 AU. (2010)								
	h 5261		21 52.2 −85 50	8.3, 8.6	197	4.9″	104	F7V	205195, 258908
	h 5306	(AaAb) B	22 03.1 −76 07	6.0, 10.6	71	35″	63	F3III	208741, 257993
	Δ 238	A B C	22 25.9 −75 01	6.2, 8.9, 13.0	80, 129	21″, 4.4′	23.1	G3V M8V	212168, 258036
	Local, solar type 2+1 triple; dark field. No orbit. AB: ps = 655 AU. AC: ps = 8,230 AU. (2010)								
	Ophiuchus		Oph					*Charts 18, 24*	
	Σ 2005	(AaAb) B C E	16 05.7 −06 17	6.5, 9.0, 11.3, 7.9	86, 232, 270	0.7″, 29″, 4.3′	87	F3V	144362, 140945
	Solar type (2)+1+1+1 quintuple, visual quadruple. MSC system mass 4.9 M⊙. Aa,Ab: P = 5.0y, e = 0.56. (2010)								
★ 5 rho	H II 19	A B (DE) [C F]	16 25.6 −23 27	5.1, 5.7, 6.8, [5.8, 13.2]	334, 252, 0, [206]	3.3″, 2.6′, 2.5′, [4.8″]	111	B2IV B2V	147933, 184382
	Young, high mass 2?,(2),2 group near an intensively studied star forming region. AB (103 measures) has both a linear and low quality orbital solution, with a small CPM divergence, so it may be orbiting edgewise or comoving; both DE (β 1115, 0.3″) 2.6′ s. p. and CF (Kou 63, 5″) 2.5′ n. are CPM or orbital binaries comoving with A. (2009)								
	Σ 2048	A (BaBb)	16 28.8 −08 08	6.6, 9.7	300	5.5″	51	F4V	148515, 141195
★ 10 lam	Σ 2055	A B C	16 30.9 +01 59	4.2, 5.2, 11.8	40, 169	1.5″, 120″	53	A0V A0V	148857, 121658
	Marfik. A type 2+1 triple. MSC gives 6.1 M⊙. AB: P = 129 y, orbit r = 48 AU, e = 0.61, periastron 2069. (802 measures; 2013)								
	Ho 407		16 32.9 −10 34	6.8, 11.8	218	15″	85	A9V	149108, 159972
	Σ 2106	A B	16 51.1 +09 24	7.1, 8.2	172	0.7″	64	F7IV	152113, 121908
	Solar type binary; faint field. AB: P = 1,270 y, orbit r = 90 AU. (2012)								
21	OΣ 315		16 51.4 +01 13	5.8, 7.3	312	0.7″	116	A2V	152127, 121911
	A type binary; low quality orbit AB: P = 1,115 y, orbit r = 195 AU. (219 measures; 2011)								
★ 24	β 1117		16 56.8 −23 09	6.3, 6.3	305	1.0″	93	A0V	152849, 184822
	A type binary; exact "Dawes" (matched m.6) arcsecond resolution test. AB: ps = 125 AU. (2011)								
	Sh 240		16 57.1 −19 32	6.6, 7.6	232	4.5″	178	B6V B7V	152909, 160180
	Σ 2114		17 02.0 +08 27	6.7, 7.6	195	1.3″	121	A4V	153914, 122023
	A type spectroscopic binary and visual double; sparse field. Frequently measured, but no visual orbit or linear solution. AB: ps = 210 AU. (309 measures; 2012)								
	Σ 2119		17 06.5 −13 56	8.2, 8.3	185	2.4″	108	F6V	154520, 160280
	Σ 2122		17 06.9 −01 39	6.4, 9.7	279	20″	91	A9V	154660, 141522
	A 1145	(AaAb) B	17 08.2 −01 05	6.3, 7.8	343	0.7″	90	A1V F3V	154895, 141528
	A type (2)+1 spectroscopic triple; four stars f. AB: P = 204 y, orbit r = 50 AU, e = 0.43, apastron 2062. (2012)								
★ 35 eta	β 1118	A B E	17 10.4 −15 44	3.1, 3.3, 11.1	235, 317	0.6″, 63″	27.1	A1IV A1IV	155125, 160332
	Sabik. Matched A type 2+1 triple. AB: P = 88 y, orbit r = 38 AU, e = 0.95, apastron 2068. AE: ps = 2,300 AU. (257 measures; 2010)								

Label	Catalog ID	Components	Coordinates (J2000)	Mag.	θ	ρ	Dist.	Spectral type(s)	HD, SAO No.
	β 125	A B	17 12.2 −27 03	6.9, 9.7	66	1.8″	141	G8IV	155363, 185137
★ 36 A	Sh 243	A B C	17 15.3 −26 36	5.1, 5.1, 6.5	142, 74	4.9″, 12.2′	5.46	K5Ve K1V	155886, 185198
			Local, very high CPM, solar type 2+1 triple. AB: P = 471 y, orbit r = 70 AU. AC: ps = 5,400 AU. (270 measures; 2012)						
41	A 2984		17 16.6 −00 27	4.9, 7.5	15	0.8″	63	K1IV	156266, 141586
	H I 35		17 17.7 −26 38	6.9, 9.1	336	5.8″	156	B9.5V	156252, 185233
	β 126	A B C	17 19.9 −17 45	6.3, 7.6, 11.2	263, 138	2.4″, 12″	115	A2V	156717, 160462
40 xi	Don 832		17 21.0 −21 07	4.4, 8.9	40	4.4″	17.4	F1III	156897, 185296
			Local, solar type double with F giant; faint field. AB: ps = 100 AU. Neglected. (1989)						
	Rag 9	A [B C]	17 22.9 −02 23	6.3, [11.9, 12.0]	146, [86]	46″, [0.8″]	19.5	G3V M3V	157347, 141642
			Local, solar type 1+2 triple with low mass binary companion; sparse field. AB: ps = 1,210 AU. (2012)						
	Σ 2166		17 27.9 +11 23	7.2, 8.6	281	27″	115	A5V	158263, 102835
	Σ 2171		17 29.3 −09 59	8.3, 8.5	61	1.5″	115	F2	158373, 141684
	β 1089		17 29.8 −05 55	6.6, 9.0	323	1.5″	79	G8IV	158463, 141691
★	Σ 2173	A B	17 30.4 −01 04	6.1, 6.2	148	0.8″	16.3	G5V	158614, 141702
			Local, matched spectroscopic binary; faint field. AB: P = 46.4 y, orbit r = 16 AU, e = 0.18, apastron 2032. (716 measures; 2013)						
V2373	OΣ 331	(AaAb) B	17 32.0 +02 49	7.7, 8.8	351	1.0″	480	B5V	158976, 122481
	Shy 726		17 33.9 +08 06	7.9, 8.6	12	3.8′	84	G0 G	159333, 122512
	Σ 2185	(AaAb) B	17 34.8 +06 01	7.5, 10.3	4	28″	91	F8	159481, 122529
			High CPM, solar type (2)+1 triple. AB: ps = 3,440 AU. (2013)						
	Σ 2186		17 35.8 +01 00	8.2, 8.4	77	3.0″	300	B8IV	159660, 122544
	Sh 251	A B	17 39.1 +02 02	6.4, 7.8	328	111″	113	K0III	160315, 122607
	Σ 2191	A B	17 39.8 −04 58	7.8, 8.5	267	26″	140	F2V	160388, 141793
61	Σ 2202	(AaAb) B	17 44.6 +02 35	6.1, 6.5	93	21″	85	A1IV	161270, 122690
	Σ 2223		17 49.0 +04 58	7.6, 9.7	211	18″	87	F0	162056, 122780
	Σ 3128	A B	17 53.0 −07 55	7.8, 10.0	51	1.2″	58	G0IV	162756, 141935
	Σ 2244		17 57.1 +00 04	6.9, 6.6	100	0.7″	118	A3V	163624, 122950
			High mass A type binary. System mass 5.1 M⊙. AB: P = 368 y, orbit r = 105 AU, e = 0.52. (218 measures; 2012)						
67	H VI 2	A C	18 00.6 +02 56	4.0, 8.1	142	54″	380	B5Ib	164353, 123013
	β 1202	E [A B]	18 01.5 +03 31	8.0, [8.7, 9.7]	318, [345]	90″, [0.6″]	250	A0	164529, 123031
			An A type 1+2 triple; unrelated double β 1202 CD (m.10, 2.8″) 2′ n.f. AB: ps = 200 AU; EA: ps = 30,400 AU. (1999)						
★ 69 tau	Σ 2262	(AaAb) B C	18 03.1 −08 11	5.3, 5.9, 11.3	286, 125	1.5″, 101″	51	F4IV F5V	164765, 142050
			Solar type (2)+1+1 spectroscopic quadruple; faint field, brightest of three stars about 40′ s.p. globular cluster NGC 6539. MSC system mass 3.1 M⊙ implies low mass components. AB: P = 257y, orbit r = 70 AU, e = 0.77, closing to periastron 2086. (684 measures; 2013)						
	Σ 2266	A B	18 04.4 +03 29	7.9, 9.6	185	8.5″	108	F5	165111, 123086
★ 70 p	Σ 2272	A B	18 05.5 +02 30	4.2, 6.2	130	5.9″	5.08	K0V K4V	165341, 123107
			Local, very high CPM, low mass binary; pretty YO/O color. Discovered by C. Mayer (1977), it has completed more than 2.5 orbits since it was first measured by W. Herschel (1779) and is always easily resolved: it is currently the most measured double star. A third component is suspected but not confirmed; other "components" in WDS are field stars. AB: P = 88 y, orbit r = 23 AU, e = 0.50, next apastron in 2028. (1,731 measures; 2013)						
	Σ 2276	(AaAb) B	18 05.7 +12 00	7.1, 7.4	256	7.0″	138	A7	165475, 103373
73	Σ 2281	A B	18 09.6 +04 00	6.0, 7.5, 12.6	287, 194	0.7″, 68″	55	F2V	166233, 123187
			Solar type double with gam Dor type variable; faint rich field. AB: P = 294 y, orbit r = 65 AU, e = 0.61, apastron 2060. (403 measures; 2011)						
	β 637	(AaAb) B C	18 09.9 +03 07	5.7, 11.7, 10.9	189, 245	6.7″, 98″	46.9	F6V	166285, 123198
	Σ 2294		18 14.6 +00 11	8.2, 8.6	93	1.3″	114	F2	167278, 123283
			Solar type binary; rich field, tiny unrelated pair (m.11, 15″) 9′ p. AB: P = 345y, orbit r = 85 AU, widening. (242 measures; 2012)						
	LDS 1012	(AaAb) B	18 26.2 +08 47	7.9, 8.4	163	10.1′	28.9	G7V	169822, 169822
			High CPM, solar type (2)+1 triple; rich field. MSC gives 1.7 M⊙. Interferometric binary Aa,Ab: P = 302d, e = 0.35. AB: ps = 23,600 AU; near 100% probability pair is physical. (2001)						
	OΣ 350	A B	18 26.9 +06 25	7.8, 9.4	166	1.8″	520*	A0III	169959, —
			Distant, probably high mass double with A type giant; difficult to find in a rich field (NGC 6633). AB: ps = 1,260 AU. (2011)						

Label	Catalog ID	Components	Coordinates (J2000)	Mag.	θ	ρ	Dist.	Spectral type(s)	HD, SAO No.
	Orion		Ori					*Chart 14*	
	β 552	(AaAb) (BaBb) C	04 51.8 +13 39	6.4, 8.9, 12.4	258, 224	0.7″, 49″	40.7	dF6	30869, 94171
		Solar type (2)+(2)+1 spectroscopic quintuple; sparse field, member of Hyades cluster. MSC 3.6 M$_\odot$. System mass 2.0 M$_\odot$. Ab,Ab: P = 143 d. AB: P = 95 y, e = 0.59, apastron 2029. AC: ps = 2,700 AU. (2012)						
	S 457		04 53.1 −01 17	7.9, 8.1	355	41″	200	A	31125, 131570
	Σ 612	A B C	04 54.3 +07 22	8.3, 8.4, 13.3	200, 264	16″, 60″	30.3	K0	31208, 112196
		High CPM, solar type 2+1 triple; Y/Y color, in close group of faint stars. AB: ps = 650 AU. (2010)							
★ V1834	OΣ 90	(AaAb) B [C D]	04 54.9 +08 36	7.0, 9.0, [12.2, 13.0]	339, 97, [320]	1.8″, 40″, [12″]	145	A0	31306, 112205
		Compact A type (2)+1+2 (?) system, visual quadruple with Algol type eclipsing binary (P = 1.5 d): it's unclear from the data if AB and CD are related. AB: ps = 350 AU; AC ps = 7,800 AU. (2012)							
	h 689		04 58.2 −02 13	6.4, 10.9	277	21″	133	A2V	31739, 131640
	Sh 49	A B	04 59.0 +14 33	6.1, 7.4	307	39″	350	B7V	31764, 94240
	S 463		05 01.8 +11 23	7.2, 10.1	29	32″	470	B8	32202, 94274
★ 14 i	OΣ 98	(AaAb) B	05 07.9 +08 30	5.8, 6.7	295	0.9″	65	Am	33054, 112440
		A type (2)+1 spectroscopic triple; W/– color, centered in faint group that includes CPM companion Σ 643 (m.9, 2.4″) 6′ s. AB: P = 197 y, orbit r = 65 AU, e = 0.18, apastron 2075. (479 measures; 2013)							
	Bvd 50		05 09.2 +11 30	8.4, 8.5	104	31″	87	F3V F3V	33221, 94353
	Σ 652		05 11.8 +01 02	6.3, 7.4	180	1.7″	195	A G2III	33646, 112509
★ 17 rho	Σ 654	(AaAb) B	05 13.3 +02 52	4.6, 8.5	63	6.8″	107	K0.5III	33856, 112528
		Solar type (2)+1 spectroscopic triple; YO/B color, brilliant in sparse field. MSC gives 6.4 M$_\odot$. Aa,Ab: P = 2.8 y, e = 0.10. (2011)							
	OΣ 517	A B C	05 13.5 +01 58	6.8, 7.0, 13.0	241, 138	0.7″, 6.5″	210	A5V	33883, 112535
		An A type, probable 2+1 triple; faint field. MSC system mass 4.3 M$_\odot$. AB: P = 987 y, orbit r = 160 AU. (2013)							
★ 19 bet	Σ 668	A (BaBbC)	05 14.5 −08 12	0.3, 6.8	205	9.5″	260	B8Iae B9V	34085, 131907
		Rigel. Distant 1+(2+1) quadruple, visual double; spectacular example of B type blue supergiant, pulsating variable with high mass (2+1) triple companion. MSC system mass 33.4 M$_\odot$. W/w color. A,BC: ps = 3,330 AU. (2013)							
	Σ 667		05 14.7 −07 04	7.2, 8.8	316	4.2″	300	K2	34121, 131910
	Σ 664		05 15.2 +08 26	7.8, 8.4	177	4.7″	120	A9IV	34081, —
	Bvd 51		05 15.4 −03 22	7.8, 9.6	318	45″	83	F7V G1V	34195, 131918
	Σ 688		05 19.3 −10 45	7.5, 7.6	95	11″	96*	F0	34750, 150333
	β 189	(AaAb) B	05 20.4 −05 22	6.4, 9.8	285	4.6″	280	B8III	34880, 132004
	OΣ 106		05 22.2 +05 24	7.1, 10.1	41	9.4″	57	F6V	35066, 112681
	Σ 700	(AaAb) B	05 23.1 +01 03	7.7, 7.9	6	4.9″	230	B9V B9.5V	35192, 112704
	Σ 701		05 23.3 −08 25	6.1, 8.1	141	6.0″	123	B8III	35281, 132053
	Wnc 2	(AaAb) (BC)	05 23.9 −00 52	6.9, 7.0	159	3.1″	56	F6V	35317, 132060
		Solar type (2)+(2) spectroscopic quadruple, visual matched binary; the middle star of six bright stars. A,BC: P = 923 y, orbit r = 162 AU. BC (A 847): P = 48 y, orbit r = 18 AU, e = 0.22, apastron 2033. (2011)							
	Σ 708		05 25.2 +01 55	7.7, 8.9	321	2.7″	540	B8V	35501, 112744
	Σ 712	A B	05 26.5 +02 56	6.7, 8.6	66	3.2″	146	B9.5V	35673, 112765
★ 32	Σ 728		05 30.8 +05 57	4.4, 5.8	43	1.4″	93	B5V	36267, 112849
		High mass double; dominant in field. Ambiguous: low quality orbit and nearly rectilinear motion, but parallel proper motion. AB: P = 614 y, orbit r = 200 AU. (233 measures; 2012)							
	Σ 726		05 30.9 +10 15	7.9, 8.6	262	1.1″	280*	B9	36263, 94602
★ 33 n 1	Σ 729	A B	05 31.2 +03 18	5.7, 6.7	27	1.9″	350	B1.5V	36351, 112861
		Distant, high mass double; sparse field. AB: ps = 900 AU. (2012)							
	Σ 734	A B	05 33.1 −01 43	6.7, 8.2	357	1.7″	320	B4Vn	36646, 132247
		Distant, high mass double; splendid field. The pair CD (β 1049, m.8, 0.5″) 30″ s.p. is unrelated. (2008)							
	Σ 743	A B	05 34.7 −04 24	7.7, 8.3	282	1.8″	315	B5V	36883, 132285
		Distant, high mass double in bright high mass group (NGC 1981); β 13 (m.7, 1.0″) 8′ s.p.; Σ 750 (see below) 13′ f. (2012)							

Label	Catalog ID	Components	Coordinates (J2000)	Mag.	θ	ρ	Dist.	Spectral type(s)	HD, SAO No.
	Σ 741		05 34.9 −00 07	7.1, 10.0	286	10″	230*	B5	36898, 132291
★ 39 lam 1	Σ 738	A B	05 35.1 +09 56	3.5, 5.5	44	4.4″	340	O8III B0.5V	36861, 112921
		Meissa. Distant, high mass, tiny OB "diamond ring," leading a line of four stars s.; YW/− color. AB: ps = 2,020 AU. (2012)							
★ 41 the 1	Σ 748	(A) (B) (C) (D) E	05 35.3 −05 23	6.6, 7.5, 5.1, 6.4, 11.1	32, 132, 96, 352	8.9″, 13″, 21″, 4.6″	430	O7 B1V O6 B0.5V	37022, 132314
		The Trapezium: a 3 million year old, (2)+(4)+(3)+(2)+1 association of high mass (O and B type) double and multiple stars illuminating the core of the Great Orion nebula (NGC 1976, M 42). Independently detected as a triple (ACD) by Galileo (1617), then as a quartet by J. Picard (1673); N.-C. de Peiresc and J. Cysat (c. 1611) first noticed the nebulosity. The quartet can be resolved by a large binocular; E,F need good seeing, F is optical. MSC system mass >107 M☉. A and B: Algol type variables, D: pre main sequence star, E: Orion type variable. Divergent true motions suggest the group is already breaking apart, as binding mass is lost in the gas and dust dispersed by the stars' intense radiation. AB: ps = 5,890 AU. AD: ps = 13,900 AU. (2013) (408 measures; 2013) See Figure 3.							
42	Dawes 4		05 35.4 −04 50	4.6, 7.5	205	1.1″	270	B1V	37018, 132320
43 the 2	Σ I 16	(AaAb) B	05 35.4 −05 25	5.1, 6.2	93	52″	470	O9.5Ve	37041, 132321
★ 44 iot	Σ 752	(AaAb) B	05 35.4 −05 55	2.8, 7.7	141	12″	710	O9III	37043, 132323
		Nair al Saif. Distant, high mass (2+1)+1 spectroscopic triple; near Σ 745/747. MSC gives 69.3 M☉. AB: ps = 11,500 AU. (2012)							
	Σ 750		05 35.5 −04 22	6.4, 8.4	60	4.2″	360	B2.5IV	37040, 132325
	Dawes 3	(AaAb) B	05 35.9 −05 38	7.3, 8.5	173	0.9″	440*	B6Ve	37115, −
		Distant, high mass (2)+1 spectroscopic triple, in s.f. rim of NGC 1976 (M 42), 5′ p. HD 37150. AB: ps = 530 AU. (2006)							
★	Σ 757	A B D	05 38.1 −00 11	8.0, 8.3, 8.5	239, 79	1.5″, 42″	270	B6V	37370, −
		High mass 2+1 triple; W/W color, in row of stars, C (m.9, 50″ f.) is unrelated. AB: ps = 550 AU. AD: ps = 15,300 AU. (2010)							
★ 48 sig	Σ 762	(AaAbB) G C D E	05 38.7 −02 36	3.8, 12.0, 8.8, 6.6, 6.3	20, 239, 85, 62	3.2″, 11″, 13″, 41″	330	O9.5V	37468, 132406
		A = V1030 Ori. High mass, distant, young septuple system; a gravitationally unstable group displayed in remarkable "planetary" alignment; Σ 761 (4′ n.p.) is apparently unrelated. Despite the attribution to F.W. Struve, D,E were first detected by C. Mayer (1777), C by W. Dawes (1831). Estimates vary, but system mass likely exceeds 60 M☉ if undetected lower mass components are included. AB: P = 157 y, e = 0.05, apastron 2078. (2013)							
	Σ 766	A B	05 40.3 +15 21	7.0, 8.4	275	10″	260	F0	37603, 94746
★ 50 zet	Σ 774	(AaAb) B	05 40.8 −01 57	1.9, 3.7	167	2.4″	230	O9.5Ibe BOIII	37742, 132444
		Alnitak. High mass (2)+1 spectroscopic triple with O supergiant and B giant; divergent proper motion suggests C (m.10, 58″ n.) is unrelated. System mass ~50 M☉. Aa,Ab: P = 7.4 y, e = 0.34. (256 measures; 2013)							
	β 1052		05 41.7 −02 54	6.7, 8.2	185	0.6″	81	A9IV	37904, 132465
		A type binary; s.f. corner of 16′ rectangle of m.8 stars. System mass 3.0 M☉. AB: P = 192 y, orbit r = 55 AU, e = 0.13. (2013)							
	Σ 788	A B C	05 44.7 +03 50	7.6, 10.1, 10.4	89, 148	7.3″, 35″	300	B9	38270, 113093
	Σ 790		05 46.0 −04 16	6.4, 9.0	89	7.1″	111	G8III	38495, 132515
52	Σ 795	(AaAb) B	05 48.0 +06 27	6.0, 6.0	222	1.2″	165	A5V	38710, 113150
	J 36		05 48.3 +03 54	7.8, 9.6	101	1.5″	58*	F8	38767, 113155
	Σ 797		05 48.5 +04 42	7.4, 9.8	19	7.2″	320	A0	38798, 113161
	Ary 39	A B	05 53.0 +20 47	7.7, 10.6	184	32″	250*	B9	39358, 77667
	OΣ 123		05 54.2 +10 15	7.3, 9.1	187	2.1″	133	G5	39612, 94975
	Σ 816		05 54.9 +05 52	6.9, 9.3	286	4.4″	240	B9	39773, 113267
59	Arn 37	(AaAb) C	05 58.4 +01 50	5.9, 6.9	293	3.0′	112	A5m	40372, 113315
		A = V1004 Ori. A type (2)+1 spectroscopic triple with del Sct type variable; sparse field. AC: ps = 27,200 AU. (2008)							
	OΣ 124		05 58.9 +12 48	6.1, 7.4	299	0.6″	240	K2III A5V	40369, 95075
	OΣ 125		05 59.7 +22 28	7.9, 8.9	0	1.4″	200*	A0	40423, —
	Σ 840	A [B C]	06 06.5 +10 45	7.2, [9.8, 10.1]	249, [126]	22″, [0.4″]	230	A0V+F0	41580, 95234
	OΣ 133		06 08.0 +21 18	7.4, 11.2	33	3.3″	103	F0	41786, 78006
★	Σ 848	A B D	06 08.5 +13 58	7.3, 8.2, 8.3	110, 122	2.6″, 28″	300	B1V B2V	41943, 95282
		V1154 Ori. Distant, high mass 2+1 triple in open cluster NGC 2169, a situation that has confused the correct attribution of component measures. Optical pair Σ 844 (m.8, 23″) 3′ n.p. (2012)							

Label	Catalog ID	Components	Coordinates (J2000)	Mag.	θ	ρ	Dist.	Spectral type(s)	HD, SAO No.
	Σ 855	A B	06 09.0 +02 30	5.7, 6.7	114	29″	200	A3Vn A0V	42111, 113507
	Σ 867	A B	06 11.7 +17 23	7.5, 8.9	159	2.2″	460*	B9.5III	42476, 95354

Pavo — Pav — Chart 29

Label	Catalog ID	Components	Coordinates (J2000)	Mag.	θ	ρ	Dist.	Spectral type(s)	HD, SAO No.
	h 4979		17 52.1 −60 24	7.5, 10.0	237	11″	108	F0IV	161918, 254059
	h 5029		18 15.1 −57 51	8.3, 8.6	82	1.9″	51	G2V	166653, 245335
	I 249		18 19.7 −63 53	6.2, 10.8	352	7.8″	23.1	F9V	167425, 254209

Local, solar type double with probable low mass companion; sparse or faint field. AB: ps = 245 AU. (2000)

Label	Catalog ID	Components	Coordinates (J2000)	Mag.	θ	ρ	Dist.	Spectral type(s)	HD, SAO No.
xi	Gale 2	(AaAb) B	18 23.2 −61 30	4.4, 8.1	156	3.4″	144	K4III	168339, 254226

High mass, long period (2)+1 spectroscopic triple, visual binary with K giant; rich field. MSC gives 10.9 M$_\odot$. Aa,Ab: P = 6.1 y, e = 0.26. AB neglected. (1988)

Label	Catalog ID	Components	Coordinates (J2000)	Mag.	θ	ρ	Dist.	Spectral type(s)	HD, SAO No.
	Skf 973	(AB) D	18 29.9 −57 31	5.9, 8.9	273	2.3′	98	K0III	169836, 245510
	MlbO 5		18 34.2 −66 17	7.0, 9.1	292	4.7″	1150	G5III A5	170407, 254286
	Skf 105	A (BaBb)	18 45.4 −64 52	4.8, 10.1	65	71″	28.6	A7V K5V	172555, 254358
	R 314	(AaAb) B	18 49.7 −73 00	6.2, 8.1	271	1.9″	220	B9.5IV	172881, 257630
	h 5065		18 51.9 −57 56	7.9, 9.9	21	23″	135	A0V	174041, 245756
	h 5075		19 04.1 −63 47	7.7, 7.7	113	1.7″	200	A0	176340, 254453
★	h 5085		19 10.6 −60 03	7.6, 9.1	239	2.7″	540	B9II	177999, 254482

Distant, high mass double with B supergiant; at rim of globular cluster NGC 6752. AB: ps = 1,970 AU. Neglected. (1991)

Label	Catalog ID	Components	Coordinates (J2000)	Mag.	θ	ρ	Dist.	Spectral type(s)	HD, SAO No.
	Gale 3		19 17.2 −66 40	6.1, 6.4	350	0.5″	92	A5V+A8V	179366, 254515

A type binary, subarcsecond resolution test. AB: P = 157 y, orbit r = 50 AU, e = 0.57, increasing θ to apastron in 2029. (2013)

Label	Catalog ID	Components	Coordinates (J2000)	Mag.	θ	ρ	Dist.	Spectral type(s)	HD, SAO No.
	I 117		19 32.2 −60 16	7.5, 8.2	185	1.0″	260	A3IV	183237, 254592
	I 119		19 42.6 −59 01	7.9, 9.0	150	2.4″	41.7	G5V	185454, 246222
	h 5132		19 44.0 −66 18	7.6, 9.7	308	22″	67	G1V	185523, 254636
	I 120	A B C	19 49.1 −61 49	8.3, 8.0, 10.5	188, 340	0.4″, 12″	56	F7V	186602, 254655

Solar type 2+1 triple; two wide m.12 pairs are 2′ and 5′ f. MSC gives 2.9 M$_\odot$. AB: P = 65 y, orbit r = 19 AU, e = 0.68, apastron 2029. (2013)

Label	Catalog ID	Components	Coordinates (J2000)	Mag.	θ	ρ	Dist.	Spectral type(s)	HD, SAO No.
	h 5140		19 49.8 −64 54	8.2, 8.3	252	1.3″	64	F7V	186632, 254659
	Shy 759	Aa Ab B	19 50.7 −59 12	5.5, 7.1, 7.1	151, 232	0.8″, 6.8′	85	A0IV A5IV	186957, 246293

A type 2+1 triple, comoving with a fourth star C (HD 188162, m.5) 1.6° n.f. The Aa,Ab pair (I 121) is included despite a linear solution, in view of the matching CPM, small ρ and few measures. AC (Shy 761): ps = 1.3 pc. Neglected. (1999)

Label	Catalog ID	Components	Coordinates (J2000)	Mag.	θ	ρ	Dist.	Spectral type(s)	HD, SAO No.
	h 5163		20 05.1 −63 04	7.7, 8.4	249	1.3″	153	A3V	189721, 254722
	h 5171	A B	20 14.6 −64 26	6.9, 9.8	305	18″	104	A2IV	191585, 254757
	h 5194		20 30.3 −69 04	7.1, 11.2	259	3.5″	110	A3V	194441, 254800
	h 5200		20 33.0 −68 22	7.5, 10.0	136	12″	141	A1V	194972, 254811
	Mug 4	A (BaBb)	20 37.9 −60 38	7.5, 12.1	175	11″	50	G3V	196050, 254837
★	Rmk 26		20 51.6 −62 26	6.2, 6.6	81	2.4″	76	A3IV	198160, 254883

A type double; prominent in dark field. AB: ps = 245 AU. (2009)

Pegasus — Peg — Charts 6, 11, 12

Label	Catalog ID	Components	Coordinates (J2000)	Mag.	θ	ρ	Dist.	Spectral type(s)	HD, SAO No.
	Σ 2767		21 10.5 +19 58	8.2, 8.5	29	2.5″	182	F4V	201672, 106894
	β 681		21 13.3 +16 55	7.5, 10.9	243	2.8″	210	K3III	202091, 106927
1	Σ II 11	A (BaBb)	21 22.1 +19 48	4.2, 9.3	311	37″	47.8	K0.5III	203504, 107073
	Cou 430	A B	21 25.2 +18 28	8.0, 9.4	234	0.6″	177	A0	203991, 107115
KP	Σ 2797	A (BaBb)	21 26.7 +13 41	7.4, 8.8	217	3.5″	220	A2V	204215, 107139
	Σ 2799	A B	21 28.9 +11 05	7.4, 7.4	261	1.8″	105	F4V	204509, 107165

Solar type binary; faint group 5′ s.f. Low quality orbit AB: P = 978 y, orbit r = 220 AU. (442 measures; 2012)

Label	Catalog ID	Components	Coordinates (J2000)	Mag.	θ	ρ	Dist.	Spectral type(s)	HD, SAO No.
	h 1647	A B	21 29.0 +22 11	6.1, 10.2	178	42″	200	M5III	204585, 89737
	β 74		21 35.2 +21 24	7.5, 9.1	335	1.0″	85	F5V	205497, 89810

Label	Catalog ID	Components	Coordinates (J2000)	Mag.	θ	ρ	Dist.	Spectral type(s)	HD, SAO No.
★ 3	Σ I 56	(AaAb) B	21 37.7 +06 37	6.2, 7.5	349	39″	88	A2V	205811, 126940
		A type (2)+1 occultation triple; dark field. AB: estimated P = 100,000 y, ps = 4,630 AU. OΣ 443 7′ north preceding. (2012)							
	OΣΣ 222		21 44.1 +07 09	7.5, 8.5	258	88″	163	F2V	206751, 127028
	β 692	A B	21 50.1 +31 51	7.5, 11.0	10	2.9″	181	K0	207703, 71749
	Σ 2834	A B	21 51.7 +19 18	6.9, 9.9	298	4.2″	92	F5IV	207859, 107450
	Σ 2833		21 51.9 +09 05	7.8, 10.2	337	9.1″	118	A5	207862, 127121
	β 75	A B	21 55.5 +10 53	8.4, 8.6	24	1.0″	47.8	G5	208348, 107510
		Solar type binary; dark field. System mass 1.7 M⊙. AB: P = 159 y, orbit r = 36 AU, e = 0.61, apastron 2047. (225 measures; 2013)							
	OΣΣ 225		21 57.5 +04 09	7.1, 8.6	287	75″	97	F5	208632, 127190
	Σ 2854		22 04.4 +13 39	7.8, 7.9	83	1.6″	72	F6V	209601, 107633
		Solar type matched double, YO/YO color. AB: ps = 155 AU, but no orbit. (231 measures; 2011)							
	Σ 2869		22 10.4 +14 38	6.3, 12.4	252	21″	136	K0III	210461, 107707
	Ho 178		22 11.5 +32 05	7.4, 11.2	223	3.4″	122	F0	210684, 72108
	Σ 2878	A B	22 14.5 +07 59	6.9, 8.1	116	1.5″	220	B9IV	211048, 127402
	Σ 2881		22 14.6 +29 34	7.7, 8.2	75	1.3″	172	F6III	211139, 90348
		Double with F giant; sparse or dark field. AB: ps = 300 AU; frequently measured, but no orbit (284 measures; 2011)							
	OΣ 467		22 14.8 +22 31	6.7, 10.7	274	24″	220	G8III	211153, 90349
	Ho 292		22 23.3 +05 39	7.7, 11.2	65	3.9″	141	A2	212317, 127491
	β 701	A B C	22 28.1 +12 15	7.3, 9.6, 12.0	186, 133	0.9″, 2.1′	66	K0V	212989, 107935
	Σ 2908		22 28.2 +17 16	7.7, 9.7	113	9.0″	460	G9III	213014, 107941
	Σ 2920	A B	22 34.5 +04 13	7.6, 8.9	143	14″	89	B9.5V	213892, 127609
	Ho 296	A B D	22 40.9 +14 33	6.1, 7.2, 14.1	60, 93	0.5″, 4.5′	33.8	G4V	214850, 108094
		High CPM, solar type 2+1 spectroscopic triple. AB: P = 21 y, orbit r = 10 AU, e = 0.73, closing. (332 measures; 2012)							
	Ho 482	A B	22 51.4 +26 23	7.3, 8.3	17	0.5″	133	A9V	216285, 90833
		A type binary; sparse field; three stars s. System mass 2.7 M⊙. AB: P = 383 y, orbit r = 75 AU, e = 0.61 (2011)							
	Ho 191	A B C	22 53.6 +30 46	7.8, 13.5, 11.8	90, 280	3.2″, 24″	125	A6V	216562, 72815
	Σ 2952	A B	22 54.2 +28 01	7.7, 10.5	138	18″	51	F8V	216632, 90864
	Cou 240		22 56.4 +22 57	7.7, 8.8	290	0.8″	186	F0	216879, 90881
	Σ 2958		22 56.9 +11 51	6.6, 9.1	15	3.9″	99	A3V	216900, 108275
	OΣΣ 241		22 58.6 +12 03	8.3, 8.4	161	84″	111	F2	217163, 108300
52	OΣ 483		22 59.2 +11 44	6.1, 7.3	359	0.5″	94	A8III	217232, 108307
		Binary with A type giant; dark field. System mass at least 3.3 M⊙. AB: P = 249 y, orbit r = 65 AU, e = 0.39. (324 measures; 2008)							
	Σ 2968		23 00.7 +31 05	6.7, 9.5	93	3.2″	136	B9	217477, 72924
	Σ 2974		23 05.0 +33 23	8.1, 8.5	165	2.7″	240	A0V A3V	218097, 72984
	OΣ 488		23 07.4 +20 35	6.7, 10.4	335	15″	149	K0III	218381, 91021
	Σ 2978		23 07.5 +32 50	6.4, 7.5	145	8.3″	163	A3V	218395, 73010
	β 78	A B	23 07.9 +31 28	7.5, 11.1	54	19″	92	A5V	218472, 73021
	Σ 2986		23 10.0 +14 26	6.6, 8.9	268	32″	24.8	G0V	218687, 108437
		Local, solar type double; very dark field. AB: ps = 1,070 AU. (2013)							
	h 5532	A B C	23 10.3 +32 29	7.4, 8.2, 9.4	85, 77	0.6″, 58″	330	B9V	218767, 73054
		Distant, high mass 2+1 triple; center star of 5 stars in dark field. AB (β 385): P = 830 y, orbit r = 205 AU. (2011)							
	β 852	(AaAb) [B C]	23 10.7 +26 31	7.2, [10.5, 11.0]	283, [322]	58″, [1.2″]	85	A7IV	218806, 91061
	Σ 3007	A B	23 22.8 +20 34	6.7, 9.8	92	5.8″	37.4	G2V dK6	220334, 91222
		High CPM double; dark field, m.10 star 1.5′ n.p. is unrelated. AB: ps = 290 AU, but no orbit. (268 measures; 2007)							
	Σ 3018	(AaAb BaBb) C	23 30.4 +30 50	7.4, 9.8	202	19″	68	F7V	221264, 73306
		Solar type (2+2)+1 spectroscopic quintuple, visual double; dark field. Aa,Ab: P = 1.9 d; Ba,Bb: P = 13.0 d; AB (β 1266): P = 48 y, orbit r = 13 AU, e = 0.43, apastron 2031. (2007)							
	Σ 3023		23 32.4 +17 24	7.2, 9.1	280	1.8″	153	F4IV	221479, 108669
★ 72	β 720	A (BaBb)	23 34.0 +31 20	5.7, 6.1	102	0.6″	168	K4III	221673, 73341
		Solar type 1+(2) interferometric triple with K giant; dark field, O/− color. Ba,Bb: P = 4.2 y, orbit r = < 4 AU. AB: P = 492 y, orbit r = 95 AU. (465 measures; 2011)							
	β 858	A B C	23 41.3 +32 34	7.8, 8.8, 12.9	220, 52	0.8″, 23″	210	A1V	222529, 73436

Label	Catalog ID	Components	Coordinates (J2000)	Mag.	θ	ρ	Dist.	Spectral type(s)	HD, SAO No.
	OΣ 503	A B	23 42.0 +20 18	8.3, 8.6	133	1.0″	108	F8	222610, 91425
	OΣ 504		23 42.5 +18 40	7.4, 10.3	176	7.6″	200	K0	222659, 108780
78	AGC 14		23 44.0 +29 22	5.1, 8.1	280	0.8″	69	G8III	222842, 91457
	OΣ 505		23 45.5 +20 25	6.8, 9.6	60	2.3″	159	G8III	222978, 91467
	Σ 3044		23 53.0 +11 55	7.3, 7.9	283	20″	72	F0	223839, 108883
★ 85	β 733	(AaAb) B	00 02.2 +27 05	5.8, 8.9	273	0.8″	12.2	G5V K5V	224930, 91669
			Local, very high CPM, solar type (2)+1 triple; visual binary; dark field, C 3′ n.p. is unrelated. System mass 1.6 M$_\odot$. AB: P = 26 y, orbit r = 10 AU, e = 0.38, minimum separation (and periastron) in 2015, then rapidly widens with increasing θ. (2009)						
	Σ 3055	A B C	00 04.0 +12 09	7.3, 10.3, 12.7	359, 32	5.5″, 2.0′	176	F0III	225161, 91683
	Σ 3058		00 05.2 +30 20	7.8, 9.2	52	13″	250	F3V F6V	4, 53627
	Σ 3061		00 05.7 +17 50	8.4, 8.5	149	7.6″	128	F5V F5V	85, 91703
	Gic 2	A B	00 09.3 +25 17	7.8, 11.5	237	30″	51	G0	471, 73776

Perseus Per Charts 3, 7

Label	Catalog ID	Components	Coordinates (J2000)	Mag.	θ	ρ	Dist.	Spectral type(s)	HD, SAO No.
★	Σ 162	(AaAb) B C	01 49.3 +47 54	6.5, 7.2, 9.2	198, 179	1.8″, 21″	128	A3V	11031, 37536
			Splendid (2)+1+1 quadruple; YW/BW color, rich field. MSC has 6.4 M$_\odot$, AC: P = 39,100 y, orbit r = 3,450 AU. (2011)						
	Sti 1797		02 10.4 +56 18	7.5, 11.9	140	9.3″	117	K2III	13149, 22997
	Σ 230		02 15.0 +58 29	7.9, 9.4	259	24″	720*	B8III	13633, 23071
★	Σ 268		02 29.4 +55 32	6.7, 8.5	130	2.8″	163	A2Shell	15253, 23369
			Close pair, delicate brightness contrast, W/WB color, rich field. CPM double Σ 270 (below) 10′ f. AB: ps = 620 AU. (2012)						
	Σ 270	A B	02 30.8 +55 33	7.0, 9.7	305	21″	55	F4V	15407, 23389
	Σ 272		02 33.1 +58 28	8.3, 8.4	216	1.9″	130	A3V	15641, 23419
★	Ary 72	A B C D E	02 36.9 +55 55	7.7, 10.0, 9.3, 9.8, 8.7	244, 101, 229, 276	35″, 69″, 2.2′, 2.3′	625*	K4II	16068, 23469
			Distant, high mass multiple star or comoving group with K supergiant, at the core of Trumpler 2, a cluster about 80 million years old. Fine visual example of the transition from natal cluster to multiple star. AB: ps = 117,200 AU. (2011)						
	h 1123		02 42.0 +42 48	8.4, 8.5	249	20″	910*	B8III	16705, 38244
★	OΣ 44	A B C	02 42.2 +42 42	8.5, 9.0, 8.3	56, 290	1.4″, 86″	250	B9V	16728, 38254
			Distant, high mass 2+1 triple; in rich field s. of a bright stellar arc (M 34). Enjoy the recognition moment when the wide pair resolves into a triple. AB and C joined by parallax, AC: ps = 29,000 AU. (2009)						
	Σ 292		02 42.5 +40 16	7.6, 8.2	212	23″	350	B9	16772, 38265
13 the	Σ 296	A B	02 44.2 +49 14	4.2, 10.0	305	21″	11.1	F7V +M1.5	16895, 38288
			Local, solar type double; rich field. AB: ps = 315 AU; a premature orbit gives P = 2,720 y. (2013)						
	β 9	A B C	02 47.1 +35 33	6.4, 8.6, 13.3	210, 127	1.0″, 30″	51	F0IV	17240, 55872
	Σ 301	(AaAb) B	02 47.6 +53 57	7.9, 8.7	17	8.2″	990	A0	17198, 23611
★ 15 eta	Σ 307	(AaAb) B	02 50.7 +55 54	3.8, 8.5	295	31″	270	K3Ib B9V	17506, 23655
			Miram. Probable high mass, (2)+1 spectroscopic triple; characteristic "giant type" pairing of an M supergiant with a high mass companion, seen at great distance, and the attractive Y/B complementary color that results. CD (WRD 1, m.12, 5″) 1′ p. is unrelated. AB: ps = 11,300 AU. (2012)						
20	Σ 318	(AB) C	02 53.7 +38 20	5.0, 9.7	237	14″	71	F4IV	17904, 55975
			Solar type (2)+1 triple, visual binary. MSC system mass 3.4M$_\odot$. AB (β 520): P = 4.3 y, e = 0.76. (222 obs.) AC: ps = 1,340 AU. (2008)						
★	Σ 331	(AaAb) (BaBb)	03 00.9 +52 21	5.2, 6.2	85	12″	141	B7V B9V	18537, 23763
			High mass (2)+(2) spectroscopic quadruple; blazing in sparse field. AB: ps = 2,280 AU. (2012)						
	Σ 336		03 01.5 +32 25	7.0, 8.3	8	8.4″	168	G5IV	18715, 56095
	Σ 337	A B	03 02.3 +41 24	8.0, 9.3	163	18″	47.8	F5	18751, 38525
	Es 558		03 06.8 +45 45	7.7, 10.6	0	8.3″	170	B9	19174, —
	Σ 352		03 08.8 +35 28	7.8, 9.7	0	3.8″	156	A0	19444, 56196
	Σ 360		03 12.2 +37 13	8.0, 8.3	126	2.8″	39.9	G0	19771, 56241
	Hu 544		03 15.8 +50 57	6.7, 8.2	102	1.6″	135	A0	20096, 23907

Label	Catalog ID	Components	Coordinates (J2000)	Mag.	θ	ρ	Dist.	Spectral type(s)	HD, SAO No.
★	Σ 369		03 17.2 +40 29	6.8, 7.7	29	3.6″	198	B9V	20283, 38700
		High mass double; dominant in sparse field. AB: ps = 960 AU. (2012)							
	OΣ 53	A B (CaCb)	03 17.7 +38 38	7.7, 8.5, 13.3	237, 180	0.6″, 107″	60	G0	20347, 56320
		Solar type 2+(2) quadruple; sparse field. MSC gives 2.4 M$_\odot$. AB: P = 114 y, orbit r = 30 AU, e = 0.76, periastron 2041. (2012)							
	Σ 382	A B	03 24.5 +33 32	5.8, 9.3	153	4.8″	134	A0V	20995, 56419
	Σ 388		03 28.7 +50 26	8.0, 9.0	214	2.8″	130	F0	21332, 24055
	AG 67	A B	03 29.0 +40 11	7.5, 10.4	349	24″	163	G5	21449, 38870
	Σ 391		03 29.2 +45 03	7.6, 8.3	96	4.0″	1100	B2IV	21448, 38873
	Σ 392		03 30.3 +52 54	7.5, 10.3	348	26″	210	K0	21488, 24068
	β 533	A B C	03 35.6 +31 41	7.6, 7.7, 9.2	221, 294	1.1″, 3.8′	86	F4V	22195, 56569
	S 430	A B	03 38.3 +44 48	7.2, 7.5	96	41″	153	A0	22428, 39005
	OΣ 59		03 40.7 +46 01	7.9, 8.9	355	2.8″	111	G5	22679, 39031
★ 38 omi	β 535		03 44.3 +32 17	3.9, 6.7	23	1.0″	340	B1III	23180, 56673
		Atik. Spectroscopic binary with B type giant, an ellipsoidal variable star; β 880 (m.9, 0.6″) 9′ s.f. in rich field (star forming region and young star cluster IC 348). AB: ps = 460 AU. (2009)							
V580	Σ 443	(AaAb) B	03 47.0 +41 26	8.2, 8.8	55	6.8″	21.9	K1V K2V	23439, 39100
		Local, very high CPM, solar type (2)+1 triple with eclipsing binary; CD (Fox 135) 2′ n.p. is unrelated. The WDS notes cite an m.13 CPM companion (G095-059) at an angular separation of more than 2°. AB: ps = 200 AU. (2011)							
	Σ 448	(AaAb) B	03 47.9 +33 36	6.7, 9.4	13	3.4″	330	B2.5V	23625, 56709
	Σ 446	A B	03 49.5 +52 39	6.9, 9.9	255	8.7″	800	B0.5III	23675, 24248
		Distant, high mass double with B giant; member of Camelopardalis OB1 association, lucida (brightest star) of star cluster NGC 1444. (2012)							
	OΣ 66		03 52.1 +40 48	8.1, 8.5	144	1.0″	240	A2	24117, 39161
44 zet	Σ 464	(AaAb) B E	03 54.1 +31 53	2.9, 9.2, 10.0	209, 186	13″, 120″	230	B1Ib	24398, 56799
		Menkib. High mass (2)+1+1 spectroscopic quadruple with variable B supergiant; W/– color, dominant in its field. Brightest member of the Perseus OB2 association, about 1 million years old. AE: ps = 37,000 AU. (2012)							
43	S 440	(AaAb) B	03 56.6 +50 42	5.3, 10.7	31	76″	37.4	F5IV	24546, 24314
	Σ 469		03 57.3 +41 53	6.9, 9.9	147	8.9″	121	A2	24689, 39212
★ 45 eps	Σ 471	(AaAb) B D	03 57.9 +40 01	2.9, 8.9, 9.3	10, 146	8.7″, 2.7′	196	B0.5V A2V	24760, 56840
		High mass (2)+1+1 quadruple with bet Lyr type primary; delicate contrast, W/W color, rich field. AD: ps = 43,100 AU. (2012)							
	OΣ 69		03 59.7 +38 49	6.6, 9.1	325	1.5″	194	A1V	24982, 56866
IQ	OΣ 68	(AaAb) B	03 59.7 +48 09	7.8, 9.2	176	39″	260	B9	24909, 39231
	Σ 483		04 04.1 +39 31	7.4, 9.4	58	1.4″	32.4	G5	25444, 56936
		Solar type binary; faint field. Low quality orbit AB: P = 496 y, orbit r = 80 AU. (2009)							
	A 1710		04 06.4 +43 25	8.2, 8.3	317	0.6″	69	G5	25693, 39312
		Solar type matched binary; pretty field. AB: P = 110 y, orbit r = 28 AU, periastron 2057, closing from cusp of orbit. (2009)							
AG	OΣ 71	(AaAb) B	04 06.9 +33 27	6.9, 8.7	230	0.8″	290	B3V B5V	25833, 56973
50	OΣ 531	E [A B]	04 07.6 +38 04	5.6, [7.3, 9.7]	280, [356]	12.4′, [2.6″]	20.4	F7V K1V	25893, 56982
		Local, solar type 1+2 triple; middle of three stars in dark field. A is BY Dra variable V491 Per, E is RS CVn variable V582 (50 Per); near 100% probability the pair AE is physical. Binary CD (β 545, m.9, 1″) 4′ s.p. is unrelated. (2012)							
	Σ 3114	A B	04 09.2 +40 10	7.7, 9.6	158	3.1″	61	F8V	26051, 39342
51 mu	OΣ 73	(AaAb) B	04 14.9 +48 25	4.2, 10.3	351	14″	280	G0II	26630, 39404
		High mass (2)+1 spectroscopic triple with bet Lyr type semi-detached binary (P = 284 d). MSC system mass 14.6 M$_\odot$. (2011)							
★	OΣ 77	A B C	04 15.9 +31 42	8.0, 8.2, 8.6	297, 43	0.5″, 56″	102	F8V	26842, 57110
		Solar type 2+1 triple; pretty, nearly matched trio beautifully displayed in dark field; m.8 O/– color star 2′ n.p. MSC system mass 3.5M$_\odot$. AB: P = 188y, orbit r = 55 AU, e = 0.45, periastron 2075. (2012) See Figure 2.							
	OΣΣ 44		04 17.3 +46 13	7.1, 8.0	323	58″	680	A2	26907, 39438
	Σ 521		04 21.8 +50 02	7.4, 9.2	259	2.0″	810	G0	27395, 39484
	β 310		04 22.0 +39 56	7.1, 12.6	173	19″	55	F8	27495, 57186

Label	Catalog ID	Components	Coordinates (J2000)	Mag.	θ	ρ	Dist.	Spectral type(s)	HD, SAO No.
V590	Σ 533	(AaAb) B	04 24.4 +34 19	7.3, 8.5	62	20″	290*	B8V	27770, 57211
56	OΣ 81	(AaAb) [Ba Bb]	04 24.6 +33 58	5.8, [9.6, 11.3]	16, [292]	4.3″, [0.6″]	40.6	F4IV	27786, 57216
★	Σ 552		04 31.4 +40 01	6.8, 7.2	117	9.0″	370	B8V	28503, 57278
		Distant, high mass double; brightest star in the sparse field. AB: 4,490 AU. (2011)							
	Σ 565	A B	04 38.1 +42 07	7.7, 9.1	167	1.3″	23*	K0	29235, 39660

Phoenix   Phe   *Chart 26*

Label	Catalog ID	Components	Coordinates (J2000)	Mag.	θ	ρ	Dist.	Spectral type(s)	HD, SAO No.
iot	Shy 834	A B C	23 35.1 −42 37	4.7, 12.8, 6.7	273, 212	6.1″, 4.9′	76	A2V	221760, 231675
	Shy 836		23 38.9 −45 37	4.7, 7.1	122	13.4′	62	A2V A7	222226, 231715
★ the	Δ 251		23 39.5 −46 38	6.5, 7.3	278	3.9″	79	A8V F0V	222287, 231719
		A type double; prominent in dark field. AB: ps = 415 AU. (2009)							
	Hd 303		23 40.0 −47 20	7.2, 9.3	66	2.0″	220	K1III	222363, 231726
	h 5416	A B [C D]	23 43.1 −46 19	6.7, 11.1, [10.8, 11.6]	74, 215, [172]	3.8″, 45″, [0.5″]	141	G8IV	222688, 231749
	Skf 1024		23 57.3 −46 07	7.8, 12.0	161	42″	56	F5V	224360, 231863
	Wg 2		00 15.3 −40 06	7.6, 9.7	2	14″	132	K1III	1116, 215021
	I 45	A B C	00 33.5 −55 20	7.7, 8.6, 9.9	205, 246	0.7″, 6.7″	186	A1V	3075, 232087
	LDS 21	(AaAb) B	00 36.6 −49 08	6.9, 8.5	273	5.5′	43.7	G3V	3405, 215165
	MlbO 1		00 42.0 −55 47	8.0, 9.0	161	6.3″	90	F7III	4001, 232154
	h 3390		00 43.3 −45 11	7.1, 10.3	313	14″	158	K0III	4130, 215208
	I 47		00 51.9 −43 43	7.5, 8.0	24	0.7″	96	F2V F5V	5042, 215270
	h 3415		01 03.9 −40 39	7.4, 8.4	138	1.0″	210	A3III	6354, 215348
★ bet	Slr 1	A B C	01 06.1 −46 43	4.1, 4.2, 11.5	93, 55	0.5″, 65″	58*	G8III	6595, 215365
		Distant, solar type 2+1 triple with G giant; dark field. Low quality orbit AB: P = 168 y, orbit r = 55 AU, widening. (2013)							
★ zet	Rmk 2	(AaAb) B C	01 08.4 −55 15	4.0, 6.8, 8.2	113, 242	0.6″, 6.7″	92	B6V B9V	6882, 232306
		High mass (2)+1+1 spectroscopic quadruple with Algol type eclipsing variable; dark field. MSC gives 10.0 M$_\odot$. AC: ps = 845 AU. (2013)							
	Slr 2	A B	01 08.6 −46 40	7.1, 8.7	182	1.2″	82	A9V	6869, 215379
	Hu 1342		01 09.4 −56 36	7.8, 8.3	329	0.4″	99	F5V	6996, 232315
		Solar type binary; unidentified m.12 double 1.5′ s.p. System mass 2.6 M$_\odot$. AB: P = 79 y, orbit r = 26 AU, slowly closing. (2009)							
	h 3428		01 20.3 −48 41	7.7, 9.7	157	21″	75	F6V	8188, 215464
	h 3430		01 20.5 −57 20	7.2, 9.5	223	3.3″	49.3	F7V	8224, 232385
	Hd 185		01 23.7 −40 57	7.4, 10.9	241	4.6″	187	K0III	8534, 215486
	I 51		01 34.3 −45 41	7.1, 9.4	8	1.6″	122	G8IV	9733, 215567

Pictor   Pic   *Chart 27*

Label	Catalog ID	Components	Coordinates (J2000)	Mag.	θ	ρ	Dist.	Spectral type(s)	HD, SAO No.
	CorO 23		04 41.9 −47 50	7.5, 9.6	228	3.7″	83	A8IV	30065, 216942
★ iot	Δ 18	(AaAb) (BaBb) C	04 50.9 −53 28	5.6, 6.2, 9.1	59, 48	13″, 5.0′	41.6	F0IV F4V K2V	31203, 233709
		Solar type (2)+(2)+1 quintuple; sparse field. MSC system mass (AB only) 5.7 M$_\odot$. AB: ps = 730 AU; AC: ps = 16,800 AU. (2009)							
★ VW	h 3715	(AaAb) B	04 59.5 −49 27	7.2, 9.1	112	9.9″	151	F3V	32278, 217101
		Solar type (2)+1 triple with W UMa type eclipsing contact binary (P = 0.43 d); sparse field. AB: ps = 2,020 AU. Neglected. (1999)							
	Cbl 124		05 01.6 −44 50	7.6, 11.2	266	51″	94	K0+G	32517, 217124
★ the	Δ 20	(AaAbB) C	05 24.8 −52 19	6.2, 6.7	288	38″	157	A0V	35860, 233965
		Splendid high mass (2+1)+1 quadruple, visual binary; dominant in faint field. MSC system mass 10.8 M$_\odot$. AB (I 345): P= 123 y, orbit r= 45, e= 0.69, apastron 2057. (2013)							
	Skf 828		05 32.9 −47 41	7.7, 10.0	145	55″	109	A3V G0	37004, 217394
	HDS 739		05 34.7 −48 57	7.3, 10.4	280	14″	140	A3m	37260, 217408
	h 3787		05 37.8 −54 34	7.8, 10.0	249	25″	112	G5III	37761, 234056

Label	Catalog ID	Components	Coordinates (J2000)	Mag.	θ	ρ	Dist.	Spectral type(s)	HD, SAO No.
	h 3784		05 38.2 −46 06	7.6, 9.4	76	4.7″	25.5	G6V	37706, 217444
	I 63	A B	05 48.2 −48 55	7.3, 8.6	17	1.0″	210	A1V	39177, 217547
	I 3		06 12.5 −61 28	7.1, 7.6	6	1.1″	230	B9.5V	43519, 249475
★ mu	h 3874		06 32.0 −58 45	5.6, 9.3	230	2.5″	230	B9Ve	46860, 234564
			High mass double; easy to find in sparse field, but a severe brightness contrast challenge. AB: ps = 775 AU. Neglected. (1991)						
AK	I 5	A B C	06 38.0 −61 32	6.3, 8.8, 9.8	101, 76	0.6″, 14′	21.3	G2V	48189, 249604
			Local, solar type 2+1 triple with BY Dra variable, comoving with D (HD 45270, m.7), a pre main sequence star 2.1° n.p. AB: P= 218 y, orbit r= 43 AU. AC: ps = 17,200 AU. AD (Shy 35): ps = 0.85 pc. (2012)						
	I 6		06 42.5 −61 45	7.7, 8.2	257	0.6″	130*	F5V	49076, 249627

Pisces   Psc   *Charts 6, 12*

Label	Catalog ID	Components	Coordinates (J2000)	Mag.	θ	ρ	Dist.	Spectral type(s)	HD, SAO No.
	OΣ 491	A [B C]	23 13.7 +02 12	7.2, [10.9, 11.1]	319, [86]	118″, [0.8″]	88	F2	219150, 128046
	Σ 2995		23 16.6 −01 35	8.2, 8.6	32	5.2″	65	G5	219542, 146605
	Σ 3009	A B	23 24.3 +03 43	6.9, 8.8	229	7.3″	220	K2III	220512, 128160
	Cbl 194		23 28.1 −02 27	8.1, 8.5	226	57″	114	F2	220966, 146717
	Σ 3031		23 41.2 +06 16	7.8, 8.6	309	15″	58*	F8	222502, 128327
	Σ 3045		23 54.4 +02 28	8.0, 9.3	272	1.7″	200	A2	224004, 128456
	A 2100		23 56.8 +04 44	7.4, 7.9	260	0.4″	112	F0	224315, 128487
			Solar type binary; sparse field. System mass 2.8 M$_\odot$. AB: P= 89 y, orbit r= 30 AU, e= 0.77, widening to apastron 2035. (2013)						
34	Σ 5		00 10.0 +11 09	5.5, 9.4	160	7.7″	94	B9Vn	560, 91750
★ 35	Σ 12	(AaAb) B	00 15.0 +08 49	6.1, 7.5	145	12″	78	A9V F3V	1061, 109087
			UU Psc. A type (2)+1 spectroscopic triple with ellipsoidal, Algol type variable star (P = 0.84 d); YW/O color. AB: ps = 1,260 AU. (2014)						
	Σ 15		00 15.9 −05 36	7.7, 9.8	199	4.9″	210	G5	1153, 128668
★ 38	Σ 22	(AB) C	00 17.4 +08 53	7.1, 7.7	235	4.0″	15.2	F7V	1317, 109111
			Local, solar type (2)+1 triple, visual binary; dark field. MSC system mass 3.7 M$_\odot$. AB (A 1803): P= 36 y, orbit r= 4 AU, e= 0.0, periastron 2022; AC: ps = 82 AU. (2008)						
	β 1093		00 20.9 +10 59	6.7, 8.6	117	0.8″	310	A0V	1663, 91858
★ 55	Σ 46		00 39.9 +21 26	5.6, 8.5	192	6.6″	127	K0III F3V	3690, 74182
			High mass double with K supergiant and complementary contrast Y/B color; dark field. AB: ps = 1,130 AU. (2014)						
	OΣ 18	A B	00 42.4 +04 10	7.9, 9.7	209	2.0″	60	F8V	3972, 109392
			Solar type binary; dark field. Low quality orbit AB: P= 387 y, orbit r= 80 AU, widening. (2011)						
66	OΣ 20	A B C	00 54.6 +19 11	6.1, 7.2, 12.8	175, 9	0.6″, 2.5′	108	A0V	5267, 92145
			A type 2+1 triple W/− color; very dark field. AB: P= 343 y, orbit r= 75 AU, e= 0.29, apastron 2076. (311 measures; 2010)						
★ 74 psi 1	Σ 88	(AaAb) B	01 05.7 +21 28	5.3, 5.5	159	29″	84	B9.5V A0V	6456, 74482
			High mass (2)+1 triple; m.11 star 2′ s.f. unrelated. Aa,Ab: P= 14.4 y, e= 0.52, periastron 2022. (2012)						
77	Σ 90	A (BaBb)	01 05.8 +04 55	6.4, 7.3	84	33″	40.6	F3V F5V	6479, 109666
sig 2	S 393	A C	01 06.2 +32 11	6.4, 11.7	234	2.3′	173	K0	6476, 54421
	OΣ 22		01 07.1 +11 33	7.3, 10.5	200	8.7″	80	A9V	6614, 92255
	β 303		01 09.7 +23 48	7.3, 7.6	292	0.6″	122	F0IV	6886, 74523
	Σ 98	(AaAb) B C	01 12.9 +32 05	7.0, 8.1, 11.9	248, 35	20″, 85″	132	A0V A3IV	7215, 54514
			Beautiful A type (2)+1+1 spectroscopic quadruple with similar stars in dark field; galaxy NGC 420 10′ p. AB: ps = 3,560 AU. (2014)						
	OΣ 26	(AaAb) B	01 13.0 +30 04	6.3, 10.5	260	11″	116	G9III G1V	7229, 74561
★ 86 zet	Σ 100	(AaAb) [(BaBb) C]	01 13.7 +07 35	5.2, [6.3, 12.2]	63, [75]	23″, [1.8″]	53	A7IV F7V	7344, 109739
			High mass (2)+(2)+1 spectroscopic and occultation quintuple; O/− color; dark field: discovered by C. Mayer (1777). BC (β 1029, Δm = 6) is the most challenging pair in this Atlas. AB: ps = 1,640 AU. (324 measures; 2012)						

Label	Catalog ID	Components	Coordinates (J2000)	Mag.	θ	ρ	Dist.	Spectral type(s)	HD, SAO No.
85 phi	Σ 99	(AaAb) B C	01 13.7 +24 35	4.7, 9.1, 12.7	221, 173	7.8″, 2.4′	137	G8III	7318, 74571
	h 636		01 14.3 +30 33	7.4, 11.8	288	20″	270	B9.5IV	7384, 54541
	β 4	A B	01 21.3 +11 32	7.1, 8.9	110	0.6″	210	F1III	8187, 92388
		High mass double with F giant; dark field, unrelated stars n.f. System mass 4.8 M☉. AB: P= 291 y, orbit r= 90 AU, widening. (2008)							
	OΣ 30	A B C	01 25.6 +31 33	8.1, 11.8, 8.1	245, 106	4.6″, 57″	64	F8	8610, 54687
	Σ 122		01 26.9 +03 32	6.7, 9.5	326	6.0″	166	B9V A8V	8803, 109895
	S 398	A (BaBb)	01 28.4 +07 58	6.3, 8.0	100	69″	115	K1III	8949, 109907
	OΣ 31		01 33.3 +08 13	6.5, 10.6	78	4.0″	148	K0III	9496, 109964
100	Σ 136	A B	01 34.9 +12 34	7.3, 8.3	77	15″	230	A6V	9656, 92521
★	Σ 138	A B	01 36.0 +07 39	7.5, 7.6	58	1.6″	76	F6V	9817, 110001
		Solar type matched pair; Y/Y color, dark field, unrelated m.10 star 1′ f. After discovering this pair in 1792, W. Herschel revisited it often: "a very beautiful minute object." AB: ps = 160 AU. (328 measures; 2011)							
	h 644	A B	01 48.7 +07 41	7.2, 11.7	278	18″	172	K0	11049, 110142
★ 113 alp	Σ 202	A B C	02 02.0 +02 46	4.1, 5.2, 8.3	263, 63	1.6″, 6.8′	42.6	A0 A3m	12447, 110291
		Alrischa. A type 2+1 triple; dark field, YW/– color, southern apex in triangle of 7th mag. stars. A and possibly B is an alp2 CVn type variable. MSC system mass 7.8 M☉. AB: P= 933 y, orbit r= 170 AU. (609 measures; 2014)							

Piscis Austrinus   PsA   *Chart 20*

Label	Catalog ID	Components	Coordinates (J2000)	Mag.	θ	ρ	Dist.	Spectral type(s)	HD, SAO No.
	Bvd 137		21 33.5 −27 53	7.7, 9.8	58	40″	49.1	G3V K1V	205067, 190437
	Cbl 184		21 37.0 −35 53	7.4, 8.6	265	79″	62	F3V	205530, 213139
		High probability CPM pair found by datamining Hipparcos astrometry and sky survey photographs (see references). AB: ps = 6,600 AU. (2000)							
	h 5311	A B	21 59.5 −29 03	7.2, 11.4	292	40″	154	K0III	208810, 190798
	Stone 56		21 59.6 −27 38	7.3, 10.5	36	11″	76	F2V	208851, 190800
12 eta	β 276		22 00.8 −28 27	5.7, 6.8	113	1.9″	250	B8Ve	209014, 190822
	β 769	(AaAb) B	22 11.6 −34 28	7.1, 8.2	356	0.9″	240	K1III	210525, 213625
	Shy 805	(AaAb) B	22 22.0 −34 31	7.5, 7.5	37	2.4′	86	F4V F2	212025, 213765
17 bet	Pz 7	A C	22 31.5 −32 21	4.3, 7.1	172	31″	43.8	A0V	213398, 213883
	H VI 119	A [B C]	22 39.7 −28 20	6.4, [7.5, 8.6]	159, [69]	86″, [3.1″]	112	K0III	214599, 191308
	h 5365	A B C	22 51.8 −35 53	7.5, 11.0, 11.3	276, 36	4.8″, 55″	101	A3	216237, 214144
22 gam	h 5367		22 52.5 −32 53	4.5, 8.2	256	4.1″	66	A0III	216336, 214153
23 del	Howe 91		22 55.9 −32 32	4.2, 9.2	249	4.9″	47.3	G8III	216763, 214189
	h 5371		22 57.8 −26 06	7.7, 9.2	343	9.2″	61	G3V	217004, 191529
	Rss 39	A B	22 57.8 −33 12	8.0, 9.6	246	59″	81	F3V	216986, 214205

Puppis   Pup   *Charts 22, 27*

Label	Catalog ID	Components	Coordinates (J2000)	Mag.	θ	ρ	Dist.	Spectral type(s)	HD, SAO No.
★	h 3834	A B E	06 04.7 −45 05	6.0, 9.0, 6.4	215, 323	5.9″, 3.4′	26.9	F5V	41742, 217706
		Solar type 2+1 triple; wide pair in glittering field. MSC system mass 3.3 M☉. AB: ps = 215 AU. AE: ps = 7,370 AU. (2010)							
	Δ 23		06 04.8 −48 28	7.3, 7.7	127	2.6″	30.0	G6V	41824, 217708
	h 3856		06 22.9 −45 38	6.9, 10.7	7	35″	240	K1III	45056, 217893
	I 156		06 25.7 −48 11	5.9, 8.1	125	1.0″	180	B9V	45572, 217922
★	Δ 30	A B [C D]	06 29.8 −50 14	6.0, 6.2, [8.0, 8.7]	258, 312, [159]	0.5″, 12″, [0.4″]	51	F2V	46273, 234539
		High mass, solar type 2+2 quadruple; a tiny gem in large aperture. MSC system mass 4.3 M☉. AB (R 65): P= 53 y, orbit r= 32 AU, e= 0.98, periastron 2022; CD (Hd 195): P= 99 y, orbit r= 28 AU, e= 0.25, apastron 2055. AB,CD: ps = 825 AU. (2013)							
	Δ 31		06 38.6 −48 13	5.1, 7.4	326	14″	146	G8III	47973, 218093
	h 5443		06 41.2 −40 21	6.1, 9.4	108	16″	360	B4V	48383, 218126
	Δ 32		06 42.3 −38 24	6.6, 7.7	277	7.9″	83	A5m	48543, 197108

Label	Catalog ID	Components	Coordinates (J2000)	Mag.	θ	ρ	Dist.	Spectral type(s)	HD, SAO No.
	h 3895		06 46.7 −47 48	7.1, 10.9	66	26″	320	K3III	49614, 218186
	h 3900		06 54.2 −34 13	7.6, 8.9	277	2.2″	169	A1IV	51042, 197334
	I 66	A B	06 58.3 −35 25	7.8, 9.3	253	1.9″	280	A	52093, 197424
★	Δ 38	(AaAb) B C	07 04.0 −43 37	5.6, 6.7, 8.8	126, 337	21″, 3.1′	16.5	G3V K0V	53705, 218421
		Local, high CPM, solar type (2)+1+1 quadruple, visual triple. MSC has 2.9 M☉. AC: ps = 4,100 AU. Neglected. (1999)							
	h 3928	A B C	07 05.5 −34 47	6.5, 7.8, 9.7	144, 291	2.7″, 38″	53	F2V	53952, 197557
	h 3931	A C	07 06.0 −42 20	7.2, 8.8	41	72″	142	A0V	54208, 218447
	β 757		07 12.4 −36 33	6.0, 8.4	69	2.5″	280	B3V	55718, 197694
★ pi	HDS 1008	Aa Ab B	07 17.1 −37 06	2.9, 6.5, 7.9	152, 213	0.7″, 67″	250	K3Ib	56855, 197795
		Tureis. Distant, high mass 2+1 triple with K supergiant; near Collinder 135, the difficult Δm needs a large aperture. Neglected. (1991)							
	I 7		07 17.5 −46 59	7.1, 8.4	203	0.7″	14.6	K3V	57095, 218611
		Local, high CPM, low mass binary. AB: P= 85 y, orbit r= 15 AU, predicted to go below 0.5″ in 2026, but if delayed this may suggest the true period is actually longer. (2012)							
	Δ 45		07 21.4 −48 32	6.8, 7.7	157	23″	240	B9V	58017, 218666
	h 3957		07 22.3 −35 55	7.1, 7.9	191	7.5″	60	F8	58038, 197907
	h 3966		07 24.8 −37 17	6.9, 6.9	322	6.9″	210	F0V F0V	58635, 197975
	h 3969		07 27.0 −34 19	7.1, 8.1	227	18″	32.0	F6V F8	59099, 198005
	Δ 49		07 28.9 −31 51	6.3, 7.0	54	9.0″	230	B3V B4V	59499, 198038
	Σ 1104	A B C	07 29.4 −15 00	6.4, 7.6, 11.8	35, 185	1.8″, 22″	35.1	F7V	59438, 152943
★	Howe 18		07 34.0 −23 42	8.1, 8.9	204	1.9″	980*	B6III	60535, —
		Distant, high mass double with B giant; mixed field, stellar group f. AB: ps = 2510 AU. Neglected. (1991)							
	H N 19	(AaAb) B	07 34.3 −23 28	5.8, 5.9	117	9.7″	31.7	F5V	60584, 174019
★	Σ 1120	A C D E F	07 36.1 −14 30	5.6, 9.7, 10.0, 37, 41, 327, 7	20″, 2.5′, 104″, 92″	389	B2Ve B6V	60855, 153118
		Distant, high mass CPM group in NGC 2422; optical (!) pair Σ 1121 (m.6, 7″) 7′ f. AD: ps = 78,700 AU. (2001)							
★ k 1,2	H III 27	A B	07 38.8 −26 48	4.4, 4.6	317	10″	106	B6V B6V	61556, 174199
		Beautiful high mass, matched double; both components suspected variable, 4 m.8 stars s.p and s.f. AB: ps = 1430 AU. (2009)							
	β 201		07 38.9 −20 16	7.8, 8.4	334	2.9″	128	A7III	61532, —
d 2	I 160		07 39.7 −38 08	5.8, 8.6	149	1.2″	175	B5Vn	61878, 198265
	I 353		07 39.7 −43 17	7.6, 8.4	41	0.7″	580	B5V	61946, 218877
	Hu 710		07 43.0 −17 04	7.0, 8.0	63	0.5″	109	G5III	62351, 153301
		Solar type binary with G giant; faint field. System mass 2.7 M☉. AB: P= 159 y, orbit r= 39 AU, e= 0.62, apastron 2031. (2013)							
	Δ 55	A C	07 44.2 −50 27	6.6, 7.6	133	52″	30.9	F9V	63008, 235458
	Skf 1444		07 44.6 −37 57	5.9, 8.6	204	2.9′	210	B7V+A3V	62893, 198379
		CPM double in bright field (NGC 2451). High mass likely prevails over wide orbit. AB: ps = 49,300 AU. (2010)							
2	Σ 1138	A (BaBb) C	07 45.5 −14 41	6.0, 6.7, 10.6	340, 228	17″, 100″	78	A2V A8V	62864, 153363
	WFC 58		07 45.7 −48 21	7.7, 11.0	178	16″	300	K2III	63255, 218966
5	Σ 1146		07 47.9 −12 12	5.7, 7.3	341	1.1″	28.6	dF5	63336, 153414
	I 186		07 49.4 −30 33	8.2, 8.5	243	1.5″	81	F3V	63785, 198483
9	β 101		07 51.8 −13 54	5.6, 6.5	290	0.5″	16.5	G1V	64096, 153500
		Local, solar type visual/spectroscopic binary; faint rich field. AB: P= 23 y, orbit r= 10 AU, e= 0.76, apastron 2020. (243 measures; 2012)							
	B 1566		07 52.0 −31 38	8.1, 9.0	100	1.3″	400	B7V	64301, 198533
212	Howe 65		07 52.3 −34 42	5.1, 8.6	267	3.8″	17.9	F5V	64379, 198540
		Local, high CPM, solar type double; faint field. AB: ps = 92 AU, infrequently measured and no orbit. (2001)							
	λ 91		07 55.8 −43 51	6.6, 6.9	344	0.7″	620	B6V	65211, 219157
	I 30		07 58.8 −47 41	7.6, 8.7	352	1.7″	390	B7V	65905, 219237
	Δ 59		07 59.2 −49 59	6.2, 6.2	48	17″	620*	B2IV	66005, 219249
	h 4032	(AB) C	07 59.8 −47 18	6.7, 9.4	351	29″	510	B8V	66079, 219265
	β 333	A B	08 01.3 −22 20	7.2, 9.2	45	1.6″	160	A1V	66097, 174989

Label	Catalog ID	Components	Coordinates (J2000)	Mag.	θ	ρ	Dist.	Spectral type(s)	HD, SAO No.
	β 202	A B	08 02.0 −27 13	6.9, 9.9	162	7.9″	1230*	B7II	66306, 175018
			Distant, high mass CPM double with B supergiant; pretty brightness contrast, splendid field, star s.f. unrelated, but large and uncertain distance makes this pair ambiguous. AB: ps = 13,100 AU. Neglected. (1999)						
QQ	I 8		08 02.5 −44 40	6.5, 9.1	303	2.3″	210	A0	66605, 219333
	h 4037	A B	08 02.6 −27 33	7.2, 8.4	242	7.3″	490	G2IV	66454, 175052
	I 487		08 04.5 −25 42	6.6, 10.0	25	1.9″	260	K4.5III	66885, 175114
	Δ 63		08 09.8 −42 38	6.6, 7.5	80	5.5″	290	B7V	68242, 219507
18	LDS 204	A (BaBb)	08 10.7 −13 48	5.5, 12.3	236	97″	22.4	F7V	68146, 153924
			Local, solar type 1+(2) triple; rich field. AB: ps = 2,930 AU. (2000)						
	λ 505		08 13.3 −24 18	7.5, 9.0	259	1.3″	165	A8III	68821, –
OS	Δ 67		08 14.0 −36 19	5.0, 6.0	174	66″	310	B1.5IV	69081, 198969
	β 454	A B	08 15.9 −30 56	6.5, 8.2	2	1.9″	98	G8III	69445, 199010
	h 4073	(AaAb) B	08 18.2 −37 22	7.2, 7.8	175	1.9″	410	A0IV	70003, 199059
	h 4087	A B C	08 22.1 −40 59	7.6, 8.0, 12.4	255, 346	1.5″, 10″	80	F5V	70764, 219839
	Rst 4888	A (BaBb)	08 25.0 −42 46	6.6, 6.8	103	0.5″	550	B3V	71302, 219890
NO	h 4093	(AaAb) (BaBb)	08 26.3 −39 04	6.5, 7.1	124	8.2″	172	B9V A2V	71487, 199222
			High mass (2)+(2) quadruple with Algol type variable (A). MSC has 10.9 M$_\odot$. Ba,Bb: P= 96 y, widening. Neglected. (1999)						

Pyxis   Pyx   *Chart 22*

Label	Catalog ID	Components	Coordinates (J2000)	Mag.	θ	ρ	Dist.	Spectral type(s)	HD, SAO No.
	I 489		08 31.5 −19 35	5.8, 6.7	290	0.4″	106	A0V	72310, 154359
			A type binary; m.7 star 5′ n.p. in faint field. AB: P= 261 y, orbit r= 55 AU, widening. (2013)						
★	β 205	A B	08 33.1 −24 36	7.1, 6.8	291	0.6″	73	A8IV	72626, 176061
			A type binary; in small group of m.10 stars. AB: P= 143 y, orbit r= 39 AU, e= 0.29, widening to apastron 2025. (2013)						
	β 206		08 35.5 −25 06	8.2, 8.5	277	1.9″	460	K1III	73071, —
	h 4115	A B	08 37.5 −33 45	6.5, 11.8	159	22″	61	F0IV	73476, 199436
	β 207		08 38.7 −19 44	6.6, 9.2	102	4.4″	360	K5III	73603, 154492
	β 208	A B	08 39.1 −22 40	5.4, 6.8	40	1.0″	19.4	G6IV	73752, 176226
			Local, high CPM, solar type binary with G subgiant. AB: P= 123 y, orbit r= 33 AU, e= 0.33; below 0.4″ during the 2020s, then widens to apastron 2048 and greatest visual separation around 2065. (2013)						
	I 314		08 39.4 −36 36	6.4, 7.9	244	0.8″	37.9	F1IV	73900, 199473
			Solar type binary; rich field. System mass 1.8 M$_\odot$. AB: P= 67 y, orbit r= 20 AU, widening to apastron around 2030. (2011)						
zet	h 4120		08 39.7 −29 34	5.0, 9.6	59	51″	75	G4III	73898, 176253
★	Skf 1307		08 46.6 −29 44	7.5, 7.8	189	110″	650	C5.5J +K2II	75021, 176458
			UZ Pyx. Striking pair: a distant carbon star and semiregular pulsating variable, comoving with K supergiant in a rich field. Culled from the legacy research literature by B. Skiff. AB: ps = 96,700 AU; high mass could sustain the wide orbit. Neglected. (1999)						
	h 4144		08 50.4 −35 56	7.0, 9.0	315	2.4″	490	B5IV	75722, 199690
★	h 4166	A (BC)	09 03.3 −33 36	7.1, 7.9	153	14″	140*	A0V	77737, 199924
			A type 1+(2) triple, visual double; Skf 1466 (see below) 8′ s.f. A,BC: ps = 2,640 AU. Neglected. (1999)						
	Skf 1466		09 03.6 −33 42	7.8, 9.4	147	74″	340	A2IV+A4V	77788, 199933
	β 410		09 09.8 −25 48	7.3, 8.8	157	1.8″	370	A0V	78876, 177055
eps	H N 96	(AaAb) (BC)	09 09.9 −30 22	5.6, 9.5	147	18″	65	A4IV	78922, 200047
	h 4200		09 20.7 −31 46	7.4, 7.9	75	3.0″	260	B9.5V	80773, 200259
	Jc 5		09 26.7 −28 47	6.4, 7.5	281	0.5″	630	B6Ve	81753, 177461

Reticulum   Ret   *Chart 27*

Label	Catalog ID	Components	Coordinates (J2000)	Mag.	θ	ρ	Dist.	Spectral type(s)	HD, SAO No.
	Δ 12	A [B C]	03 15.2 −64 27	6.7, [9.5, 9.8]	104, [25]	19″, [0.7″]	110	F5	20586, 248764
★ zet 1,2	Alb 1	A B	03 18.2 −62 30	5.3, 5.6	217	5.2′	12.0	G2V	20807, 248774
			Local, very high CPM, solar type double; pretty binocular pair, near 100% probability it is physical. AB: ps = 5,050 AU. (2007)						

Label	Catalog ID	Components	Coordinates (J2000)	Mag.	θ	ρ	Dist.	Spectral type(s)	HD, SAO No.
	Δ 14		03 38.2 −59 47	7.0, 8.3	272	57″	74	F3V F5	22989, 233197
★	h 3592		03 44.6 −54 16	6.5, 9.3	17	5.0″	76	K1III	23697, 233252
		Double with K type giant; sparse field, north corner of three m.7 stars. AB: ps = 515 AU. Neglected. (1993)							
	Pol 2		04 02.9 −59 35	8.3, 8.6	104	1.9″	270	F3III	25903, 233379
the	Rmk 3		04 17.7 −63 15	6.0, 7.7	5	3.9″	141	B9III	27657, 248986
	h 3670		04 33.6 −62 49	5.9, 9.3	100	32″	45.2	K1III	29399, 249054

Sagitta   Sge   *Chart 11*

Label	Catalog ID	Components	Coordinates (J2000)	Mag.	θ	ρ	Dist.	Spectral type(s)	HD, SAO No.
V338	β 139	(AaAb) B	19 12.6 +16 51	7.1, 8.0	135	0.7″	240	B9IV	179588, 104602
		V338 Sge. High mass (2)+1 triple Algol type variable; rich field, stars 30″ f. and 2′ p. are unrelated. AB: *P*= 587 y, orbit *r*= 130 AU. (2009)							
	Σ 2484		19 14.3 +19 04	7.9, 9.5	240	2.1″	67	F8	180054, 104635
	Σ 2504	A B	19 21.0 +19 09	7.0, 9.0	282	8.6″	63	F5V	181752, 104753
	OΣ 375		19 34.6 +18 08	7.7, 8.9	186	0.6″	230	G5	184591, 104991
	H IV 99	A C	19 50.0 +17 57	8.0, 9.1	256	69″	480*	B7V B9V	187566, 105324
	Webb 12		20 07.8 +19 50	8.4, 8.4	77	41″	290	F0	191140, 105724
	Σ 2634	A B	20 09.6 +16 48	7.8, 9.9	14	4.1″	23.7	G9V	191499, 105765
		Local, solar type binary; s.f. of two stars in field strewn with faint close pairs. AB: ps = 130 AU. (2011)							
17 the	Σ 2637	A B	20 09.9 +20 55	6.6, 8.9	331	12″	45.3	F3V	191570, 88276

Sagittarius   Sgr   *Chart 25*

Label	Catalog ID	Components	Coordinates (J2000)	Mag.	θ	ρ	Dist.	Spectral type(s)	HD, SAO No.
★	Pz 6	A B	17 59.1 −30 15	5.4, 7.0	104	5.7″	310	M1Ib G8II	163755, 209553
		Distant, colorful and extremely rare CPM double of M and G supergiants; rich field. AB: ps = 2,400 AU. (2013)							
V1647	h 5000	(AaAb) B	17 59.2 −36 56	7.1, 9.0	101	7.3″	123	A3III	163708, 209552
	Howe 88	A (BC)	18 05.7 −36 35	7.9, 8.9	4	3.2″	510	B8	165063, 209691
	β 244		18 08.6 −27 52	7.6, 9.1	265	2.2″	280	G6III	165723, 186598
★	β 245		18 10.1 −30 44	5.8, 8.0	352	4.1″	138	K1II	166023, 209803
		Albireo colors in an m.6 system with K supergiant primary. AB: ps = 760 AU. (2007)							
	β 132	(AaAb) B	18 11.2 −19 51	7.0, 7.1	188	1.4″	102	A4V	166393, 161153
11	h 5030		18 11.7 −23 42	5.1, 11.5	286	43″	78	K0III	166464, 186437
eta	β 760	A B	18 17.6 −36 46	3.3, 8.0	8	3.6″	44.7	M3.5III	167618, 209957
21	Jc 6		18 25.4 −20 32	5.0, 7.4	284	1.7″	126	K1III A1V	169420, 186794
	β 133		18 27.7 −26 38	6.6, 8.5	243	0.9″	78	A8V F2V	169851, 186837
	H N 125		18 28.9 −25 03	8.2, 8.5	107	2.5″	62	G2V	170121, 186864
	WNO 6		18 29.0 −26 35	6.7, 8.0	181	42″	132	A3III	170141, 186863
	Howe 43	A B	18 31.1 −32 59	5.4, 9.8	185	3.5″	89	A3m	170479, 210257
U	β 966	A [Ba Bb] M Q	18 31.9 −19 08	6.9, [9.1, 9.9], 10.4, 10.2	252, [95], 317, 219	66″, [0.9″], 2.8′, 2.6′	1510*	G1.5Ib	170764, 161571
		Distant, high mass CPM group with G type supergiant, a Cepheid variable; in IC 4725 (M25). (2012)							
	Stone 62		18 34.5 −34 49	7.6, 7.8	132	2.3″	84	F0V	171119, 210321
	λ 361		18 49.7 −33 36	8.4, 8.5	233	1.2″	130	F3V	173968, 210606
★ 38 zet	Hd 150	A B	19 02.6 −29 53	3.3, 3.5	271	0.5″	27.0	A2III	176687, 187600
		Ascella. A type binary; dominant in rich field. AB: *P*= 21 y, orbit *r*= 13 AU, *e*= 0.21, apastron in 2016, drops below 0.4″ in 2020, then reappears in 2033. A bright, matched resolution challenge. (2013)							
	h 5082	A B C	19 03.1 −19 15	6.2, 9.0, 10.8	89, 11 2	7.6″, 20″	610	G5II	176884, 162130
	H N 129		19 04.2 −22 54	6.9, 9.2	309	8.1″	300	A0V	177120, 187632
	H N 126		19 04.3 −21 32	7.9, 8.1	187	1.3″	56	F8V	177166, 187634
		Solar type binary; very tiny in a pretty field. AB: *P*= 500y, orbit *r*= 90 AU. (2013)							

Label	Catalog ID	Components	Coordinates (J2000)	Mag.	θ	ρ	Dist.	Spectral type(s)	HD, SAO No.
	S 710		19 06.9 −16 14	6.1, 8.4	0	6.3″	260	B8IV	177817, 162201
	CorO 233	A B	19 08.7 −33 48	8.2, 9.0	256	12″	220	F5	178079, 210980
	S 715		19 17.7 −15 58	7.1, 7.9	17	8.4″	133	A3V	180562, 162417
		A type double; rich field, S 716 (m.8, 5″) 6′ f. AB: ps = 1,510 AU. (2007)							
	Shy 753		19 22.1 −29 31	7.2, 7.2	227	7.3′	51	G0IV G0IV	181544, 188025
	λ 375		19 26.7 −26 19	7.5, 12.0	166	12″	230	A0V	182649, 188125
	Scj 22		19 28.2 −12 09	8.1, 8.7	284	1.1″	36.6	G8V	183063, 162662
		Solar type binary; faint field. System mass 1.6 M☉. AB: P= 165 y, orbit r= 37 AU, e= 0.57, apastron 2067. (2013)							
	H N 119		19 29.9 −26 59	5.6, 8.8	145	7.6″	72	K2III	183275, 188192
	S 722	(AaAb) B	19 39.2 −16 54	7.2, 7.5	236	9.9″	74	A8III	185344, 162853
54 e1	h 599	(AaAb) C	19 40.7 −16 18	5.4, 7.7	41	46″	74	K2III F8V	185644, 162883
	β 467		19 46.5 −21 31	7.7, 9.9	132	3.3″	97	A8 F3	186666, 188551
	I 122		19 50.7 −41 52	7.8, 10.1	337	5.2″	167	A8V	187211, 229890
	Hd 294		20 01.2 −38 35	8.1, 9.1	31	1.2″	68	F3V	189386, 211734
	Σ 2625		20 06.8 −12 56	7.8, 10.2	5	12″	450	K0III	190723, 163267
	h 5168		20 07.4 −29 43	6.9, 10.3	80	18″	230	K2IV	190727, 188927
	h 5178	A B C	20 13.7 −34 07	7.1, 8.2, 10.8	9, 16	2.8″, 82″	163	K1III	191957, 211916
	Stone 64		20 16.9 −32 36	8.1, 8.5	296	1.7″	39.5	G5V	192614, 211959
	Δ 230		20 17.8 −40 11	7.4, 7.7	112	9.9″	80	F8	192724, 230134
★	h 5188	A C	20 20.5 −29 12	6.7, 7.6	320	27″	136	A3III A2V	193281, 189164
		Striking group of two A type doubles and a triple – physical h 5188 AC, optical Glp 18 and ambiguous Wie 1. Parallax or proper motion indicate they are probably unrelated. Neglected. (1998)							
	R 321		20 26.9 −37 24	6.6, 8.1	126	1.6″	37.3	K2IV K1V	194433, 212126
		Solar type binary, delicate brightness contrast. AB: P= 178 y, orbit r= 39 AU, widening. (2013)							
	Rss 36		20 28.0 −42 37	8.2, 8.8	267	35″	99	F6V	194570, 230226

Scorpius Sco *Chart 24*

Label	Catalog ID	Components	Coordinates (J2000)	Mag.	θ	ρ	Dist.	Spectral type(s)	HD, SAO No.
2	β 36		15 53.6 −25 20	4.7, 7.0	268	2.1″	154	B2.5Vn	142114, 183896
	I 977		15 55.7 −26 45	8.0, 8.5	253	0.5″	76	F7V	142456, 183933
		Solar type binary; subarcsecond resolution challenge. System mass 2.3 M☉. AB: P= 275 y, orbit r= 55 AU, widening. (2010)							
	β 38		16 02.9 −25 01	7.2, 9.5	345	4.3″	270	B9.5V	143715, 184058
★ 51 xi	Σ 1998	A B C	16 04.4 −11 22	5.2, 4.9, 7.3	0, 47	1.0″, 6.9″	25.3	F7V	144069, 159665
		Local, solar type triple, joined by CPM with Σ 1999 (see below) as a 2+1+2 quintuple; regal in sparse field. MSC system mass 5.0 M☉. AB: P= 46 y, orbit r = 17 AU, e= 0.74, apastron 2020. (630 measures) AC: P= 1,514 y, orbit r= 196 AU. (380 measures) A(Σ 1998)A(Σ 1999): 4.7′, ps = 9,520 AU. Also a CPM pair (Shy 287) with binary ups Oph (Rst 3949, m.5, 0.8″) 6.6º n.f. (2012)							
★	Σ 1999	A B	16 04.4 −11 27	7.5, 8.1	98	12″	25.3	G8+K3	144087, 159668
		Local, solar type double; related by CPM to Σ 1998 (above). AB: ps = 410 AU. (2013)							
	Kou 49	(AC) B	16 05.7 −21 50	7.8, 12.4	165	9.0″	162	A0V	144254, 184106
11	β 39	A B	16 07.6 −12 45	5.8, 9.8	263	2.8″	112	B9.5V	144708, 159715
	BrsO 11		16 09.5 −32 39	6.7, 7.2	84	8.0″	121	K2III	144927, 207396
★ 14 nu	β 120	(AaAbAc) B [C (DaDb)]	16 12.0 −19 28	4.4, 5.3, [6.6, 7.2]	3, 336, [56]	1.4″, 41″, [2.3″]	145	B2IV	145502, 159764
		Jabbah. High mass (2+1)+1+[1+(2)] septuple, visual quadruple; faint field. MSC 35.6 M☉. (A)B: ps = 275 AU. (2013)							
12	h 4839	(AaAb) B	16 12.3 −28 25	5.8, 8.1	71	3.8″	93	B9V	145483, 184217
	Σ 2019	(AB) C	16 14.3 −10 25	7.4, 9.8	153	23″	79	F7V	145996, 159803
	λ 268		16 14.7 −39 08	8.4, 8.5	183	1.8″	127	A9V	145840, 207482
		A type double; unrelated m.11 star 1′ n.f. AB: ps = 310 AU. Last measured by Hipparcos satellite, since then neglected. (1991)							
	Skf 1313		16 14.9 −25 29	6.1, 9.9	37	46″	149	B7IV	146001, 184258

Label	Catalog ID	Components	Coordinates (J2000)	Mag.	θ	ρ	Dist.	Spectral type(s)	HD, SAO No.
	Sh 225		16 20.1 −20 03	7.4, 8.1	333	47″	163	B9	147010, 159860
	High mass double; in field with optical pair Sh 226. AB: ps = 10300 AU. (2013)								
20 sig	H IV 121	(Aa1Aa2) Ab B	16 21.2 −25 36	3.1, 5.2, 8.4	244, 273	0.5″, 20″	210	B1III B1V	147165, 184336
	Al Niyat. High mass (2)+1+1 triple with bet Cep type variable (Aa); faint field, good resolution test. Aa1,Aa2: P = 33 d, e= 0.32, with system mass of 30 M$_\odot$, orbit r = 0.6 AU. Aa,Ab: ps = 140 AU. Aa,B: ps = 5,670 AU. (2013)								
	h 4843		16 21.4 −33 18	7.4, 11.5	267	12″	150	F3V	147149, 207586
	β 624		16 22.9 −23 07	7.8, 9.3	319	1.0″	131	A2.5V	147432, 184350
	h 4845		16 23.8 −41 15	8.1, 8.5	128	1.8″	84	F2IV	147435, 226715
	h 4848	A B	16 23.9 −33 12	6.9, 7.3	152	6.2″	110	A0V A0V	147553, 207625
	Rag 8		16 24.0 −39 12	5.4, 11.0	248	5.8′	12.8	G1V DA2	147513, 207622
	Local, solar type wide double; B seems lost in sparse, faint field, but near 100% probability pair is physical. AB: ps = 6,010 AU. (2004)								
★	H N 39		16 24.7 −29 42	5.9, 6.6	355	4.1″	33.1	G0IV G0V	147723, 184369
	Solar type double; unequal pair isolated in rich field. AB: ps = 185 AU. (2012)								
★ 21 alp	Gnt 1		16 29.4 −26 26	1.0, 5.4	274	2.5″	170	M1.5I B4V	148478, 184415
	Antares. Prototype of a high mass (~30 M$_\odot$) binary where one component has entered the supergiant phase, here expanding to a radius of ~3.4 AU; the O/BG color often reported for this pair is illusory, created by complementary contrast of a nearly white (B type) star against the bright Y or O hue. Poor quality, edgewise orbit suggests AB: P= 1,220 y, orbit r= 360 AU, inclination ~80°, closing. Compare with Albireo, Almach, Izar and Rasalgethi, similar high mass "giant type" systems. Neglected. (1997)								
	HDS 2335		16 31.5 −39 01	7.4, 10.5	221	4.1″	24.5	K1V+G9V	148704, 207733
	Local, high CPM, solar type spectroscopic double; isolated in faint field near NGC 6139. AB: ps = 135 AU. Neglected. (1991)								
	I 95		16 33.0 −33 32	7.5, 9.6	0	1.8″	500	A0V	148950, 207759
	λ 283	A B C	16 40.4 −39 55	7.7, 12.9, 10.0	149, 269	7.4″, 22″	310	F0IV	150070, 207902
	Solar type 2+1 triple; rich field, optical pair TOB 8 (m.8, 22″) 5′ f. AB: ps = 3,100 AU. Neglected. (1999)								
	R 283		16 42.5 −37 05	7.0, 7.8	250	0.8″	95	G3III	150420, 207943
	β 1116	A B	16 44.3 −27 27	6.6, 10.2	13	2.0″	96	A2V	150768, 184591
	I 99		16 49.5 −43 57	8.0, 8.7	65	1.0″	179	A0V	151473, 227264
	h 4889		16 51.0 −37 31	6.2, 7.8	5	6.6″	340	B9	151771, 208089
	WFC 181	A C	16 51.9 −38 03	3.0, 9.4	257	80″	154	B1.5 +B6.5	322326, 208102
★	CorO 289	(AaAb) B C D	16 56.9 −40 31	7.3, 13.0, 9.5, 9.6	129, 253, 237	7.3″, 7.7″, 15″	400*	O6.5III	152723, 227479
	Distant, high mass, probably unstable quintet with O giant; the star cluster Trumpler 24, at 1100 pc, is far behind it. AD: ps = 8,090 AU. (2014)								
	λ 318		17 06.2 −38 38	8.4, 8.7	352	1.0″	97	F3V	154287, 208399
	Hd 266	A B C	17 06.3 −37 14	5.7, 10.0, 12.5	78, 186	5.8″, 41″	142	A2IV	154310, 208406
	Tok 310		17 06.4 −37 06	7.9, 10.0	213	97″	58	F8V	154337, 208407
	CorO 208		17 07.4 −44 27	7.1, 9.2	138	4.8″	280	A0V	154410, 227637
	Howe 86		17 13.9 −38 18	6.9, 9.0	146	2.7″	103	F5V	155536, 208556
	I 408		17 16.3 −42 20	7.0, 8.9	173	1.7″	550	B2V	155896, 227768
★ 4	MlbO 4	A B	17 18.9 −34 59	6.4, 7.4	185	1.3″	6.84	K3V+K5V	156384, 208670
	Local, very high CPM, low mass binary. AB: P= 42 y, orbit r= 12 AU, e= 0.58, visible through periastron in 2018. (242 measures; 2010)								
	Δ 217		17 29.0 −43 58	6.3, 8.5	168	13″	570	B5III	158042, 228010
	B 342		17 29.4 −38 31	6.8, 7.6	109	0.4″	118	A2V	158156, 208870
	Howe 87		17 31.3 −39 01	7.4, 9.0	231	3.3″	100	F6V	158468, 208903
	Hld 136	A B C	17 31.7 −41 02	7.8, 8.1, 8.5	107, 250	1.1″, 64″	270	B7IV	158531, 228078
	I 603		17 33.2 −45 31	7.2, 9.4	79	1.2″	430	B9.5III	158747, 228107

Label	Catalog ID	Components	Coordinates (J2000)	Mag.	θ	ρ	Dist.	Spectral type(s)	HD, SAO No.
35 lam	Δ 218	(AaAb) C	17 33.6 −37 06	2.1, 9.2	330	94″	175	B2IV B	158926, 208954
		Shaula. High mass (2)+1 triple, visual binary; faint field, primary is bet Cep type (high mass and rapidly pulsating) variable star. Aa,Ab: *P*= 2.9 y, *e*= 0.12. AC ps = 22,200 AU. (2000)							
	I 247	(AaAb) B	17 37.9 −37 52	7.0, 8.7	97	0.6″	47.6	G8V	159704, 209047
	CorO 222		17 56.8 −39 56	7.9, 8.4	127	3.7″	810	K1III	163195, 209497
	R 306	A B	17 57.9 −36 00	6.8, 9.5	15	3.4″	152	A0III	163482, 209525
	I 1013		17 58.0 −39 08	6.5, 8.2	138	1.1″	127	A0IV	163433, 209524

Sculptor   Scl   *Chart 20*

Label	Catalog ID	Components	Coordinates (J2000)	Mag.	θ	ρ	Dist.	Spectral type(s)	HD, SAO No.
	Howe 92		23 13.0 −32 19	7.7, 10.7	266	6.9″	169	K0III	219034, 214386
	Howe 63		23 24.0 −27 17	7.8, 10.7	268	6.7″	179	A1V	220455, 191872
	λ 489		23 33.8 −36 16	7.3, 11.0	145	20″	56	F3V	221609, 214624
	I 693		23 37.0 −36 48	8.0, 9.2	89	1.4″	82	F6V	221982, 214657
	Howe 93		23 37.1 −31 52	6.7, 9.9	250	5.7″	300	K1III	222004, 214659
	h 5417		23 44.5 −26 15	6.3, 9.4	320	7.9″	69	F7V	222872, 192116
del	h 3216	A C	23 48.9 −28 08	4.6, 9.4	297	74″	42.1	A0V	223352, 192167
	h 5423		23 49.8 −25 20	6.4, 11.6	304	13″	98	A3V	223466, 192180
★ phi	Lal 192		23 54.4 −27 03	6.8, 7.4	272	6.6″	81	A2V F2V	223991, 192231
		Pretty, A type visual and spectroscopic binary; sparse field. A pair actually discovered by J. Dunlop (Δ 253). AB: ps = 720 AU. (2013)							
	Skf 760		23 55.3 −31 55	6.1, 6.8	0	2.2′	270	B6V B8V	224113, 214860
	Lal 193		23 59.5 −26 31	8.1, 8.3	170	11″	102	F0	224641, 192295
kap 1	β 391	A B	00 09.4 −27 59	6.1, 6.2	259	1.4″	77	F4IV	493, 166083
	h 3375		00 33.7 −35 00	6.6, 8.5	171	4.9″	33.1	G3IV	3074, 192609
		High CPM, solar type double; dark field, pretty YO/− color. AB: ps = 220 AU. Neglected. (1998)							
	h 1991		00 38.8 −25 06	6.6, 9.7	95	47″	270	K0	3605, 166443
	h 1992		00 38.9 −25 36	7.8, 8.9	247	46″	123	A7V	3622, 166446
lam 1	Hd 182		00 42.7 −38 28	6.6, 7.0	21	0.7″	145	A0V	4065, 192690
	Stone 60		01 04.5 −33 32	6.6, 10.2	218	8.5″	164	K0III	6403, 192907
	h 3436		01 27.1 −30 14	6.9, 9.6	128	9.6″	430	K1III	8887, 193123
	Arg 4		01 32.3 −26 33	8.0, 9.1	73	18″	270	Am G2V	9451, 167119
tau	h 3447		01 36.1 −29 54	6.0, 7.4	185	0.8″	69	F2V	9906, 193201
★ eps	h 3461	A B	01 45.6 −25 03	5.4, 8.5	20	5.1″	28.1	F2V	10830, 167275
		Solar type double; delicate brightness contrast, sparse or dark field. AB: ps = 195 AU, low quality orbit *P* = 1,200 y. (2009)							

Scutum   Sct   *Chart 18*

Label	Catalog ID	Components	Coordinates (J2000)	Mag.	θ	ρ	Dist.	Spectral type(s)	HD, SAO No.
	Σ 2313		18 24.7 −06 36	7.5, 8.7	195	5.7″	72	G0IV	169392, 142290
	Σ 2325		18 31.4 −10 48	5.8, 9.3	257	12″	330	B2V	170740, 161569
	Σ 2373		18 45.9 −10 30	7.4, 8.4	337	4.4″	270	F2	173457, 161805
	Rst 4596		18 46.8 −14 28	7.4, 11.5	154	5.8″	54	F5V	173614, 161816
	Σ 2391	A B	18 48.7 −06 00	6.5, 9.6	332	38″	230	A2II	174005, 142640

Serpens   Ser   *Charts 10, 17, 18*

Label	Catalog ID	Components	Coordinates (J2000)	Mag.	θ	ρ	Dist.	Spectral type(s)	HD, SAO No.
	Σ 1919		15 12.7 +19 17	6.7, 7.4	10	23″	27.2	G1V G5V	135101, 101437
	Ho 547		15 16.4 +16 48	7.8, 11.5	291	5.5″	41.9	G0	135792, 101471
	β 943		15 18.4 +00 56	6.7, 10.9	92	3.0″	178	K1III	136027, 120938
	Σ 1931	A B	15 18.7 +10 26	7.2, 8.1	167	13″	50	F7V G3V	136160, 101480
★ 5	Σ 1930	A B	15 19.3 +01 46	5.1, 10.1	35	11″	25.4	F8V	136202, 120946
		MQ Ser. Local, high CPM, solar type double; primary is BY Dra type variable star. AB: ps = 380 AU. (2013)							

Label	Catalog ID	Components	Coordinates (J2000)	Mag.	θ	ρ	Dist.	Spectral type(s)	HD, SAO No.
6	β 32		15 21.0 +00 43	5.5, 8.8	22	3.4″	73	K2III	136514, 120955
	Σ 1950		15 30.0 +25 31	8.1, 9.2	91	3.4″	490	K4III	138232, 83852
★ 13 del	Σ 1954	A B	15 34.8 +10 32	4.2, 5.2	173	3.9″	70	F0IV	138918, 101624
			Solar type binary with del Sct type variable star; YW/YW color, sparse field. In large apertures, forms charming "reflection" quadruple with CD (m.14, 4″) 1′ n., which may be related. AB: P= 1,038 y, orbit r= 270 AU, at cusp. (476 measures; 2013)						
28 bet	Σ 1970	A B C	15 46.2 +15 25	3.7, 10.0, 8.2	264, 255	30″, 27′	47.6	A2IV	141003, 101725
V382	Eis 1	A B	15 48.2 +01 34	7.5, 9.5	357	18″	21.3	G8V	141272, 121196
			Local, solar type double with BY Dra type variable; 11′ f. m.11 K giant star in sparse field. AB: ps = 515 AU, estimated P = ~7,000 y. (2006)						
	Skf 1311		15 50.3 +02 12	5.3, 10.3	300	75″	84	G8III+K0	141680, 121215
	Σ 1985		15 55.9 −02 10	7.0, 8.7	353	5.9″	36.9	F8V	142661, 140842
	Σ 1988		15 56.8 +12 29	7.6, 7.8	249	1.9″	96	F1V	142910, 101829
	Σ 1987		15 57.2 +03 24	7.3, 8.7	322	11″	179	A0V	142930, 121277
	OΣ 303	A B	16 00.9 +13 16	7.7, 8.1	173	1.6″	97	F7V	143597, 101874
			Solar type double; dark field. AB: ps = 210 AU. Frequently measured, but no orbit or linear solution. (296 measures; 2012)						
	Σ 2003		16 03.7 +11 26	7.3, 10.5	171	14″	133	K3III	144064, 101898
	Σ 2031	A B	16 16.3 −01 39	7.2, 11.0	229	21″	47.4	F7V	146433, 141069
	Σ 2041		16 21.8 +01 13	7.5, 10.5	1	2.6″	187	K0	147411, 121534
	Hu 189		17 53.1 −13 39	7.4, 9.0	254	1.5″	63	F8V	162739, 160930
	Hld 139	A B C	17 54.9 −11 38	7.0, 11.1, 12.9	144, 97	3.2″, 66″	68	F3V	163117, 160899
	h 2814	A B	17 56.3 −15 49	5.9, 9.2	157	21″	74	A1V	163336, 160915
	Σ 2303		18 20.1 −07 59	6.6, 9.3	240	1.6″	60	F2V	168459, 142229
	AC 11		18 25.0 −01 35	6.7, 7.2	355	0.9″	134	A9III+F6III	169493, 142294
			Solar type binary; faint field. AB: P= 248 y, orbit r= 84 AU, e= 0.38, at cusp, periastron 2116. (258 measures; 2013)						
59 d	Σ 2316	(AaAb) B	18 27.2 +00 12	5.4, 7.6	320	3.7″	144	G0III+A6V	169985, 123497
	OΣ 360		18 38.7 +04 51	6.9, 9.1	282	1.7″	142	K2III	172190, 123756
★	Σ 2375	(AaAb) (BaBb)	18 45.5 +05 30	6.3, 6.7	120	2.6″	189	A1V	173495, 123886
			Matched A type (2)+(2) interferometric quadruple; visual binary. The "Tweedledum and Tweedledee" system: two similar binaries, first detected with an eyepiece interferometer by W. Finsen (1953), so alike they are frequently mismeasured. Aa,Ab: P= 27 y, e= 0.779, periastron 2021. Ba,Bb: P= 38.6y, e= 0.87, apastron 2025. AB: ps = 670 AU. (322 measures; 2013)						
★ 63 the 1,2	Σ 2417	A B	18 56.2 +04 12	4.6, 4.9	104	22″	47.4	A5V A5Vn	175638, 124068
			Lovely wide, bright, matched A type pair; discovered by C. Mayer (1777). AB: ps = 1,410 AU, no orbit. (218 measures; 2012)						

Sextans   Sex   *Chart 15*

Label	Catalog ID	Components	Coordinates (J2000)	Mag.	θ	ρ	Dist.	Spectral type(s)	HD, SAO No.
	Σ 1377		09 43.5 +02 38	7.5, 10.5	137	4.2″	154	F7V	84184, 117871
8 gam	AC 5	A B C	09 52.5 −08 06	5.4, 6.4, 12.3	49, 333	0.5″, 37″	85	A1V	85558, 137199
			A type 2+1 triple. MSC has 3.3 M⊙. AB: P= 78 y, orbit r= 34 AU, e= 0.74, drops below 0.4″ in 2025. (330 measures; 2011)						
	Σ 1401		10 00.2 +06 15	7.7, 10.5	21	22″	60	F5	86683, 118045
	Hjl 1056	A B	10 02.8 −01 04	6.8, 10.6	32	109″	110	K0	87095, 137312
	Σ 1440		10 29.8 −03 55	7.8, 9.2	346	15″	118	G0	90934, 137604
	Σ 1441	(AaAb) B	10 31.0 −07 38	6.5, 8.8	167	2.8″	350	K5III F6V	91106, 137614
	A 556	(AaAb) B	10 37.0 −08 50	7.2, 9.7	199	0.9″	39.9	G0V K5V	91962, 137678
	Σ 1457		10 38.7 +05 44	7.7, 8.2	333	1.8″	65	F5V	92184, 118410
			Solar type binary; very dark field. AB: ps = 160 AU. (203 measures; 2013)						
34	A 2768		10 42.6 +03 35	6.9, 8.5	247	0.6″	71	F7V	92749, 118443
			Solar type binary; very dark field. System mass 2.1 M⊙. AB: P= 81 y, orbit r= 28 AU, e= 0.55, apastron in 2017. (2013)						
★ 35	Σ 1466	(AaAb) B	10 43.3 +04 45	6.2, 7.1	240	6.7″	169	K2III+K1III	92841, 118449
			(2)+1 spectroscopic triple, a rare pair of two K giants; OY/OY color, dark field. AB: ps = 1,530 AU. (2012)						
40	Σ 1476		10 49.3 −04 01	7.1, 7.8	16	2.4″	86	A2IV	93742, 137808

Label	Catalog ID	Components	Coordinates (J2000)	Mag.	θ	ρ	Dist.	Spectral type(s)	HD, SAO No.
	Taurus		Tau						*Charts 7, 13*
	β 879		03 28.6 +11 23	6.8, 12.8	71	25″	118	G9III	21524, 93448
	Σ 406		03 30.8 +05 09	7.6, 9.3	127	9.1″	182	F0	21788, 111233
	Σ I 7		03 31.1 +27 44	7.4, 7.8	233	44″	120	B9	21700, 75964
★	Σ 401	A B	03 31.3 +27 34	6.6, 6.9	269	11″	97	A2V	21743, 75970
			A type double; W/W color, with Σ I 7 (above) 10′ n.p., spectacular in a sparse field. AB: ps = 1,440 AU. (2013)						
	AG 68		03 32.2 +11 33	6.8, 9.9	248	17″	220	A1V	21915, 93479
★ 7	Σ 412	A B C	03 34.4 +24 28	6.6, 6.9, 9.9	353, 54	0.8″, 22″	132	A3V+A3V	22091, 75999
			A type 2+1 triple; YW/– color, sparse field. AB: system mass 3.5 M$_\odot$, P= 522 y, orbit r= 85 AU, e= 0.68, widening. (298 measures; 2012)						
★ V711	Σ 422	(AaAb) B	03 36.8 +00 35	6.0, 8.9	274	7.1″	30.7	G8V K6	22468, 111291
			Solar type (2)+1 spectroscopic triple with RS Cvn variable; Y/W color, 12′ n. of 10 Tau. Aa,Ab: P = 2.8 d. AB: ps = 290 AU. (200 meaures; 2013)						
	Σ 427		03 40.6 +28 46	7.4, 7.8	208	7.0″	113	A1V A2V	22766, 76071
	Σ 435		03 43.1 +25 41	7.2, 8.9	3	14″	66*	F3V	23075, 76094
	Hjl 1024	(AaAb) B	03 43.7 +23 39	8.0, 9.6	175	3.3′	132	A9V	23157, 76103
	Hjl 1025	(AaAb) B	03 45.6 +24 20	7.2, 9.9	164	2.9′	150*	A1V F5V	23387, 76152
			A type (2)+1 CPM triple, wide visual double; in Pleiades (M 45), unrelated to 20 Tau 3′ n.f. AB: ps = 34,000 AU. (2011)						
	Σ 444	A B	03 45.8 +23 09	6.9, 10.1	332	3.8″	121	A0V	23410, 76156
★ 21	Hjl 1026	A B [D E]	03 45.9 +24 33	5.8, 6.4, [12.7, 13.0]	130, 74, [162]	2.5′, 2.9′, [18″]	114	B8V A0V	23432, 76159
			Asterope. High mass 1+1+2 quadruple or comoving group, in Pleiades (M 45); B is 22 Tau, E may be optical. Wide, fragile pairs may originate as stars released in parallel trajectories from the natal cluster: linked by CPM, parallax and radial velocity, this group is perhaps an example. (Compare with Hjl 1024 and OΣΣ 40.) AB: ps = 17,100 AU. (2013)						
	OΣΣ 40	(AaAb) B	03 49.4 +24 23	6.6, 7.5	309	87″	122	B9.5V	23873, 76236
	OΣ 64	(AaAb) B [Ca Cb]	03 50.0 +23 51	6.8, 10.2, [9.1 11.1]	236, 235, [259]	3.2″, 10″, [0.6″]	147	B9.5V	23964, 76251
			High mass (2)+1+2 spectroscopic quintuple. MSC system mass 5.5 M$_\odot$. Aa,Ab: P = 16.7 d. Ca,Cb (Bov 25): ps = 120 AU, requires large aperture. (2012)						
	OΣ 65		03 50.3 +25 35	5.7, 6.5	195	0.2″	57	A2V A5V	23985, 76256
			A type binary; faint field. System mass 3.8 M$_\odot$. AB: P = 61 y, orbit r = 25 AU, e = 0.63, ps = 34 AU, reappears in 2016. (236 measures; 2008)						
	A 1830	A B C	03 51.3 +26 21	8.0, 8.0, 11.1	194, 330	0.4″, 61″	64*	F8	24104, 76270
V479	OΣΣ 41	A (BC)	03 54.5 +05 10	7.5, 8.9	358	59″	158	F3II	24550, 111492
	Σ 479	A B C	04 00.9 +23 12	6.9, 7.8, 9.5	127, 242	7.2″, 57″	330	B9V	25201, 76388
	Σ 495		04 07.7 +15 10	6.1, 8.8	222	3.7″	43.6	F3V	26015, 93775
	OΣ 72		04 08.0 +17 20	6.1, 9.7	328	4.7″	123	K5III	26038, 93777
	Σ 494		04 08.9 +23 06	7.5, 7.7	188	5.3″	105	A8IV A8IV	26128, 76476
	Σ 510		04 12.2 +00 44	6.7, 10.1	303	11″	146	G8III	26573, 111659
47	β 547	A B C	04 13.9 +09 16	5.1, 7.3, 13.2	341, 227	1.2″, 29″	102	G5IV A8V	26722, 111674
	OΣΣ 45	A B	04 15.5 +06 11	6.4, 7.0	316	64″	21.3	G0IV G3V	26923, 111698
			Local, solar type double with BY Dra variable; striking pair in dark field; near 100% probability pair is physical. AB: ps = 1,840 AU. (2012)						
	Σ 517		04 16.0 +00 27	7.4, 9.3	8	3.3″	117	A1V	26991, 111705
	Ho 328		04 17.0 +19 41	7.4, 9.1	2	0.5″	85	F5V	27028, 93840
			Solar type binary; dark field, LDS 5536 (m.10, 12″) 8′ n.f. AB: P = 63 y, orbit r = 31 AU, at cusp, periastron 2035. (2008)						
	LDS 5535		04 17.3 +20 35	5.0, 9.6	118	3.0′	28.9	A3	27045, 76532
55	OΣ 79		04 19.9 +16 31	7.3, 8.6	358	0.5″	46.5	F9V	27383, 93870
			Solar type spectroscopic binary; sparse field, member of Hyades cluster. AB: P = 90 y, orbit r = 27 AU, e = 0.60, apastron 2031. (212 measures; 2012)						

Label	Catalog ID	Components	Coordinates (J2000)	Mag.	θ	ρ	Dist.	Spectral type(s)	HD, SAO No.
	β 87		04 22.4 +20 49	6.2, 8.6	167	1.9″	710	B3V K3II	27639, 76571
	OΣ 82	(AaAb) B (C)	04 22.7 +15 03	7.3, 8.6	333	1.3″	43.6	F9V	27691, 93896
		colspan-note							

Solar type (2)+1+(1) spectroscopic quadruple (C is m.16), visual binary; YW/B color, member of Hyades cluster. Aa,Ab: P = 4.0 d. AB: P = 241 y, orbit r = 50 AU, e = 0.26. (211 measures; 2012)

Label	Catalog ID	Components	Coordinates (J2000)	Mag.	θ	ρ	Dist.	Spectral type(s)	HD, SAO No.
	Σ 535		04 23.3 +11 23	7.0, 8.3	271	1.1″	87	A5III	27762, 93899

Double with A type giant; delicate brightness contrast, dark field. AB: ps = 130 AU. (231 measures; 2013)

Label	Catalog ID	Components	Coordinates (J2000)	Mag.	θ	ρ	Dist.	Spectral type(s)	HD, SAO No.
★ 65,67 kap 1,2	Σ 541	(AaAb) B	04 25.4 +22 18	4.2, 5.3	174	5.8′	47.2	A7V	27934, 76601

K Tau. A type (2)+1 CPM triple; dark field, both members of the Hyades cluster: A is an occultation binary, B is a del Sct type variable. The AB pair brackets CD, a tiny (m.10, 6″) matched and unrelated pair. AB: ps = 22,000 AU. (2011)

Label	Catalog ID	Components	Coordinates (J2000)	Mag.	θ	ρ	Dist.	Spectral type(s)	HD, SAO No.
68 del 3	H VI 101	(AaAb) B C D	04 25.5 +17 56	4.3, 7.9, 11.1, 9.1	341, 236, 40	1.8″, 77″ 6.9′	45.5	A2IV	27962, 93923

A = V776 Tau. A type (2)+1+1+1 spectroscopic quintuple with alp2 CVn type variable; sparse field. AB: ps = 110 AU. AD: ps = 25,300 AU. (2006)

Label	Catalog ID	Components	Coordinates (J2000)	Mag.	θ	ρ	Dist.	Spectral type(s)	HD, SAO No.
	Σ 546		04 27.0 +19 07	7.9, 9.2	181	6.9″	65	G0	28139, 93940
	Σ 545	(AaAb) B	04 27.1 +18 12	6.9, 8.8	58	19″	107	A0V	28150, 93942
★ 78 the 2	Σ I 10	(AaAb) B	04 28.7 +15 52	3.4, 3.9	347	5.7′	46.1	A7III	28319, 93957

A type (2)+1 triple with del Sct variable; sparse field, member of Hyades cluster. LDS 2246 (see below) 30′ n.f. AB: ps = 21,300 AU. Neglected. (1998)

Label	Catalog ID	Components	Coordinates (J2000)	Mag.	θ	ρ	Dist.	Spectral type(s)	HD, SAO No.
V921	Bgh 2		04 29.5 +17 52	7.0, 9.1	9	110″	46.4	G5	28406, 93963
80	Σ 554	(AaAb) B	04 30.1 +15 38	5.7, 8.1	14	1.6″	45.8	F0V G0V	28485, 93970

Solar type (2)+1 spectroscopic triple; dark field, member of Hyades cluster. Aa,Ab: P = 30.5d. AB: P= 180 y, orbit r = 46 AU, closing. (2013)

Label	Catalog ID	Components	Coordinates (J2000)	Mag.	θ	ρ	Dist.	Spectral type(s)	HD, SAO No.
81	β pm 62		04 30.6 +15 42	5.5, 8.9	339	2.7′	44.9	Am	28546, 93978
	LDS 2246		04 30.6 +16 12	4.8, 6.5	130	4.2′	43.2	A6IV	28527, 93975
	OΣ 84		04 31.1 +06 47	7.2, 8.1	256	9.4″	84	K0IV	28630, 111863
★	Σ 559		04 33.6 +18 01	7.0, 7.0	276	3.1″	134	B9IVn	28867, 94002

High mass double; possible T Tauri (pre main sequence star), dark field. AB: ps = 560 AU. (2012)

Label	Catalog ID	Components	Coordinates (J2000)	Mag.	θ	ρ	Dist.	Spectral type(s)	HD, SAO No.
	h 5461	A B C	04 34.6 +28 58	5.9, 10.3, 11.7	102, 134	25″, 50″	149	B9	28929, 76654
	Σ 562		04 34.8 +22 42	6.8, 9.9	285	1.9″	159	F5III	28976, 76658
88	Sh 45	(Aa1Aa2Ab1Ab2) (BaBb)	04 35.7 +10 10	4.3, 7.8	300	71″	47.9	A6m +G3V	29140, 94026

A type (2+2)+(2) spectroscopic sextuple, visual binary with alp2 CVn type variable; dark field. MSC system mass 6.9 M⊙. Aa,Ab: P= 18.0 y, orbit r = 12 AU, e = 0.07. (2008)

Label	Catalog ID	Components	Coordinates (J2000)	Mag.	θ	ρ	Dist.	Spectral type(s)	HD, SAO No.
	OΣ 86		04 36.6 +19 46	8.7, 7.7	1	0.5″	270	A2	29193, 94031

A type binary; dark field. AB: P= 828 y, orbit r = 165 AU. (2008)

Label	Catalog ID	Components	Coordinates (J2000)	Mag.	θ	ρ	Dist.	Spectral type(s)	HD, SAO No.
	OΣΣ 53		04 37.4 +00 33	7.6, 7.6	353	78″	67	G5	29355, 111935
	h 346	A B	04 41.3 +28 37	5.7, 10.7	54	43″	117	A2V	29646, 76707
	LDS 2266	A B	04 42.9 +18 43	7.2, 10.2	102	2.4′	43.5	G5+K3	29836, 94078
	Σ 623		04 59.9 +27 20	7.0, 8.7	206	21″	110	B7V A0V	31806, 76880
	S 461	(AB) (CaCb)	05 01.7 +26 40	6.9, 8.3	159	79″	40.6	G2V	32092, 76903

Solar type (2)+(2) spectroscopic quadruple, visual binary. MSC system mass 3.9 M⊙. AB: P = 25 y, e = 0.29, periastron 2025. (2013)

Label	Catalog ID	Components	Coordinates (J2000)	Mag.	θ	ρ	Dist.	Spectral type(s)	HD, SAO No.
	OΣ 95		05 05.5 +19 48	7.0, 7.6	296	0.9″	138	A5m	32642, 94306

A type binary; dark field. AB: P = 760 y; given the spectral type, the orbit r = ~140 AU. (210 measures; 2012)

Label	Catalog ID	Components	Coordinates (J2000)	Mag.	θ	ρ	Dist.	Spectral type(s)	HD, SAO No.
	Σ 645	(AaAb) (BC)	05 09.8 +28 02	6.0, 9.1	28	11″	53	A5m	33204, 76990
	Bgh 21		05 10.1 +27 33	7.0, 9.3	353	5.3′	40.0	F5 G5	33252, 76998
	Σ 670	A (BaBb)	05 16.7 +18 26	7.7, 8.3	164	2.6″	199	B3V	34251, 94431
CD	Σ 674	(AaAb) B	05 17.5 +20 08	6.8, 9.7	149	9.9″	81	F7V dK2	34335, 77084
	Σ 680		05 19.2 +20 08	6.2, 9.7	203	8.9″	120	K0III	34579, 77098
	OΣ 108		05 29.3 +18 22	6.8, 10.4	130	3.2″	187	A2	35985, 94586

Label	Catalog ID	Components	Coordinates (J2000)	Mag.	θ	ρ	Dist.	Spectral type(s)	HD, SAO No.
★ 118	Σ 716	(AaAb) (BaBb)	05 29.3 +25 09	5.8, 6.7	209	4.7″	130	B8.5V	35943, 77201
		High mass (2)+(2) quadruple; WB/W color, dark field. MSC system mass 8.6M$_\odot$. AB: ps = 820 AU. (2012)							
	h 3275	(AB) [C D]	05 29.8 +18 25	7.7, [8.2, 12.0]	21, [254]	56″, [1.4″]	174	A0	36073, 94589
★	Σ 730	(AaAb) B	05 32.2 +17 03	6.1, 6.4	141	9.6″	920	B7IIIe	36408, 94630
		Distant, high mass (2)+1 spectroscopic triple, visual double with B giant; lovely matched YW/YW color. AB: ps = 11,900 AU. (2013)							
	Σ 742	(AaAb) B	05 36.4 +22 00	7.1, 7.5	274	4.1″	67	F8	37013, 77313
		Solar type (2)+1 triple; ambiguous Σ 740 (m.9 22″) 50′ s. MSC system mass 2.6 M$_\odot$. AB: ps = 370 AU. (206 measures; 2012)							
★	Σ 749	A B	05 37.1 +26 55	6.5, 6.6	320	1.2″	155	B9IV	37098, 77322
		High mass binary; pretty field, CD (Bow 4, m.11, 4″) 3′ n.p. is unrelated. AB: P = 987 y, orbit r = 160 AU. (256 measures; 2012)							
	Σ 787	A B D	05 46.0 +21 19	8.3, 8.8, 8.1	57, 160	0.7″, 2.1′	158	F2	38363, 77527
	OΣΣ 67	(AB) C	05 48.4 +20 52	6.0, 8.3	162	76″	168	B9Vn	38670, 77578
	Ku 23		05 50.8 +14 27	7, 8.6	107	0.8″	470	B9	39098, 94914

Telescopium   Tel   *Chart 29*

Label	Catalog ID	Components	Coordinates (J2000)	Mag.	θ	ρ	Dist.	Spectral type(s)	HD, SAO No.
	h 5033	A B	18 15.4 −48 51	6.7, 9.8	114	17″	91	G8III	166949, 228845
★	h 5034		18 16.2 −46 01	7.5, 8.6	98	2.2″	180*	A5IV	167153, 228857
		A type double; faint rich field. AB: ps = 535 AU. As with many southern hemisphere stars: neglected. (1994)							
QW	Δ 220	(AaAb) B	18 22.2 −55 34	8.1, 8.5	177	31″	91	G0	168292, 245426
	h 5041		18 25.8 −53 38	7.3, 9.2	260	2.9″	132	Fm	169058, 245461
	h 5045		18 30.9 −48 01	6.7, 9.7	21	7.9″	69	F6V	170283, 229077
	Skf 1147		18 31.5 −50 11	7.2, 10.4	173	85″	51	F0III	170338, 245531
	h 5053	(AB) C	18 43.4 −55 46	7.7, 10.2	197	33″	62	F7V	172447, 245656
		Solar type (2)+1 triple; visual binary, faint field. AB: P = 73 y, orbit r = 17 AU, widening. (2013)							
	I 112	A B	18 54.0 −47 16	7.1, 9.1	190	1.8″	69	F5V	174691, 229354
★	R 317	A B C	19 03.1 −45 43	8.0, 8.8, 8.8	284, 211	1.5″, 19″	560*	B9III	176555, 229458
		Distant, high mass 2+1 triple; a mini bet Mon, similar brightness to stars in rich field. AC was detected by J. Herschel during his observing expedition in South Africa (1834). AB: ps = 1,100 AU, AC: ps = 14,400 AU. Neglected. (1991)							
	Δ 225	A B	19 12.4 −51 48	7.2, 8.4	250	70″	440	K5III	178734, 245970
	h 5092		19 13.9 −47 22	8.1, 8.4	350	18″	340*	G2III	179211, 229568
eta	Tok 331	A B C	19 22.9 −54 25	5.0, 11.9, 7.1	170, 171	4.2″, 6.9′	48.2	A0Vn+M7	181296, 246055
	CorO 238		19 42.3 −52 57	7.7, 9.3	48	3.3″	141	A6V	185559, 246224
	Tok 335		19 48.0 −56 22	5.3, 9.3	333	102″	49.6	A7IV	186543, 246271
	Δ 227		19 52.6 −54 58	5.8, 6.4	148	23″	172	K0III A2V	187420, 246311
		A type double with K giant; mixed field, stars n. unrelated. AB: ps = 5,340 AU. (2013)							
	Δ 229		19 58.3 −51 54	7.7, 8.2	242	80″	88	F0	188557, 246357
	I 256		20 01.4 −47 24	7.1, 9.6	189	0.8″	102	F2IV	189307, 229991
	λ 415		20 20.9 −45 33	7.3, 12.0	95	11″	67	G6III	193213, 230165
	Skf 376		20 21.7 −50 00	6.3, 12.8	300	21″	31.0	F9V	193307, 246546

Triangulum   Tri   *Chart 7*

Label	Catalog ID	Components	Coordinates (J2000)	Mag.	θ	ρ	Dist.	Spectral type(s)	HD, SAO No.
★	Σ 183	(AB) C	01 55.1 +28 48	7.7, 9.3	163	5.6″	166	F2	11671, 75020
		Solar type (2)+1 triple, visual binary; sparse field. MSC has 3.9 M$_\odot$. AB: P = 333 y, orbit r = 80 AU, e = 0.52, widening. (2008)							
	Frk 2		01 56.4 +30 26	8.0, 9.1	307	54″	220	F0	11813, 55115
	A 819	(AB) C	01 57.0 +31 01	7.8, 10.0	271	67″	86	F5	11849, 55122
		Solar type (2)+1 triple, visual binary; sparse field. MSC system mass 3.4 M$_\odot$. AB: P = 164 y, e = 0.46, apastron 2092. (2007)							
★ 6 iot	Σ 227	(AaAb) (BaBb)	02 12.4 +30 18	5.3, 6.7	69	3.8″	89	G0III	13480, 55347
		A = TZ Tri. Bright (2)+(2) spectroscopic quadruple, visual double with G giant, an ellipsoidal variable; YW/BW color, faint field. Aa,Ab: P = 14.7 d. Ba,Bb: P = 2.2 d. AB: ps = 460 AU. (244 measures; 2012)							

Label	Catalog ID	Components	Coordinates (J2000)	Mag.	θ	ρ	Dist.	Spectral type(s)	HD, SAO No.
	Σ 239		02 17.4 +28 45	7.1, 7.8	212	14″	34.5	F7V F9V	14082, 75265
	Σ 246		02 18.7 +34 29	7.8, 9.3	122	9.6″	90	G0IV	14202, 55446
	Ho 216		02 27.0 +31 17	7.9, 9.4	3	1.3″	79	F6V	15128, 55566
	Σ 269		02 28.2 +29 52	7.6, 9.0	345	1.7″	192	G0	15256, 75383
	h 653		02 33.6 +31 25	7.6, 12.7	43	23″	370	K3Ib	15832, 55658
	Σ 285		02 38.8 +33 25	7.5, 8.1	163	1.7″	220	K2III+	16396, 55748
	Σ 286		02 39.8 +33 57	7.8, 10.2	258	3.0″	430	K0III+	16511, 55761
	Σ 310		02 49.4 +33 56	7.4, 10.4	92	2.2″	139	A2	17497, 55911

Triangulum Australe — TrA — Chart 29

Label	Catalog ID	Components	Coordinates (J2000)	Mag.	θ	ρ	Dist.	Spectral type(s)	HD, SAO No.
	I 332		15 20.7 −67 29	6.4, 8.2	107	1.1″	310	B3V	135737, 253115
	Rss 26		15 29.9 −67 29	7.3, 9.3	243	21″	450	B3IV	137384, 253183
	Rmk 20	A B	15 47.9 −65 27	6.2, 6.4	147	1.8″	112	A5III	140484, 253297
	Δ 194	A B	15 54.9 −60 45	6.4, 8.1	95	1.1″	420	B9II	141913, 253344
	Rss 29		16 03.7 −60 30	7.1, 8.1	180	53″	650	B3IV	143448, 253406
	I 336		16 31.9 −62 17	7.8, 8.1	199	1.1″	230	B9V	148431, 253588
	HDS 2352		16 37.8 −64 15	7.7, 10.3	148	11″	91	K1IV	149237, 253631

Tucana — Tuc — Chart 26

Label	Catalog ID	Components	Coordinates (J2000)	Mag.	θ	ρ	Dist.	Spectral type(s)	HD, SAO No.
	I 20		22 18.0 −62 49	7.4, 8.4	188	0.6″	122	F5V	211299, 255187
del	h 5334	(AaAb) B	22 27.3 −64 58	4.5, 8.7	282	7.0″	77	B9.5V	212581, 255222
	Δ 245		23 08.6 −59 44	7.5, 9.4	289	14″	79	F5V	218392, 255413
	h 5402		23 31.0 −69 05	7.2, 9.1	198	36″	30.7	G0V	221231, 255503
	Gli 289		00 00.6 −66 41	7.7, 9.2	273	3.8″	65	G2V	224783, 255620
★ bet 1,2	Lcl 119	(AB) (CD) E	00 31.5 −62 57	4.3, 4.5, 5.1	168, 118	27″, 9.2′	41.4	B9V	2884, 248201

High mass, matched (2)+(2)+1 quintuple, visual triple, in 2° wide comoving group with bet 3 Tuc, 9′ s.f. MSC system mass 15.3 M⊙. CD (I 260): P = 45 y, e = 0.74, reappears in 2024, apastron 2035. AE (Shy 114): ps = 30,700 AU. (2009)

Label	Catalog ID	Components	Coordinates (J2000)	Mag.	θ	ρ	Dist.	Spectral type(s)	HD, SAO No.
	CorO 3		00 44.5 −62 30	6.3, 8.0	65	2.3″	88	F5III	4294, 248243
lam 1	Δ 2		00 52.4 −69 30	6.7, 7.4	82	20″	63	F7IV	5190, 248269
	h 3416	A B	01 03.3 −60 06	7.6, 7.7	129	5.2″	80	F5V	6334, 248309
★ kap	h 3423	(AaAb) B [C D]	01 15.8 −68 53	5.0, 7.7, [7.8, 8.4]	316, 310, [313]	4.7″, 5.3′, [1.1″]	20.9	F6IV	7788, —

Local, high CPM, solar type (2)+1+2 quintuple. MSC system mass 4.3 M⊙. AB: P = 857 y, orbit r = 135 AU. CD (I 27): P = 85 y, orbit r = 31 AU, e = 0.04, apastron 2046. (2013)

Label	Catalog ID	Components	Coordinates (J2000)	Mag.	θ	ρ	Dist.	Spectral type(s)	HD, SAO No.
	h 3426		01 17.1 −66 24	6.4, 8.3	330	2.4″	94	A0V	7916, 248350

Ursa Major — UMa — Charts 4, 9

Label	Catalog ID	Components	Coordinates (J2000)	Mag.	θ	ρ	Dist.	Spectral type(s)	HD, SAO No.
	Σ 1192	A B C	08 15.8 +60 23	6.5, 10.1, 10.4	256, 224	2.7″, 49″	151	A7Vm	68457, 14479
	Shy 199	A [B C]	08 21.1 +65 27	8.1, [8.5, 12.0]	227, [246]	4.3′, [2.8″]	36.5	G0 G5	69433, 14509
	Σ 1258		08 43.4 +48 52	7.7, 7.9	331	10″	143	F0	74010, 42512
	A 1584		08 53.1 +54 57	9.0, 7.7	90	0.7″	52	G0	75553, 27027

Solar type binary; dark field. System mass 2.1 M⊙. AB: P = 71 y, orbit r = 22 AU, e = 0.71, apastron 2020. (2013)

Label	Catalog ID	Components	Coordinates (J2000)	Mag.	θ	ρ	Dist.	Spectral type(s)	HD, SAO No.
★ 9 iot	h 2477	(AaAb) [B C]	08 59.2 +48 03	3.1, [9.9, 10.1]	82, [40]	2.4″, [0.7″]	14.5	A7IV	76644, 42630

Talitha Borealis. Local, high CPM, A type (2)+2 spectroscopic quadruple with del Sct type variable; dark field. Aa,Ab: P = 11.0 y. A,BC: P = 803 y, orbit r = 170 AU. BC (OΣ 196): P = 39 y, orbit r = 10 AU, e = 0.35, apastron 2018. (2012)

Label	Catalog ID	Components	Coordinates (J2000)	Mag.	θ	ρ	Dist.	Spectral type(s)	HD, SAO No.
13 sig 2	Σ 1306	A B C	09 10.4 +67 08	4.9, 8.9, 10.3	349, 148	4.3″, 3.3′	20.4	F7V	78154, 14788

Local, solar type 2+1 triple. AB: P = 1141 y, orbit r = 130 AU. AC: ps = 5,450 AU. (221 measures; 2012)

Label	Catalog ID	Components	Coordinates (J2000)	Mag.	θ	ρ	Dist.	Spectral type(s)	HD, SAO No.
	Σ 1315		09 12.8 +61 41	7.3, 7.7	27	25″	100	A3IV	78767, 14808

Label	Catalog ID	Components	Coordinates (J2000)	Mag.	θ	ρ	Dist.	Spectral type(s)	HD, SAO No.
★	Σ 1321	A B	09 14.4 +52 41	7.8, 7.9	98	17″	5.81	M0V M0V	79210, 27178

Local, very high CPM, rare low mass matched binary; A is a flare star, both may be spectroscopic binaries. AB: *P* = 975 y, orbit *r* = 114 AU Note, at just 6 pc distance, the weak brightness of low mass dwarf K/M type stars. (447 measures; 2013)

	Arn 71	A B D	09 20.7 +51 16	6.2, 10.0, 7.9	138, 52	5.6″, 3.9′	27.5	F5V	80290, 27215
	Σ 1340	A B C	09 22.5 +49 33	7.1, 9.0, 12.5	319, 83	6.2″, 2.3′	230	B9.5V	80608, 42825
	OΣ 200	(AaAb) B	09 24.9 +51 34	6.5, 8.6	337	1.2″	129	G0IV	81025, 27246
	Σ 1346	A B	09 25.6 +54 01	7.7, 8.6	314	5.7″	196	A2V	81104, 27249
	Σ 1349		09 31.2 +67 32	7.5, 9.0	166	19″	550	A3V	81787, 14903
★ 23 h	Σ 1351	A B	09 31.5 +63 04	3.7, 9.2	269	23″	23.8	F0IV	81937, 14908

Local solar type double with del Sct type variable; a dark field accents the brightness contrast. AB: ps = 740 AU. (2003)

	Σ 1350	A B C	09 34.3 +66 48	8.3, 8.3, 9.2	249, 213	10″, 2.0′	230	F4V F6V	82285, 14923
	Σ 1363		09 35.2 +60 54	7.3, 10.6	356	11″	76	F0	82569, 14931
	Cbl 38		09 50.5 +45 05	7.5, 11.7	11	53″	101	F3Vn	85039, 43063
★ 30 phi	OΣ 208		09 52.1 +54 04	5.3, 5.4	298	0.4″	156	A2V	85235, 27408

Matched A type binary; dark field. System mass 11.6 M_☉. AB: *P* = 105 y, orbit *r* = 55 AU, *e* = 0.45, widening to apastron in 2040. (316 measures; 2012)

	OΣ 209		09 53.3 +50 37	7.4, 10.3	309	4.9″	197	G8IV	85439, 27416
	OΣ 522		09 53.9 +64 47	7.5, 11.2	123	14″	220	K0	85360, 15013
	Σ 1415	A B	10 17.8 +71 04	6.7, 7.3	168	17″	400	A7m	88849, 7099
	Σ 1427		10 22.0 +43 54	8.2, 8.5	214	9.0″	105	F5V	89686, 43306
	Σ 1428		10 26.0 +52 37	8.0, 8.4	87	2.8″	87	F6V	90204, 27639
	Kui 50		10 28.1 +48 47	6.4, 7.4	22	4.0″	22.9	F9V	90508, 43351

Local, high CPM, solar type double. AB: ps = 120 AU. (2012)

	OΣ 219		10 30.2 +51 00	7.6, 11.2	297	13″	158	A7IV	90806, 27664
36	LDS 2863	A B	10 30.6 +55 59	4.9, 8.9	303	2.0′	12.8	F6V	90839, 27670

Local, solar type wide double; isolated in dark field. AB: ps = 2,070 AU. (2012)

	Σ 1462	A B	10 42.9 +50 48	7.4, 10.1	175	8.0″	129	A8IV	92668, 27744
	Sma 75	A (BaBb)	10 43.5 +46 12	5.2, 7.4	88	4.8′	37.1	F5III	92787, 43444
	Σ 1469	A B	10 47.7 +65 28	7.7, 10.4	324	11″	87	F8V	93270, 15292
	OΣ 229		10 48.0 +41 07	7.6, 7.9	260	0.6″	166	A5IV	93457, 43475

A type binary; preceding of two m.7 stars in dark field. AB: *P* = 320 y, orbit *r* = 105 AU. (219 measures; 2013)

	Skf 59		10 54.7 +36 46	7.4, 11.5	58	34″	99	F2V	94456, 62308
	UC 2059	(AaAb) Ba Bb	11 01.8 +36 41	7.5, 11.2, 11.5	46, 38	47″, 0.6″	75	F0	95485, 62360
50 alp	β 1077	A B (CaCb)	11 03.7 +61 45	2.0, 5.0, 7.2	5, 204	0.7″, 6.3′	37.7	G9III	95689, 15384

Dubhe. 2+(2) quadruple, visual triple with G giant; dominant in dark field, brightness contrast challenge. AB: *P* = 44 y, orbit *r* = 22 AU, *e* = 0.44, decreasing θ to apastron 2024; AC: ps = 19,200 AU. (2012)

51	Hzg 8	A B C	11 04.5 +38 14	6.0, 11.6, 7.6	111, 83	7.7″, 2.5′	79	A3III	95934, 62387
	Σ 1510		11 08.0 +52 49	7.7, 9.0	330	5.4″	56	F8V	96527, 27918
	Eng 45	A B	11 11.8 +42 50	7.2, 8.3	247	2.2′	46.8	F8 G5	97194, 43641
	Ho 50		11 13.7 +41 05	6.5, 8.4	35	3.0″	114	K2III	97501, 43649
	Σ 1520		11 16.1 +52 46	6.5, 7.8	343	12″	34.1	F6V F9V	97855, 27970
★ 53 xi	Σ 1523	(AaAb) (BaBb)	11 18.2 +31 32	4.3, 4.8	188	1.6″	10.4	F9V G9V	98230, 62484

Alula Australis. Local, high CPM, solar type (2)+(2) spectroscopic quadruple, visual binary; dark field. MSC system mass 2.4 M_☉. Discovered by W. Herschel (1780), first to have its orbit computed (Savary, 1830); components are astrometric (Aa,Ab, *P* = 1.8 y, orbit *r* = 1.6 AU) and spectroscopic (Ba,Bb, *P* = 4.0 d) binaries. Bright, nearly matched, never closer than 1.4″, a beautiful system in all apertures, mildly challenging in small ones. AB: *P* = 60 y, orbit *r* = 36 AU, *e* = 0.40, apastron 2025. (1,625 measures; 2013)

★ 54 nu	Σ 1524	A B D	11 18.5 +33 06	3.5, 10.1, 8.9	149, 268	7.4″, 4.7′	122	K3III	98262, 62486

Alula Borealis. Probable solar type 2+1 triple with K giant; dark field. AB: ps = 1,220 AU. (2005)

	Σ 1542		11 27.9 +44 34	6.9, 9.7	265	3.5″	89	F2V	99607, 43750
57	Σ 1543	(AaAb) B	11 29.1 +39 20	5.4, 10.7	355	5.4″	65	A2V	99787, 62572

159

Label	Catalog ID	Components	Coordinates (J2000)	Mag.	θ	ρ	Dist.	Spectral type(s)	HD, SAO No.
AW	OΣΣ 111	(AaAbAc) B	11 30.1 +29 58	7.0, 9.5	33	67″	69	A9n	99946, 62579
		A type (2+1)+1 spectroscopic quadruple, visual binary; Aa,Ab is a W UMa type (contact) eclipsing variable (P = 0.4 d), situated between two identical stars in dark field. MSC system mass 3.7 M$_\odot$. Aa,Ac: P = 17 y, e = 0.46. (2006)							
	OΣ 234	(AaAb) B	11 30.8 +41 17	7.5, 8.1	175	0.5″	75	F6V	100018, 43789
		Solar type (2)+1 spectroscopic triple; dark field. AB: P = 87 y, e = 0.36, slips below 0.4″ in 2020 to periastron 2055. (256 measures; 2012)							
	Σ 1546		11 32.3 +56 06	7.5, 10.3	347	11″	57	F8	100214, 28050
	OΣ 235	A B	11 32.3 +61 05	5.7, 7.6	32	0.8″	28.0	F8V	100203, 15542
		Solar type binary. System mass 2.3 M$_\odot$. AB: P = 73 y, orbit r = 22 AU, e = 0.40, apastron 2018. (333 measures; 2013)							
	Σ 1553		11 36.6 +56 08	7.7, 8.2	166	6.2″	47.0	G5	100831, 28071
	Σ 1561	A (BaBb)	11 38.7 +45 07	6.5, 8.2	246	9.1″	23.3	G0V	101177, 43841
		Local, high CPM, solar type 1+(2) spectroscopic triple; m.9 C 3′ f. is unrelated. AB: ps = 290 AU. Ba,Bb: P = 23.5 d. (2013)							
	Σ 1559		11 38.8 +64 21	6.8, 8.0	324	2.0″	210	A5IV	101150, 15580
	Arg 101		11 51.2 +33 23	6.3, 9.3	273	46″	65	Am	102942, 62731
★ 65	Σ 1579	(Aa1Aa2AbB) C (DaDb)	11 55.1 +46 29	6.7, 8.3, 7.0	42, 114	3.7″, 63″	210	A3Vn	103483, 43945
		Aa = DN UMa. Remarkable (2+1+1)+1+(2) spectroscopic septuple, visual triple with Algol type variable (P = 1.7 d); sparkling in very dark field. MSC system mass 10.4 M$_\odot$. Aa,Ab: P = 640 d. AB: P = 118 y, e = 0.50. AC: ps = 1,050 AU. AD: ps = 17,800 AU. (2007)							
★	OΣ 241		11 56.3 +35 27	6.8, 8.7	145	1.8″	176	F3V	103659, 62763
		Solar type double; delicate gem, n.f. of two equal stars in dark field. AB: ps = 430 AU. (2011)							
	Σ 1600		12 05.6 +51 56	7.6, 8.3	93	7.9″	370	G8III	105031, 28241
	Hu 1136		12 05.7 +62 56	6.3, 10.2	217	1.9″	118	K1III	105043, 15710
	Σ 1603		12 08.1 +55 28	7.8, 8.3	83	22″	49.1	F8V F9V	105421, 28253
★	Σ 1695	(AaAb) B	12 56.3 +54 06	6.0, 7.8	280	3.8″	87	A5m	112486, 28572
		A type (2)+1 spectroscopic triple; dark field. MSC system mass 5.4 M$_\odot$. AB: ps = 450 AU. (2014)							
78	β 1082		13 00.7 +56 22	5.0, 7.9	108	1.0″	25.5	F2V	113139, 28601
		Local solar type binary; dark field. AB: P = 105 y, orbit r = 31 AU, e = 0.39, increasing θ, dips to 0.5″ at periastron 2026. (2012)							
	OΣΣ 121		13 09.8 +62 14	6.5, 10.6	9	108″	230	A1V	114504, 15999
★ 79 zet, 80	Σ 1744	(AaAb) (BaBb) (CaCb)	13 23.9 +54 56	2.2, 3.9, 4.0	152, 70	15″, 12′	26.3	A1V	116656, 28737
		Mizar and Alcor. Nearby, A type (2)+(2)+(2) sextuple (second closest to the Sun), visual triple, comoving with a dozen stars that follow within 4 pc (the UMa star cluster, Collinder 285). Striking pure white trio, second closest sextuple to the Sun (after Castor), best viewed with low power. First double identified (Castelli, 1617), first double photographed (G. P. Bond, 1857), first double line spectroscopic binary (E. C. Pickering, 1889). Aa,Ab: P = 20.5 d; Ba,Bb: P = 175 d; Ca,Cb: ps = 28 AU; AB,C: ps = 25,500 AU. (473 measures; 2013) See Figure 2.							
	S 649	C [A B]	13 28.5 +59 57	5.5, [8.2, 9.9]	110, [108]	3.0′, [1.0″]	73	F8	117433, 16083
	Σ 1770		13 37.7 +50 43	6.9, 8.2	126	1.8″	430	K3III	118741, 28819
	β 802		13 48.6 +48 21	7.6, 11.8	224	3.7″	117	A8IV	120475, 44759
	Σ 1831	A B	14 16.1 +56 43	7.2, 9.6	139	5.6″	156	A7IV	125229, 29074
		A type double; dark field, s.f. m.7 C and m.10 EF (optical pair Σ 1830) is unrelated. AB: ps = 1,180 AU. (2013)							

Ursa Minor UMi *Chart 1*

Label	Catalog ID	Components	Coordinates (J2000)	Mag.	θ	ρ	Dist.	Spectral type(s)	HD, SAO No.
★ 1 alp	Σ 93	(AaAb) B	02 31.8 +89 16	2.0, 9.1	233	18″	133	F8Ib	8890, 308
		Polaris. (2)+1 spectroscopic triple, visual binary with supergiant Cepheid variable. MSC has 6.1 M$_\odot$. Aa,Ab: P = 30 y, e = 0.61. (2013)							
	β 799	A B C	13 04.8 +73 01	6.6, 8.5, 11.1	266, 17	1.4″, 92″	120	A7IV	113889, 7741
	Skf 1229	(AB) C	13 49.5 +75 34	7.7, 9.2	199	100″	113	F0	121128, 7888
	Σ 1798		13 55.0 +78 24	7.7, 9.7	11	7.5″	125	F2	122189, 7912

Label	Catalog ID	Components	Coordinates (J2000)	Mag.	θ	ρ	Dist.	Spectral type(s)	HD, SAO No.
	Σ 1840	A B	14 19.9 +67 47	7.0, 10.1	222	27″	143	B9V	126028, 16342
	Σ 1915		14 33.3 +85 56	7.3, 9.7	318	2.4″	180	K0	131616, 2433
	Hu 908	A B	14 53.1 +78 11	6.8, 8.9	237	1.5″	68	K0IV	132698, 8111
★	Σ 1972	(AaAb) (BaBb)	15 29.2 +80 27	6.6, 7.3	79	31″	21.9	G0IV	139777, 2556
		Local, solar type (2)+(2) spectroscopic quadruple system; dark field; doubles 6′ s.f. and 10′ n. AB: ps = 920 AU. (2010)							
18 pi 2	Σ 1989		15 39.6 +79 59	7.3, 8.2	23	0.7″	120	F2V	141652, 2588
		Solar type binary; faint field. System mass 4.7 M⊙. AB: P = 172 y, e = 0.96, closing to periastron 2076. (2011)							
	Σ 2034		15 48.7 +83 37	7.7, 8.0	108	1.1″	380	A3	144463, 2625
	OΣΣ 143	(AaAb) B	16 04.8 +70 16	6.9, 8.8	83	47″	130	A0	145309, 8415
	Ku 1	A B	16 43.1 +77 31	6.1, 10.2	179	2.6″	36.1	F5IV	152303, 8612

Vela   Vel   *Charts 27, 28*

Label	Catalog ID	Components	Coordinates (J2000)	Mag.	θ	ρ	Dist.	Spectral type(s)	HD, SAO No.
	Jc 12		08 05.2 −45 25	8.4, 8.6	18	27″	580	B7V	67269, 219390
	HDS 1162		08 09.3 −51 01	7.6, 9.0	179	17″	123	A0V	68276, 235778
★ gam 1,2	Δ 65	(AaAb) (BaBb) C	08 09.5 −47 20	1.8, 4.1, 7.3	221, 152	40″, 62″	340	WC8 B1IV	68273, 219504
		Regor. Distant, high mass (2)+(2)+1 quintuple, likely member of Vela OB2 association; the primary (gam 2) is massive Wolf-Rayet+O7.5e binary, probably the nearest to the Sun. MSC gives 155 M⊙ for AB only. Aa,Ab: estimated orbit r = 1.2 AU. AB: ps = 18,300 AU. (2009)							
	λ 96	(AaAb) B	08 12.5 −46 16	6.2, 7.7	275	0.6″	320	B5V	68895, 219602
		High mass (2)+1 triple, visual binary; rich field. MSC system mass 15.8 M⊙. Aa,Ab: P = 26 y, e = 0.61. (2012)							
B	I 67		08 22.5 −48 29	5.1, 6.1	136	0.7″	530	B2III	70930, 219848
	h 4104	(AaAb) B	08 29.1 −47 56	5.5, 7.2	250	3.0″	650	B2IV	72108, 219985
		High mass, distant (2)+1 triple, visual binary; rich field. MSC system mass 35.2 M⊙. Aa,Ab: P = 340 y, e = 0.75. (2013)							
	Δ 70	(AaAb) B	08 29.5 −44 43	5.2, 7.0	349	4.3″	800	B2IV	72127, 219996
	I 313	A B	08 30.8 −41 31	6.8, 9.6	214	3.6″	184	K1III	72348, 220026
	h 4107	(AaAb) B C	08 31.4 −39 04	6.5, 8.2, 9.1	330, 99	4.3″, 30″	460	B4V	72436, 199329
		Distant, high mass, probable (2)+1+1 spectroscopic quadruple; middle of three star row in mixed field. AC: ps = 18,600 AU. Neglected. (1999)							
	Slr 8		08 32.1 −53 13	6.1, 7.1	285	0.8″	230	K0III A3	72737, 236062
	I 195		08 34.5 −37 37	6.6, 8.9	42	1.9″	410	K5III	72993, 199389
	h 4119		08 37.2 −49 26	7.4, 9.7	226	10″	187	A2m	73609, 220195
	CorO 268	A B	08 38.4 −46 47	7.8, 9.4	255	14″	60*	B5V	73813, 220225
	CorO 74		08 40.3 −40 16	5.2, 9.1	66	4.1″	86	B9V	74067, 220252
★ HY	BrsO 18	A [B (CaCb) D]	08 42.4 −53 07	4.8, [5.6, 7.9, 9.9]	311, [153, 266]	76″, [0.6″, 60″]	149	B3IV B8	74560, 236205
		High mass 1+(1+2)+1 spectroscopic quintuple; set in quadrangle of m.8 stars in IC 2391. A is HY Vel, an ellipsoidal variable; B is KT Vel, an alp2 CVn type variable. MSC system mass 7.0 M⊙. AB: ps = 15,300 AU. Neglected. (1998)							
del	I 10	(AaAb) B F	08 44.7 −54 43	2.0, 5.6, 5.8	263, 99	0.4″, 94′	24.7	A1Va +F8V	74956, 236232
		Local, A type (2)+1 triple with Algol type variable, bound or comoving with F; faint rich field. MSC has 6.8 M⊙ (AB only). AB: P = 147 y, orbit r = 50 AU, widening. AF (Shy 49): ps = 136,000 AU, near 100% probability pair is physical. (2013)							
	Jc 13		08 46.6 −42 34	7.2, 9.2	311	2.2″	380	B3III	75126, 220439
	I 70		08 47.7 −38 56	7.1, 9.1	111	1.4″	230	B2V	75271, 199628
	CapO 9		08 52.7 −52 08	6.6, 8.2	83	2.9″	124	A0V	76230, 236362
	h 4150		08 53.9 −41 50	7.3, 10.0	265	18″	480*	B5V	76323, 220634
H	R 87	(AaAb) B	08 56.3 −52 43	4.7, 7.7	335	2.6″	108	B5V	76805, 236417
	Gli 104		09 00.0 −49 33	7.1, 9.7	307	9.0″	720	K1II	77321, 220735
	h 4165		09 01.7 −52 11	5.6, 6.6	136	0.7″	113	B9	77653, 236518
	Rst 3620		09 04.5 −56 20	7.2, 9.6	132	0.7″	140*	B5V	78190, 236568

Label	Catalog ID	Components	Coordinates (J2000)	Mag.	θ	ρ	Dist.	Spectral type(s)	HD, SAO No.
	h 4188	(AaAb) B	09 12.5 −43 37	6.0, 6.8	281	2.9″	210	B8V	79416, 220952
	h 4191		09 14.4 −43 14	5.3, 9.2	14	5.7″	188	B4V	79735, 220978
★	I 11		09 15.2 −45 33	6.6, 7.7	293	0.8″	560	B8V	79900, 220998
			Distant, high mass double; in rich field with small group p. One of the challenging subarcsecond systems catalogued by R.T.A Innes. AB: ps = 605 AU. (2010)						
	CorO 83		09 25.6 −53 15	7.0, 10.2	154	19″	46.7	F7V	81734, 236956
	Δ 77	(AaAb) (BaBb)	09 29.3 −44 32	7.1, 7.0	77	108″	42.2	F8V F8V	82241, 221213
★ psi	Cop 1		09 30.7 −40 28	3.9, 5.1	112	1.0″	18.8	F0IV F3IV	82434, 221234
			Local F subgiant binary; set in sparse group. System mass 3.0 M☉. AB: P = 34 y, orbit r = 16 AU, e = 0.43, apastron 2020. (2013)						
★	h 4220		09 33.7 −49 00	5.5, 6.2	217	1.6″	250	B5III	82984, 221288
			High mass double with B giant; rich faint field. AB: ps = 540 AU. Neglected. (1996)						
IM	R 125		09 36.4 −48 45	6.3, 10.3	188	3.0″	71	A	83368, 221339
	λ 115		09 37.2 −53 40	6.1, 6.3	9	0.7″	72	A3V	83520, 237149
	R 129		09 42.7 −55 50	7.9, 8.2	296	3.1″	43.0	G5V G	84330, 237248
	Δ 80	A B	09 45.1 −49 29	8.1, 8.2	250	19″	24.9	F8 G2V	84612, 221459
			Local, solar type matched double; attractive field, several unidentified faint doubles s.f. AB: ps = 640 AU. (2012)						
	h 4245		09 46.1 −45 55	6.8, 9.6	217	9.3″	310	G8II	84774, 221480
	CorO 92		09 50.2 −49 37	8.1, 9.2	24	5.9″	96	F5V	85409, 221531
	Δ 81		09 54.3 −45 17	5.8, 8.2	240	5.2″	310	B5V	85980, 221592
	h 4283		10 04.5 −51 48	7.3, 8.4	181	7.8″	200	A0V	87580, 237664
	h 4284		10 05.1 −45 54	7.4, 9.5	66	6.6″	210	K0III	87640, 221758
	I 173		10 06.2 −47 22	5.3, 7.1	7	1.0″	74	K1IV G5V	87783, 221773
			Solar type binary; rich field. AB: P = 203 y, orbit r = 45 AU, widening. (2010)						
	I 361		10 12.9 −47 29	8.4, 11.0	125	5.6″	39.8	G8V	88746, 221866
			Solar type double; CPM double CD (Dam 521) 30″ p. appears unrelated. AB: ps = 300 AU. (2001)						
★	R 140		10 19.0 −56 01	7.5, 8.2	281	3.2″	167	A4	89613, 237907
			A type double; n.f. of 3 m.8 stars in rich field with a possible small stellar group. AB: ps = 720 AU. Neglected. (1998)						
J	Rmk 13	A B C	10 20.9 −56 03	4.5, 7.2, 9.2	102, 191	7.1″, 36″	350	B3IIIe	89890, 237959
	Δ 86	A B	10 31.2 −42 14	7.3, 8.0	292	83″	182	A	91239, 222113
	CapO 10		10 31.9 −52 14	7.4, 9.2	346	2.3″	270	B5V	91370, 238152
	Pz 3		10 32.0 −45 04	5.6, 6.0	219	14″	167	B8	91355, 222126
	h 4330	(AaAb) B	10 32.9 −47 00	5.2, 8.6	163	40″	400	K4III	91504, 222136
	Hld 106	A [B C]	10 33.3 −55 23	6.8, [7.8, 8.1]	30, [253]	26″, [1.4″]	280	A1V	300791, 238177
	h 4332		10 33.5 −46 59	7.1, 9.8	162	28″	430	A	91590, 222145
	Δ 95	A [B C]	10 39.3 −55 36	4.4, [6.1, 11.9]	105, [174]	52″, [20″]	260	G3Ib	92449, 238309
★ mu	R 155		10 46.8 −49 25	2.8, 5.7	56	2.3″	35.9	G5III G2V	93497, 222321
			Probable high mass binary with G giant; a "count the pairs" rich field. AB: P = 149y, orbit r = 85 AU, widening. (2013)						

Virgo　Vir　　　　　　　　　　　　　　　　　　　*Charts 16, 17*

Label	Catalog ID	Components	Coordinates (J2000)	Mag.	θ	ρ	Dist.	Spectral type(s)	HD, SAO No.
	Σ 1560		11 38.4 −02 26	6.4, 9.4	280	4.9″	117	G9III	101154, 138314
	Σ 1575		11 52.0 +08 50	7.4, 7.9	212	31″	97	K0	103047, 119084
	Shy 588	A C	12 02.7 −10 43	8.6, 7.5	115	5.5′	55	G3V	104576, 157055
	Σ 1616	A B	12 14.5 +08 47	7.6, 9.7	296	23″	55	G0	106423, 119282
	h 204		12 15.0 −01 20	7.6, 11.8	56	35″	230	K0	106498, 138670
	Σ 1627		12 18.2 −03 57	6.6, 6.9	197	21″	87	F2V F3V	106976, 138704
17	Σ 1636		12 22.5 +05 18	6.5, 9.3	337	21″	29.9	F8V	107705, 119360
	Σ 1648		12 30.6 +03 30	7.5, 9.8	40	8.3″	340	G8III	108877, 108877
	Σ 1647		12 30.6 +09 43	8.1, 8.4	248	1.3″	93	F2	108875, 119436
			Solar type binary; dark field. AB: P = 4273 y, orbit r = 400 AU. (274 measures; 2012)						
	Sh 146	A B C	12 31.2 +01 20	7.7, 8.7, 12.0	290, 338	50″, 3.8″	153	A5	108959, 119447

Label	Catalog ID	Components	Coordinates (J2000)	Mag.	θ	ρ	Dist.	Spectral type(s)	HD, SAO No.
	Σ 1649		12 31.6 −11 04	8.0, 8.4	196	15″	88	A7	91169, 157339
	Σ 1668		12 40.9 +08 50	7.8, 8.1	187	1.1″	121	F5V	110280, 119530
★ 29 gam	Σ 1670	A B	12 41.7 −01 27	3.5, 3.5	7	2.0″	11.7	F0V F0V	110379, 138917
			Porrima. Local, high CPM, perfectly matched solar type binary; Y/Y color. System mass 2.8 M$_\odot$. Discovered by J. Bradley (1718), the separation is less than 0.4″ at periastron, but now is an easy split in all apertures, and widening. A summer favorite. AB: P = 169 y, orbit r = 43 AU, e = 0.88, apastron 2090. (1,558 measures; 2013) See Figure 1.						
31	β 924		12 42.0 +06 48	5.6, 10.1	40	3.8″	72	A2V	110423, 119538
	Σ 1677		12 45.3 −03 53	7.3, 8.1	347	16″	260	A9IV	110886, 138952
	Σ 1690		12 56.3 −04 52	7.2, 9.0	148	5.8″	147	A0V	112372, 139049
	OΣ 256		12 56.4 −00 57	7.3, 7.6	101	1.1″	370	F7V	112398, 139053
			Distant, solar type double; near center of 1/2° m.8 quadrangle. AB: ps = 550 AU. (225 measures; 2013)						
	Σ 1701		12 59.3 +06 30	7.6, 9.9	306	21″	95	G8IV	112815, 119696
	β 341		13 03.8 −20 35	6.3, 6.5	132	0.6″	28.2	F8V	113415, 181357
			Solar type binary; system mass 2.4 M$_\odot$. AB: P = 59 y, orbit r = 21 AU, e = 0.99, closing to periastron 2023, reappearing 2037. (2010)						
48	β 929		13 03.9 −03 40	7.1, 7.7	198	0.6″	155	F0V	113459, 139131
			Solar type binary; sparse field. AB: P = 438 y, orbit r = 120 AU, e = 0.0, disappears below 0.4″ in 2029. (2007)						
	Σ 1719	A B	13 07.3 +00 35	7.6, 8.2	358	7.4″	65	F5V F9V	113984, 119774
★ 51 the	Σ 1724	(AaAb) B C	13 10.0 −05 32	4.4, 9.4, 10.4	342, 300	6.4″, 71″	97	A0IV	114330, 139189
			A type (2)+1+1 quadruple; sparse field. MSC system mass 5.5 M$_\odot$. AC: ps = 840 AU. (2012)						
GZ	Σ 1731	(AaAb) B	13 13.2 −02 33	7.7, 10.1	303	9.4″	55	F8	114842, 139227
★ 54	Sh 161	(AaAb) (BaBb)	13 13.4 −18 50	6.8, 7.2	33	5.3″	194	A0V A1V	114846, 157798
			A type (2)+(2) spectroscopic quadruple; A is a W UMa type contact binary (P = 1.0 d); sparse or dark field. AB: ps = 1,390 AU. (2010)						
	Hu 740		13 19.7 −11 40	7.3, 11.4	274	3.6″	250	A2	115813, 157860
	Σ 1734		13 20.7 +02 57	6.8, 7.3	174	1.1″	136	A3V	115995, 119889
			A type double; pretty in a dark field. AB: ps = 200 AU. (204 measures; 2013)						
★	Σ 1740		13 23.7 +02 43	7.1, 7.4	74	26″	15.5	G5V G5V	116442, 119909
			Local, solar type matched double; dark field, attractive in all apertures. AB: ps = 540 AU. (2011)						
	β 610		13 24.0 −20 55	6.6, 10.1	17	4.2″	156	G8III	116429, 181615
	Σ 1742		13 24.3 +01 24	7.8, 8.2	356	0.9″	110	A2	116542, 119913
	Sh 165		13 32.4 −12 40	7.6, 8.6	78	48″	142	F3II	117733, 157992
	Σ 1757	A B	13 34.3 −00 19	7.8, 8.8	135	1.8″	26.6	K4III	118036, 139416
			Solar type binary with K giant; two stars f. are unrelated. AB: P = 461 y, orbit r = 73 AU, increasing θ. (385 measures; 2011)						
	β 114		13 34.3 −08 37	8.1, 8.2	166	1.3″	49.2	F8V	118024, 139415
	β 932	A B	13 34.7 −13 13	6.3, 7.3	62	0.4″	148	A0V	118054, 158021
			A type binary; dark field, optical pair S 650 (m.8, 56″) 15′ s. System mass 9.4 M$_\odot$. AB: P = 178 y, orbit r = 80 AU, slowly widening. (2009)						
	h 228		13 35.6 +10 12	6.6, 9.0	15	70″	120	K0	118266, 100630
81	Σ 1763	A B	13 37.6 −07 52	7.8, 8.1	39	2.7″	196	K0III	118511, 139447
	Σ 1764	A B	13 37.7 +02 23	6.8, 8.6	6	16″	630	K2III	118578, 120042
84	Σ 1777	(AaAb) B	13 43.1 +03 32	5.6, 8.3	226	2.7″	73	K1III	119425, 120082
	Σ 1775	(AaAb) B	13 43.5 −04 16	7.2, 10.1	336	28″	320	K2III F7V	119461, 139507
86	β 935	A B [C D]	13 45.9 −12 26	5.7, 8.5, [11.9, 13.1]	306, 163, [272]	1.2″, 27″, [2.4″]	125	G8III	119853, 158152
HT	Σ 1781	(AaAb) (BaBb)	13 46.1 +05 07	7.9, 8.1	193	1.0″	46.4	F8V F0	119931, 120102
			Solar type (2)+(2) spectroscopic quadruple, visual binary with W UMa type eclipsing binary; dark field. MSC system mass 3.7 M$_\odot$. AB: P = 262 y, orbit r = 45 AU, e = 0.64, apastron 2106. (290 measures; 2013)						
	LDS 3101		13 47.0 +06 21	6.4, 10.2	106	8.1′	31.7	G1IV	120066, 120108
			High CPM, solar type double; sparse field. AB: ps = 20,800 AU; near 100% probability pair is physical. (2003)						

Label	Catalog ID	Components	Coordinates (J2000)	Mag.	θ	ρ	Dist.	Spectral type(s)	HD, SAO No.
★	Σ 1788	A B	13 55.0 −08 04	6.7, 7.3	100	3.6″	33.1	F8V G0	121325, 139618
		Solar type double; a replica Castor, twice as far as the original. AB: ps = 160 AU. (239 measures; 2012)							
	Σ 1802		14 08.1 −12 56	8.1, 9.0	276	6.0″	63	K0V	123453, 158372
★	Σ 1819	A B	14 15.3 +03 08	7.7, 7.9	172	0.9″	40.6	G0V	124757, 120370
		Solar type binary; dark field. System mass 2.1 M⊙. AB: P = 224 y, orbit r = 45 AU, e = 0.20, decreasing θ. (489 measures; 2013)							
	Hld 18	A B	14 19.4 −18 31	7.4, 10.7	356	3.2″	127	A3m F2	125379, 158495
	Σ 1833	A B C	14 22.6 −07 46	7.5, 7.5, 12.9	175, 198	5.8″, 98″	39.4	G0V G0V	125906, 139897
105 phi	Σ 1846	A (BaBb)	14 28.2 −02 14	4.9, 10.0	112	5.3″	36.3	G2III	126868, 139951
	Skf 911		14 29.8 +00 50	6.0, 9.3	331	2.3′	79	A5IV F8	127167, 120499
	Σ 1852		14 30.0 −04 15	7.1, 10.6	267	25″	57	F2V	127168, 139969
	β 1443		14 30.8 +04 46	6.2, 10.6	195	55″	210	gK4	127337, 120504
	A 1109	A B	14 42.8 +06 35	7.4, 9.4	85	1.8″	920	F8	129517, 120616
	Σ 1881		14 47.1 +00 58	6.7, 8.8	0	3.4″	133	B9.5V	130256, 120657
	Σ 1883		14 48.9 +05 57	7.0, 9.0	278	1.0″	56	F6V	130604, 120673
		Solar type binary; dark field. System mass 2.2 M⊙. AB: P = 216 y, orbit r = 45 AU, e = 0.61, widening to apastron 2073. (339 measures; 2013)							
1	H VI 51	A B	14 57.6 −00 10	5.6, 10.4	224	86″	114	K1III	132132, 120758
	β 348		15 01.8 −00 08	6.1, 7.5	108	0.5″	40.5	M0.5II	132933, 120798
	Σ 1904		15 04.1 +05 30	7.2, 7.4	348	10″	76	F0V	133408, 120822
	β 349		15 09.0 +01 41	7.6, 10.9	35	3.0″	89	F1V	134285, 120863

Volans — Vol — Chart 30

Label	Catalog ID	Components	Coordinates (J2000)	Mag.	θ	ρ	Dist.	Spectral type(s)	HD, SAO No.
★ gam 1,2	Δ 42	A (BaBb)	07 08.7 −70 30	3.9, 5.4	296	14″	42.9	K0III F2V	55865, 256374
		Possible high mass 1+(2) spectroscopic triple with K giant; faint field. AB: ps = 810 AU. (2002)							
	h 3997	(AaAb) B	07 35.4 −74 17	7.0, 7.1	306	1.9″	360	B9IV B9IV	62153, 256428
zet	Δ 57		07 41.8 −72 36	4.1, 9.3	123	16″	43.2	K0III	63295, 256438
★ eps	Rmk 7	(AaAb) B	08 07.9 −68 37	4.4, 7.3	23	6.0″	172	B6IV	68520, 250128
		High mass (2)+1 spectroscopic triple; rich field. MSC system mass 11.9 M⊙. AB: ps = 1,390. (2010)							
	I 9		08 14.7 −73 48	7.3, 7.5	103	0.8″	106	A8III	70270, 256489
kap 1,2	BrsO 17	(AaAb) B C	08 19.8 −71 31	5.3, 5.6, 7.7	60, 50	64″, 100″	133	B9III	71046, 256497
	Hrg 19		08 48.4 −65 26	7.3, 10.0	181	4.4″	159	K0III	75795, 250332
	h 4164		08 57.4 −66 12	7.7, 9.5	145	11″	112	K0III	77187, 250396

Vulpecula — Vul — Chart 11

Label	Catalog ID	Components	Coordinates (J2000)	Mag.	θ	ρ	Dist.	Spectral type(s)	HD, SAO No.
	Σ 2445	A B	19 04.6 +23 20	7.3, 8.6	262	12″	480	B2Ve	177648, 86774
	Σ 2457		19 07.1 +22 35	7.5, 9.5	200	10″	95	A7IV	178277, 86828
2	β 248	A B	19 17.7 +23 02	5.4, 8.8	128	1.7″	370	B0.5IV	180968, 87036
	Σ 2525	A B	19 26.6 +27 19	8.2, 8.4	289	2.2″	65	F8	183032, 87213
		Solar type binary; faint rich field. AB: P = 850 y, orbit r = 115 AU. (478 measures; 2013)							
	Σ 2523	A B	19 26.8 +21 10	8.0, 8.1	148	6.3″	850*	B3V B7V	183014, 87218
	Σ 2540	A B	19 33.3 +20 25	7.5, 9.2	146	5.1″	153	A3	184360, 87342
	Σ I 48	A (BaBb)	19 53.4 +20 20	7.1, 7.3	147	42″	146	A0	188211, 87874
13	Dju 4		19 53.5 +24 05	4.6, 7.4	245	1.4″	103	B9.5III	188260, 87883
★ 16	OΣ 395		20 02.0 +24 56	5.8, 6.2	126	0.8″	68	F2III	190004, 88098
		Binary with F giant; faint rich field, n.f. of two stars. AB: P = 1,201 y, orbit r = 185 AU. (355 measures; 2013)							
	Σ 2653		20 13.7 +24 14	6.7, 9.2	274	2.8″	91	A1m	192342, 88377
	Σ 2769	(AaAb) B	21 10.5 +22 27	6.7, 7.4	299	18″	240	A1V	201671, 89505

Appendix B: Double star formulas

Basic quantity	Definition
AFOV	*Eyepiece apparent field of view (degrees)*
D	*Telescope aperture (mm)*
f_e	*Eyepiece focal length (mm)*
f_o	*Telescope focal length (mm)*
M	*Absolute magnitude*
m	*Apparent magnitude*
NELM	*Naked eye limit magnitude*
π	*Geometric parallax (arcseconds)*
ρ	*Angular separation (arcseconds)*

Symbol	Formula	Calculated quantity (units)
		Star systems
b1/b2	$= 2.512^{\Delta m}$	Brightness ratio of two magnitudes (b2 is brightness of the fainter star)
d	$= 1/\pi$	System distance (parsecs)
	$(= 10^{((m-M)/5)+1})$	(Spectroscopic parallax, from m and M, in parsecs)
e	$= \sqrt{1 - b^2/a^2}$	Orbital eccentricity (a is semi-major axis, b is semi-minor axis)
M	$= m - 5(\log(d)-1)$	Absolute magnitude (from d and m)
Δm	$= m2 - m1$	Magnitude difference (m2 is magnitude of the fainter star)
P	$= \sqrt{r^3/(M_1+M_2)}$	Period, for stars of masses M_1 and M_2 (in solar units) (years)
ps	$= \rho \cdot d$	Projected separation (AU)
	$(= d \cdot 10^{(\log(\rho)+0.13)})$	(Projected separation in AU, with Couteau correction for foreshortening)
q	$= 10^{-[\Delta m/10]}$	Mass ratio (main sequence stars only)
r_a	$= r(1 + e)$	Apastron distance (AU)
r_p	$= r(1 - e)$	Periastron distance (AU)
		Telescope optics
d_e	$= f_e/N$	Exit pupil
M	$= f_o/f_e$	Object magnification
m_L	$= NELM+5\log(D)-4$	Magnitude limit (for averted vision; increases with higher magnification)
N	$= f_o/D$	Telescope relative aperture (focal ratio)
R_D	$= 116/D$	Dawes resolution criterion ($1.08R_o$ arcseconds)
R_o	$= 113/D$	Abbe resolution limit (arcseconds)
R_R	$= 138/D$	Rayleigh resolution criterion ($1.22R_o$ arcseconds)
R_S	$= 108/D$	Sparrow resolution criterion ($0.95R_o$ arcseconds)
TFOV	$= AFOV \cdot 60/M$	True field of view (arc minutes)

Appendix C: Double star orbits

This table illustrates the wide range in double star orbits, conventionally described using the orbital period in log days. It assumes two solar mass stars ($2P^2 = r^3$), and values of period and radius have been rounded for clarity.

Period (log d)	Period (days/years)	Orbital radius (AU)	Percent of WDS stars	Label or description
10^{-1}	*0.23*/0.0006	0.009	.	CONTACT
10^{0}	*1.02*/0.0028	0.025	.	
10^{1}	*8.2*/0.022	0.10	.	INTERACTING
10^{2}	*91*/0.25	0.50	0.1	*(Mercury orbit r = ~ 0.4 AU)*
10^{3}	2.8	2.5	1.1	CLOSE *(Asteroid belt r = ~ 3 AU)*
10^{4}	22	10	7.4	*(Saturn orbit r = ~ 10 AU)*
10^{5}	250	50	21.9	MEDIAN *(Limit of Kuiper belt r = ~ 50 AU)*
10^{6}	2,800	250	25.6	*(Heliosphere r = ~ 120 AU)*
10^{7}	22,000	1,000	26.1	WIDE
10^{8}	250,000	5,000	14.0	*(limit of permanent orbits?)*
10^{9}	2,800,000	25,000	3.7	FRAGILE
10^{10}	.	100,000	0.1	*(comoving CPM pairs?)*

Appendix D: Double star catalogs

In the 1960s the Aitken Double Star Catalog (ADS) was merged with southern hemisphere catalogs to form a single resource (the Index Catalog of Double Stars, IDS), parent database of the Washington Double Star Catalog (WDS). At that time, many nineteenth century catalog abbreviations were shortened and the use of Greek letter sigla was discontinued, but the numbering of systems within a catalog remained the same (for example,

Σ 274 became STF 274). All modern publications now use these standard abbreviations. This table includes the old and modern abbreviations for selected catalogs, the name of the astronomer or observatory, the approximate period of double star discoveries, the number of pairs credited in the WDS (including optical and "lost" systems), and the number of systems in the target list and Atlas charts.

Catalog ID		Observer	Active period	No. pairs in WDS (in this Atlas)
Obsolete	Modern			
H I to VI, H N	H 1 to 6, H N	W. Herschel	1777–1821	269 (42)
Σ	STF	F. Wilhelm von Struve	1814–1843	4,294 (864)
Σ I, Σ II	STFA, STFB	*(Supplemental catalogs)*		114 (29)
H	HJ	J. Herschel	1820–1837	5,932 (294)
Sh	SHJ	J. South and J. Herschel	1821–1840	118 (32)
S	S	J. South	1822–1825	220 (37)
Rmk	RMK	C. Rumker	1825–1837	24 (16)
Δ	DUN	J. Dunlop	1825–1846	200 (77)
Lal	LAL	J.-J. de Lalande	1825–1877	5 (4)
OΣ	STT	O. von Struve	1832–1878	754 (191)
OΣΣ	STTA	*(Supplemental catalog)*		232 (37)
BrsO	BSO	*Brisbane Observatory (AUS)*	1834–1851	29 (14)
Jc	JC	W. S. Jacob	1836–1879	38 (12)
Dawes	DA	W. Dawes	1841–1858	14 (3)
Gli	GLI	J. M. Gilliss	1850–1852	108 (12)
Knott	KNT	G. Knott	c. 1865	7 (1)
Hd	HDO	*Harvard Observatory (USA)*	1866–1911	291 (11)
Stone	STN	O. Stone	1867–1882	59 (9)
R	R	H. C. Russell	1870–1882	138 (27)
β	BU	S. Burnham	1870–1911	2274 (231)
β pm	BUP	*(Proper motion pairs)*		372 (2)
Ho	HO	G. W. Hough	1873–1897	791 (23)
MlbO	MLO	*Melbourne Observatory (AUS)*	1873–1898	98 (6)
Howe	HWE	H. A. Howe	1875–1879	111 (21)
CapO	CPO	*Cape Observatory (ZAF)*	1880–1910	773 (14)
Es	ES	Rev. T. E. Espin	1882–1931	3,164 (7)
CorO	COO	*Cordoba Observatory (ESP)*	1883–1920	305 (29)
J	J	R. Jonckheere	1885–1945	3,257 (1)

(*Appendix D, cont.*)

Catalog ID		Observer	Active period	No. pairs in WDS (in this Atlas)
Obsolete	**Modern**			
Slr	SLR	R. P. Sellors	1890–1896	26 (11)
–	A	R. Aitken	1891–1926	3,489 (17)
–	I	R. T. A. Innes	1891–1927	1,666 (94)
λ	SEE	T. J. J. See	1892–1897	464 (15)
Hu	HU	W. J. Hussey	1892–1914	1,671 (19)
δ	DAW	B. H. Dawson	1904–1924	225 (3)
–	B	W. van den Bos	1903–1935	3,101 (8)
–	LDS	W. J. Luyten (*includes PM catalog*)	1914–1987	5,986 (16)
Rst	RST	R. A. Rossiter	1920–1946	5,575 (7)
φ	FIN	F. W. Finsen	1920–1961	441 (2)
Cou	COU	P. Couteau	1957–1991	2705 (4)
Mlr	MLR	P. Muller	1966–1989	702 (1)
–	TOK	A. Tokovinin	1979–	417 (7)
–	HDS	*Hipparcos Catalog*	c.1991	3,383 (7)
–	TDS, TDT	*Tycho Catalog*	c.1991	14,187 (1)
–	BVD	R. Benavides	1994–	341 (10)
–	CBL	R. Caballero	1997–	548 (10)
–	SKF	B. A. Skiff	2000–	2,408 (35)
–	SHY	E. Shaya and R. Olling	2011	226 (20)
–	UC	*USNO Astrographic Catalog*	2013	5,228 (8)

Appendix E: The Greek alphabet

Symbol	Name	Abbreviation
α	Alpha	alp (alf)
β	Beta	bet
γ	Gamma	gam
δ or Δ	Delta	del
ε	Epsilon	eps
ζ	Zeta	zet
η	Eta	eta
θ	Theta	the (tet)
ι	Iota	iot
κ	Kappa	kap
λ	Lambda	lam
μ	Mu	mu
ν	Nu	nu
ξ	Xi	xi (ksi)
o or O	Omicron	omi
π	Pi	pi
ρ	Rho	rho
σ or Σ	Sigma	sig
τ	Tau	tau
υ	Upsilon	ups
φ or φ	Phi	phi
χ	Chi	chi
ψ	Psi	psi
ω	Omega	ome